Wireless Communications: Propagation and Smart Antennas

Wireless Communications: Propagation and Smart Antennas

Editor: Kathryn Davidson

MURPHY & MOORE
www.murphy-moorepublishing.com

www.murphy-moorepublishing.com

ⓂⓂ MURPHY & MOORE

Cataloging-in-Publication Data

Wireless communications : propagation and smart antennas / edited by Kathryn Davidson.
   p. cm.
Includes bibliographical references and index.
ISBN 978-1-63987-743-0
1. Wireless communication systems. 2. Radio wave propagation. 3. Adaptive antennas. I. Davidson, Kathryn.
TK5103.2 .W57 2023
621.384--dc23

Murphy & Moore Publishing
1 Rockefeller Plaza,
New York City,
NY 10020, USA

ISBN 978-1-63987-743-0

Contents

Permissions

List of Contributors

Index

Preface

It is often said that books are a boon to mankind. They document every progress and pass on the knowledge from one generation to the other. They play a crucial role in our lives. Thus I was both excited and nervous while editing this book. I was pleased by the thought of being able to make a mark but I was also nervous to do it right because the future of students depends upon it. Hence, I took a few months to research further into the discipline, revise my knowledge and also explore some more aspects. Post this process, I begun with the editing of this book.

An antenna is the interface between the radio waves propagating through space and the electric currents flowing in metal conductors. It is utilized in conjunction with a transmitter or receiver. It can be built to transmit and receive radio waves preferentially in one direction or equally in all horizontal directions. Antennas are necessary for wireless communication systems to function properly at both the transmitter and receiver ends. The rapid development of wireless communication systems has in turn led to the development of innovative antenna technologies such as meta-material antennas, diversity antennas, smart antennas, and various software-defined antenna systems that are making their way onto the market to support modern wireless communication. Smart antenna is an antenna array with digital signal processing algorithms, which identify spatial signatures, which in turn are used by the smart antenna to calculate beamforming vectors that are used to track and locate the antenna beam on a mobile or target. This book aims to shed light on the concept of smart antennas and wave propagation for wireless communication. It also elucidates new techniques and their applications. This book aims to equip students and experts with the advanced topics and upcoming concepts in this area.

I thank my publisher with all my heart for considering me worthy of this unparalleled opportunity and for showing unwavering faith in my skills. I would also like to thank the editorial team who worked closely with me at every step and contributed immensely towards the successful completion of this book. Last but not the least, I wish to thank my friends and colleagues for their support.

Editor

A Reconfigurable CMOS Inverter-Based Stacked Power Amplifier with Antenna Impedance Mismatch Compensation for Low Power Short-Range Wireless Communications

Dong-Myeong Kim, Dongmin Kim, Hang-Geun Jeong * and Donggu Im *

Division of Electronic Engineering, Jeonbuk National University, Jollabuk-do 561-756, Korea;
sshd1222@naver.com (D.-M.K.); aksd5736@naver.com (D.K.)
* Correspondence: hgjeong@jbnu.ac.kr (H.-G.J.); dgim@jbnu.ac.kr (D.I.)

Abstract: A reconfigurable CMOS inverter-based stacked power amplifier (PA) is proposed to extend impedance coverage, while maintaining an output power exceeding the specific power level under the worst antenna impedance mismatch conditions. The adopted process technology supports multi-threshold metal-oxide-semiconductor field-effect transistor (MOSFET) devices, and therefore, the proposed PA employs high threshold voltage (V_{th}) MOSFETs to increase the output voltage swing, and the output power under a given load condition. The unit cell of the last PA stage relies on a cascode inverter that is implemented by adding cascode transistors to the traditional inverter amplifier. By stacking two identical cascode inverters, and enabling one or both of them through digital switch control, the proposed PA can control the maximum output voltage swing and change the optimum load R_{opt}, resulting in maximum output power with peak power added efficiency (*PAE*). The cascode transistors mitigate breakdown issues when the upper cascode inverter stage is driven by a supply voltage of $2 \times V_{DD}$, and decrease the output impedance of the PA by changing its operation mode from the saturation region to the linear region. This variable output impedance characteristic is useful in extending the impedance coverage of the proposed PA. The reconfigurable PA supports three operation modes: cascode inverter configuration (CIC), double-stacked cascode inverter configuration (DSCIC) and double-stacked inverter configuration (DSIC). These show R_{opt} of around 100, 50 and 25 Ω, respectively. In the simulation results, the proposed PA operating under the three configurations showed a saturated output power (P_{sat}) of +6.1 dBm and a peak *PAE* of 41.1% under a 100 Ω load impedance condition, a P_{sat} of +4.5 dBm and a peak *PAE* of 44.3% under a 50 Ω load impedance condition, and a P_{sat} of +5.2 dBm and a peak *PAE* of 37.1% under a 25 Ω load impedance condition, respectively. Compared to conventional inverter-based PAs, the proposed design significantly extends impedance coverage, while maintaining an output power exceeding the specific power level, without sacrificing power efficiency using only hardware reconfiguration.

Keywords: antenna impedance mismatch; breakdown; cascode inverter; CMOS; impedance coverage; power added efficiency; re-configurability; stacked power amplifier

1. Introduction

As wireless electronic devices continue to be miniaturized, and further support multi-functionality, antenna size reduction is required because most mobile platforms have a limited space for all of the necessary antennas. The small radiator size of a miniaturized antenna greatly reduces its bandwidth, and this leads to a high quality factor (Q-factor) for the antenna.

As a result, the antenna impedance becomes more sensitive to environmental changes, leading to several important considerations and challenges in designing RF front-end circuits, such as low noise amplifiers (LNAs), power amplifiers (PAs) and RF switches. One major concern is the preservation of the peak output power and power-added efficiency (PAE) of a PA in the worst antenna impedance mismatch conditions. Because most PAs are designed to drive a standard 50 Ω impedance load using a dedicated output matching network, antenna impedance mismatch significantly affects peak output power and PAE performance. For instance, it was reported by Keerti et al. that the output power of a PA decreases as much as 5 dB for a load mismatch with a voltage standing-wave ratio (VSWR) of 10:1 [1].

The most common method used to compensate for antenna impedance mismatch, and to preserve the RF performance of a PA, is to use a tunable matching network (TMN) between the PA and the antenna [2–4]. To adapt to changes in antenna impedance, the TMN adjusts its impedance to match the overall impedance of the TMN and antenna to 50 Ω. In implementing the TMN, it is very important to build high Q-factor passive devices (inductors and capacitors) with good linearity in order to minimize power loss and avoid signal distortion. In addition, to compensate the antenna impedance mismatch over a wide range of the Smith Chart, it is better to increase the tuning range of the integrated tunable capacitor, while keeping a high Q-factor and good linearity. Unfortunately, on-chip spiral inductors fabricated on silicon substrates suffer from poor Q-factors, and the use of off-chip inductors with high Q-factors is not desirable, because of cost and limited printed circuit board (PCB) size. Regarding integrated tunable capacitors in CMOS, the most common topology of the digitally controlled switched capacitor array inevitably faces a fundamental trade-off between the Q-factor and the tuning range [5], and its performance degrades as the operating frequency increases. Some emerging technologies, such as micro-electro-mechanical systems (MEMS) and barium strontium titanate (BST), have been proposed for use in the implementation of low-loss, wide-tuning-range tunable capacitors [6,7]. However, these technologies require a high tuning/switching voltage (>30 V), which of course is unsuitable in battery-driven mobile handsets. Additionally, they have not been proven in high-volume production.

In this paper, a reconfigurable CMOS inverter-based stacked PA for 2.4 GHz low power short-range wireless communications is designed to maintain nearly constant output power in the presence of antenna impedance variation. Through only hardware reconfiguration, and without a TMN, impedance coverage was significantly extended compared to the conventional inverter-based PAs, maintaining an output power exceeding the specific power level without sacrificing power efficiency.

2. Circuit Design

2.1. Conventional Inverter-based PA

All modern RF transceivers for low power short-range wireless communication technologies, such as Bluetooth Low Energy (BLE), IEEE802.15.4 (ZigBee), Z-wave, Thread and IEEE802.15.6 (Medical Body-Area Networks, MBAN), are integrated with highly reconfigurable digital baseband modems to provide single-chip solutions, and keep costs down [8–10]. By making use of the technology scaling available with the deep-submicron CMOS process, single chip integration of RF/analog circuits and digital circuits with micro-processor cores on a common CMOS system-on-chip (SoC) platform has become a development trend, and provides the critical advantages of low cost and low power consumption. However, with this approach, the design of traditional RF/analog circuits is difficult. Furthermore, some special masks needed to implement metal–insulator–metal (MIM) capacitors and thick metal layers are not acceptable from a fabrication cost standpoint. To compensate for the increasing mask costs of scaled CMOS technologies, RF circuitry, including PAs, should be fabricated on a small silicon area, and be compatible with a standard, low cost digital CMOS process, where there is no burden on the back-end-of-line (BEOL), resulting from the stringent Q-factor requirements of on-chip inductors (or transformers). In addition, it is better to design digital-intensive and digital-oriented

RF/analog circuits to exploit many advantages of technology scaling. For these reasons, inverter-based PAs have been widely adopted for low power short-range wireless technologies.

Figure 1a shows a schematic of a conventional three-stage inverter-based PA. The hardware configuration is very simple, because there are no inductors or transformers, and as a result, it is easily implemented in a low cost digital CMOS process The previous works of Paidimarri et al. [11], Kiumarsi et al. [12], and Van Langevelde et al. [13] presented the inverter-based, push-pull topologies for PA applications, but there was no study for compensating the antenna impedance mismatch effect with the same RF performances. The first and second stages are drive amplifiers (DAs) for boosting the weak input signal with sufficiently high gain and extremely low power consumption, and the last stage is the output stage for obtaining maximum output power under the given load condition. Instead of using a dc biasing method with a shunt feedback configuration, a replica biasing circuit is adopted to provide a stable dc operating point to the inverter stage, and to ensure a rail-to-rail output swing. By adjusting the channel width ratio between the NMOS and PMOS transistors, the dc operating point of the inverter stages is around half of V_{DD}. Compared to a common-source topology with an inductor load, the output voltage swing of the inverter-based PA is reduced by almost half. As shown in Figure 1b, its output voltage swing is expressed as

$$V_{dsN,sat} \leq v_{out} \leq V_{DD} - V_{sdP,sat} \tag{1}$$

$$(V_{gs,N} - |V_{th,N}|) \leq v_{out} \leq V_{DD} - (V_{sg,P} - |V_{th,P}|) \tag{2}$$

where $V_{gs,N}$ and $V_{sg,P}$ are the gate-to-source and source-to-gate dc bias voltages applied to the NMOS and PMOS transistors, and $V_{th,N}$ and $V_{th,P}$ are the threshold voltages of the NMOS and PMOS transistors. Typically, $V_{gs,N}$ and $V_{sg,P}$ are set to be equal to half of V_{DD} to maximize the output voltage swing. From (1) and (2), it can be seen that it is desirable to increase the threshold voltage (V_{th}) of the MOSFET for the maximum output voltage swing. Modern deep-submicron CMOS technologies support multiple-V_{th} MOSFETs to reduce leakage power by assigning a high V_{th} to some transistors in non-critical paths. Therefore, for a simple hardware configuration, in the circuit design for the last stage of the inverter-based PA, high-V_{th} MOSFETs are used instead of applying a body bias voltage to increase V_{th} in regular-V_{th} MOSFETs. Figure 2 shows the simulated output power, the *PAE* and the power contour of the three-stage inverter-based PA using regular-V_{th} MOSFETs and high-V_{th} MOSFETs under a 50 Ω load impedance condition. Both PAs were designed to consume the same static dc current of 2.3 mA. The saturated output power (P_{sat}) and peak *PAE* are +1.2 dBm and 32.6% for the PA with the regular-V_{th} MOSFETs and +2.8 dBm and 41.4% for the PA with the high-V_{th} MOSFETs. As predicted, using the high-V_{th} MOSFETs increases the output voltage swing without any additional dc current, enhances P_{sat} and peak *PAE*, and extends the impedance coverage while maintaining an output power of greater than 0 dBm.

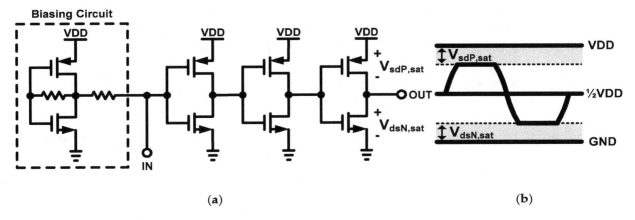

(a)   (b)

Figure 1. (a) Schematic of a conventional three-stage inverter-based power amplifier (PA) and (b) its output voltage swing limitation.

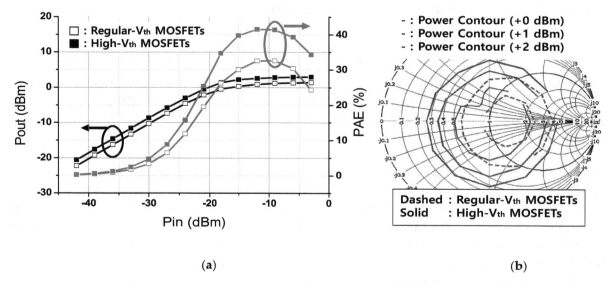

(a)   (b)

Figure 2. Simulation results of the three-stage inverter-based PA using regular-V_{th} metal-oxide-semiconductor field-effect transistors (MOSFETs) and high-V_{th} MOSFETs: (a) saturated output power (P_{sat}), PAE, and (b) the power contour.

2.2. Proposed Reconfigurable Cascode Inverter-based Stacked PA

The most efficient output power ($P_{out,eff}$) of a PA is determined by the optimum load impedance R_{opt} and maximum (zero-to-peak) output voltage swing $v_{out,max}$. This relationship is given as

$$P_{out,eff} \simeq \frac{v^2_{out,max}}{2R_{opt}}. \tag{3}$$

Equation (3) implies that the PA can extend the impedance coverage while maintaining an output power exceeding the specific power level by varying $v_{out,max}$ through hardware reconfiguration. Figure 3 shows the proposed reconfigurable cascode inverter-based stacked PA driven by a supply voltage of $2 \times V_{DD}$ with three operation modes. The first and second stages are DA stages, while the last stage is a PA stage. In designing the DA, to alleviate the breakdown issues resulting from operating at a supply voltage of $2 \times V_{DD}$, two identical inverters are stacked in series, and a bypass capacitor is placed between them to short ac signals to ground at the operating frequency. For all stages, high-V_{th} MOSFETs are employed to enhance P_{sat} and peak PAE. All transistors in the proposed reconfigurable PA are thin gate oxide MOSFETs for high frequency operation. The unit cell of the last PA stage is based on a cascode inverter, which is implemented by adding cascode transistors (M_{N2} and M_{P2}) to the traditional inverter amplifier (M_{N1} and M_{P1}). The cascode transistors mitigate breakdown issues when the upper cascode inverter stage is driven by a supply voltage of $2 \times V_{DD}$ alone, and also decrease the output impedance of the PA by changing its operation mode from the saturation region to the linear region. This variable output impedance characteristic is useful in extending the impedance coverage of the reconfigurable PA. By stacking two identical cascode inverters, and enabling one or both of them through digital switch control, the proposed PA can control the value of $v_{out,max}$ in (3), and eventually change R_{opt}, yielding the maximum output power with peak PAE. The resistance value of R_G (10 kΩ) is high enough to block ac signals, and only pass dc bias voltage. The capacitance value of C_G, C_O and C_B are 2 pF, 10 pF and 10 pF, respectively, in this design. Similarly, their capacitance value is high enough to block dc bias voltage and only pass ac signals. It was verified through simulation that their side effects on the overall performance were negligible.

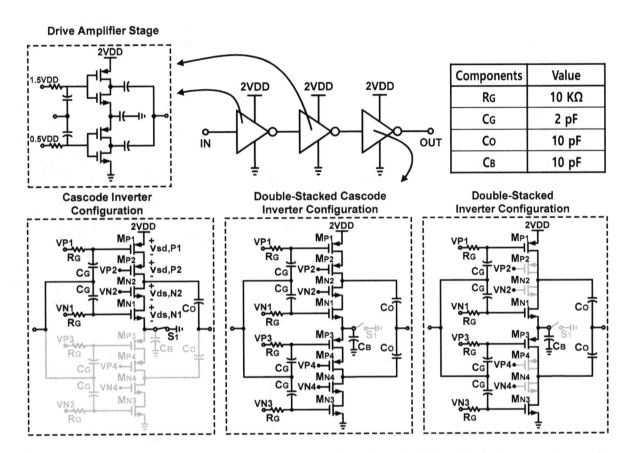

Figure 3. Proposed reconfigurable cascode inverter-based stacked PA with three operation modes: cascode inverter configuration, double-stacked cascode inverter configuration and double-stacked inverter configuration.

In the cascode inverter configuration (on the far left in Figure 3), the middle spot between the two cascode inverters is directly connected to the physical ground through switch S_1, and the upper cascode inverter, which is composed of transistors M_{N1}, M_{N2}, M_{P1} and M_{P2} operated in the saturation region, is only enabled. The output voltage swing in this mode is given as

$$V_{dsN1,sat} + V_{dsN2,sat} \leq v_{out} \leq 2V_{DD} - V_{sdP1,sat} - V_{sdP2,sat} \tag{4}$$

$$(V_{gs,N1} - |V_{th,N1}|) + (V_{gs,N2} - |V_{th,N2}|) \leq v_{out} \leq 2V_{DD} - (V_{sg,P1} - |V_{th,P1}|) - (V_{sg,P2} - |V_{th,P2}|) \tag{5}$$

where $V_{gs,N1(2)}$ and $V_{sg,P1(2)}$ are the gate-to-source and source-to-gate dc bias voltages applied to $M_{N1}(M_{N2})$ and $M_{P1}(M_{P2})$, and $V_{th,N1(2)}$ and $V_{th,P1(2)}$ are the threshold voltages of $M_{N1}(M_{N2})$ and $M_{P1}(M_{P2})$. Assuming the overdrive voltage $V_{dsN1,sat}$, $V_{dsN2,sat}$, $V_{sdP1,sat}$ and $V_{sdP2,sat}$ are all 0.2 V, the nominal supply voltage V_{DD} is 1.2 V, and a maximum output power of around +6 dBm is delivered to the load by the PA. The value of $v_{out,max}$ and R_{opt} are calculated to be 0.8 V_{op} and 80 Ω using Equation (3). Figure 4 shows the load-pull simulation results of the proposed reconfigurable PA for the three modes. The simulated R_{opt} yielding a maximum output power of around +6 dBm is about 100 Ω for the cascode inverter configuration. This closely matches the calculated result. Because the upper cascode inverter is driven by a supply voltage of $2 \times V_{DD}$ (greater than the nominal supply voltage V_{DD}), it is important to ensure in the design that the transistor's drain-to-source voltage swing does not exceed the pre-specified value of its drain-to-source breakdown voltage (BV_{DS}). This prevents the permanent damage and gradual degradation in device performance over time caused by gate-oxide breakdown and hot carrier degradation. Figure 5 presents the simulated output voltage waveform, source-to-drain voltage waveforms driven by M_{P1} and M_{P2} ($v_{sd,P1}$ and $v_{sd,P2}$), and drain-to-source voltage waveforms driven by M_{N2} and M_{N1} ($v_{ds,N2}$ and $v_{ds,N1}$) when the proposed PA in the cascode inverter configuration generates an output power of +3 dBm under a 100 Ω load impedance.

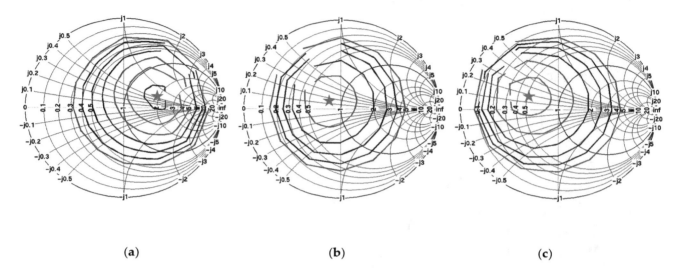

Figure 4. Load-pull simulation results of the proposed reconfigurable PA for the three modes: (**a**) cascode inverter configuration, (**b**) double-stacked cascode inverter configuration, and (**c**) double-stacked inverter configuration. The contour starts from an output power of 0 dBm (red circle), and the contour levels are in steps of 1 dB.

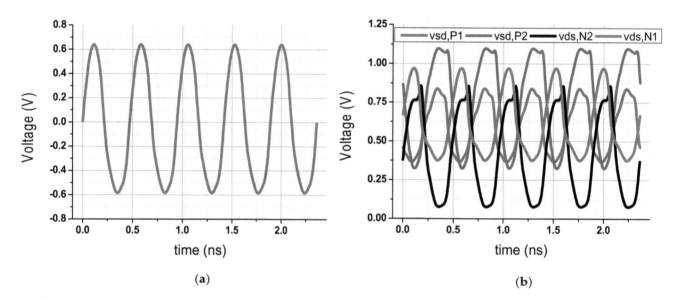

Figure 5. Simulated (**a**) output voltage waveform, (**b**) source-to-drain voltage waveforms driven by M_{P1} and M_{P2} ($v_{sd,P1}$ and $v_{sd,P2}$), and drain-to-source voltage waveforms driven by M_{N2} and M_{N1} ($v_{ds,N2}$ and $v_{ds,N1}$) when the proposed PA in the cascode inverter configuration generates an output power of +3 dBm under a 100 Ω load impedance.

It can be seen that the cascode transistors M_{P2} and M_{N2} share the burden of the over-voltage stress and mitigate the breakdown issues of the thin gate oxide MOSFETs. Because all source-to-drain and drain-to-source voltage swings do not exceed the BV_{DS} of 1.2 V provided by the adopted process technology, and because this is ac stress instead of dc stress, the cascode inverter configuration of the proposed PA does not experience device breakdown.

In the double-stacked cascode inverter configuration, both the upper and lower cascode inverters are enabled. Switch S_1 is open, the middle spot between the two cascode inverters is biased at half of the $2 \times V_{DD}$ supply voltage, and this node is ac grounded through the large bypass capacitor C_B. Ideally, this should yield almost the same maximum output power as the aforementioned cascode inverter configuration because the total output current is doubled by combining the two output currents from the upper and lower cascode inverters, while the final output voltage swing is reduced by half. This implies that the value of R_{opt}, which yields the maximum output power from the double-stacked cascode inverter configuration, is lower than that of the cascode inverter configuration. As shown in Figure 4a,b, the double-stacked cascode inverter configuration provides a simulated R_{opt} of around 50 Ω, which is approximately half the value of the simulated R_{opt} of the cascode inverter configuration. Unfortunately, because the two output currents generated from the upper and lower cascode inverters are not completely identical due to mismatches, a slight degradation (1–2 dB) of the maximum output power is unavoidable.

The last mode of the proposed reconfigurable PA is the double-stacked inverter configuration of the circuit schematic on the far right in Figure 3. Unlike in the double-stacked cascode inverter configuration, all cascode transistors M_{P2}, M_{P4}, M_{N2} and M_{N4} operate in the linear region as switches by applying a gate-to-source (source-to-gate) voltage of V_{DD} to the NMOS (PMOS) transistors M_{N2} and M_{N4} (M_{P2} and M_{P4}), and completely turning on the MOSFETs. Because the cascode configuration typically provides a high output impedance with a slightly reduced output voltage swing, the output impedance of the double-stacked inverter configuration (DSIC) is lower than that of the double-stacked cascode inverter configuration (DSCIC) at an operating frequency of 2.4 GHz, as shown in Figure 6. For maximum power transfer, it is desirable to decrease the output impedance of the reconfigurable PA to shift the impedance coverage to the lower impedance region to maintain an output power exceeding the specific power level. As shown in Figure 4c, the double-stacked inverter configuration provides a simulated R_{opt} of around 25 Ω, which is approximately half the value of the simulated R_{opt} for the double-stacked cascode inverter configuration, and covers a low impedance of less than 5 Ω for an output power greater than 0 dBm. In addition, the maximum output power is increased by 1 dB in comparison with the double-stacked cascode inverter configuration, because there is no additional voltage headroom for the cascode transistors M_{P2}, M_{P4}, M_{N2} and M_{N4}.

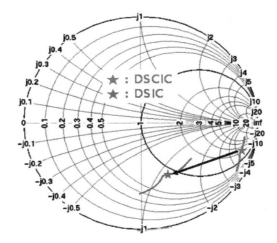

Figure 6. Simulated output impedance of the PA with double-stacked cascode inverter configuration (DSCIC) and double-stacked inverter configuration (DSIC).

3. Simulation Results

The proposed reconfigurable PA was designed using a 65-nm CMOS process. Because of a standalone circuit pattern, a simple input matching network composed of one series inductor and one shunt capacitor was placed in front of the first DA stage of the proposed PA. Figure 7 shows the simulated forward transmission coefficient (S_{21}) and input reflection coefficient (S_{11}) of the completed PA operated in the cascode inverter configuration (CIC), DSCIC and DSIC. Simulated value for S_{21} ranges from 30 dB to 35 dB, and the simulated value for S_{11} is lower than −10 dB at an operating frequency of 2.4 GHz for all three configurations. Figure 8a presents the simulated P_{sat} and PAE of the complete PA in each operation mode.

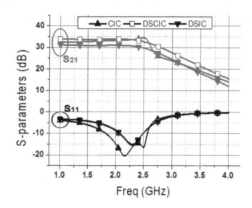

Figure 7. Simulated S-parameters of the proposed reconfigurable PA operated in the cascode inverter configuration (CIC), DSCIC, and DSIC.

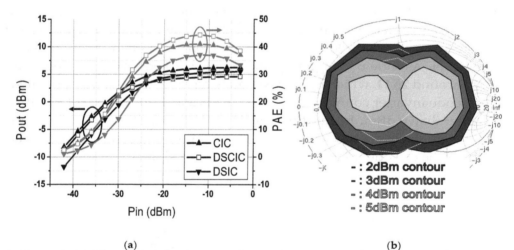

(a)   (b)

Figure 8. Simulated (**a**) P_{sat} and PAE of the complete PA in each operation mode and (**b**) total impedance coverage of the proposed reconfigurable PA while maintaining an output power of greater than +2, +3, +4, and +5 dBm.

In the simulation results, the proposed PA operated in CIC, DSCIC and DSIC shows a P_{sat} of +6.1 dBm and a peak PAE of 41.1% under a 100 Ω load impedance, a P_{sat} of +4.5 dBm and a peak PAE of 44.3% under a 50 Ω load impedance, and a P_{sat} of +5.2 dBm and a peak PAE of 37.1% under a 25 Ω load impedance, respectively. Figure 8b shows the simulated total impedance coverage of the proposed reconfigurable PA, while maintaining the output power of greater than +2, +3, +4 and +5 dBm. Compared to conventional inverter-based PAs, impedance coverage is significantly extended, while output power that exceeds the specific power level is maintained without sacrificing power efficiency using only hardware reconfiguration. The proposed PA can compensate for antenna impedance mismatch without any inductors or transformers, and can have the same effect as a TMN at the output stage.

Figure 9a,b show the simulated output voltage waveforms of the complete PA according to the change of circuit configuration under a 200 Ω load impedance and a 5 Ω load impedance, respectively. The RF sinusoidal signal with a power of −10 dBm was applied in order to achieve an output power close to P_{sat}. As predicted, the CIC yields the highest output voltage swing among three configurations when the complete PA directly drives a load impedance greater than 100 Ω, and similarly the DSIC generates the highest output voltage swing among three configurations when it directly drives a load impedance less than 25 Ω. This implies that the proposed reconfigurable PA can change its operation mode automatically over the load impedance variation, by detecting the output voltage swing through a RF envelope detector, and selecting the operation mode to maximize the detector output voltage. Figure 10 shows the possible design for the fully integrated reconfigurable PA, including an automatic tuning circuit. Many researches of [14–16] reported an extremely low power RF envelope detector circuit in CMOS at 2.4 GHz frequency band, and most of them adopted the common source-based envelope detector circuit due to high RF-to-DC conversion characteristic. Because the common source-based envelope detectors provide relatively high input impedance, and their power consumption is around microwatt level or below, their effects on the overall RF performance of the proposed PA can be negligible.

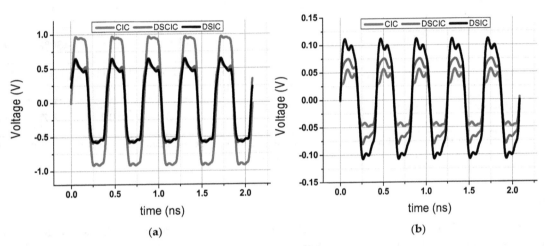

Figure 9. Simulated output voltage waveforms of the complete PA according to the change of circuit configuration under (**a**) a 200 Ω load impedance and (**b**) a 5 Ω load impedance. The RF sinusoidal signal with a power of −10 dBm was applied in order to achieve an output power close to P_{sat}.

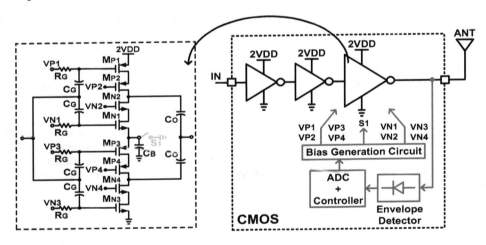

Figure 10. Possible design for the fully integrated reconfigurable PA including an automatic tuning circuit.

Table 1 summarizes and compares the performances of the proposed reconfigurable PA with previously published reports of PAs for ultra-low power (ULP) radios. The simulated *PAE* at P_{sat} of the proposed PA under all load conditions (100, 50 and 25 Ω) is higher than seen in previously reported PAs, and is achieved without the use of the on-chip matching network at the output stage. The proposed PA can minimize the degradation of the output power under the worst antenna impedance mismatch conditions, and will become a good candidate for miniaturized high-efficiency PAs for ULP short-range wireless communication systems.

Table 1. Comparison of performances of the proposed reconfigurable PA with previously published reports of PAs for ultra-low power (ULP) radios.

PA	[1] Proposed PA			[2] [17]	[2] [18]	[2] [19]
	CIC	DSCIC	DSIC			
CMOS Technology (nm)	65			55	40	130
Supply Voltage (V)	2.4			0.9–3.3	1.1	1
Output Matching Network	No			Yes	Yes	Yes
Saturated Power (P_{sat}) (dBm)	+6.1	+4.5	+5.2	0	0	1.6
PA Efficiency @ P_{sat} (%)	41.1%	44.3%	37.1%	30.0%	<30%	26.8%
Strongest Harmonic Emission (dBm)	[3] HD3 −27	[3] HD3 −20	[3] HD3 −31	[4] HD3 −52	[4] HD3 −48	[4] HD3 −32
Power Consumption @ P_{sat} (mW)	9.8	6.2	8.8	[5] 10.1	[5] 7.7	[5] 5.9

[1] Simulation results, [2] Measurement results, [3] Tested at the output power of +2 dBm, [4] Tested at the saturated output power, [5] Total power consumption of whole transmitter (Tx).

4. Conclusions

The reconfigurable CMOS inverter-based stacked power amplifier (PA) is implemented using a 65-nm CMOS process technology, and extends impedance coverage, while maintaining an output power that exceeds the specific power level in the worst antenna impedance mismatch conditions. The proposed PA can minimize the degradation of output power, and can have the same effect as a TMN at the output stage, without the use of inductors or transformers. Because the proposed PA can be easily implemented using a low cost digital CMOS process, while exploiting the advantage of technology scaling, it will become a good solution for miniaturized high-efficiency PAs for ULP radio applications.

Author Contributions: Conceptualization, D.-M.K., H.-G.J. and D.I.; Investigation, D.-M.K., D.K. and D.I.; Supervision, H.-G.J. and D.I.; Writing—original draft, D.-M.K. and D.I. All authors have read and agreed to the published version of the manuscript.

Acknowledgments: This work was supported by research funds of Jeonbuk National University in 2015. This work was also supported by Institute for Information and Communications Technology Promotion (IITP) grant funded by the Korea Government (MSIT) (No. 2018-0-01461) and Basic Science Research Program through the National Research Foundation of Korea (NRF) funded by the Ministry of Education (2018R1A2B6008816). The CAD tool was supported by IDEC.

References

1. Keerti, A.; Pham, A.-V.H. RF characterization of SiGe HBT power amplifiers under load mismatch. *IEEE Trans. Microw. Theory Tech.* **2007**, *55*, 207–214. [CrossRef]
2. Yoon, Y.; Kim, H.; Kim, H.; Lee, K.; Lee, C.; Kenney, J.S. A 2.4-GHz CMOS power amplifier with an integrated antenna impedance mismatch correction system. *IEEE J. Solid State Circuits* **2014**, *49*, 608–621. [CrossRef]
3. Song, H.; Bakkaloglu, B.; Aberle, J.T. A CMOS adaptive antenna-impedance-tuning IC operating in the 850MHz-to-2GHz band. In Proceedings of the IEEE International Solid-State Circuits Conference—Digest of Technical Papers, San Francisco, CA, USA, 8–12 February 2009; pp. 384–385.
4. Jeon, J.; Kang, M. A ruggedness improved mobile radio frequency power amplifier module with Dynamic impedance correction by software defined atomization. *Electronics* **2019**, *8*, 1317. [CrossRef]
5. Sjöblom, P.; Sjöland, H. An adaptive impedance tuning CMOS circuit for ISM 2.4-GHz band. *IEEE Trans. Circuits Syst. I Reg. Pap.* **2005**, *52*, 1115–1124.

6. Shen, Q.; Barker, S. Distributed MEMS tunable matching network using minimal-contact RF-MEMS varactors. *IEEE Trans. Microw. Theory Tech.* **2006**, *54*, 2646–2658. [CrossRef]

7. Schmidt, M.; Lourandakis, E.; Leidl, A.; Seitz, S.; Weigel, R. A comparison of tunable ferroelectric Π- and T-matching networks. In Proceedings of the European Microwave Conference, Munich, Germany, 9–12 October 2007; pp. 98–101.

8. Retz, G.; Shanan, H.; Mulvaney, K.; O'Mahony, S.; Chanca, M.; Corwley, P.; Billon, C.; Khan, K.; Quinlan, P. A Highly Integrated Low-Power 2.4GHz Transceiver Using a Direct-Conversion Diversity Receiver in 0.18 μm CMOS for IEEE 802.15.4 WPAN. In Proceedings of the IEEE International Solid-State Circuits Conference—Digest of Technical Papers, San Francisco, CA, USA, 8–12 February 2009; pp. 414–415.

9. Zolfaghari, A.; Said, M.E.; Youssef, M.; Zhang, G.; Liu, T.T.; Cattivelli, F.; Syllaios, Y.I.; Khan, F.; Fang, F.Q.; Wang, J.; et al. A Multi-Mode WPAN (Bluetooth, BLE, IEEE 802.15.4) SoC for Low-Power and IoT Applications. In Proceedings of the Symposium on VLSI Circuits, Kyoto, Japan, 5–8 June 2017; pp. 74–75.

10. Yang, W.; Hu, D.Y.; Lam, C.K.; Cui, J.Q.; Soh, L.K.; Song, D.C.; Zhong, X.W.; Hor, H.C.; Heng, C.L. A +8 dBm BLE/BT Transceiver with Automatically Calibrated Integrated RF Bandpass Filter and −58 dBc TX HD2. In Proceedings of the IEEE International Solid-State Circuits Conference—Digest of Technical Papers, San Francisco, CA, USA, 11–15 February 2017; pp. 136–137.

11. Paidimarri, A.; Nadeau, P.; Mercier, P.; Chandrakasan, A. A 2.4 GHz Multi-Channel FBAR-based Transmitter with an Integrated Pulse-Shaping Power Amplifier. *IEEE J. Solid State Circuits* **2013**, *48*, 1042–1054. [CrossRef]

12. Kiumarsi, H.; Mizuochi, Y.; Ito, H.; Ishihara, N.; Masu, K. A Three-stage inverter-based stacked power amplifier in 65 nm complementary metal oxide semiconductor process. *Jpn. J. Appl. Phys.* **2012**, *51*, 02BC01. [CrossRef]

13. Van Langevelde, R.; van Elzakker, M.; Van Goor, D.; Termeer, H.; Moss, J.; Davie, A.J. An ultra-low-power 868/915 MHz RF transceiver for wireless sensor network applications. Proceedings of IEEE Radio Frequency Integrated Circuits Symposium, Boston, MA, USA, 7–9 June 2009.

14. Cheng, K.; Chen, S. An ultralow-power wake-up receiver based on direct active RF detection. *IEEE Trans. Circuits Syst. I Reg. Pap.* **2017**, *64*, 1661–1672. [CrossRef]

15. Chen, S.; Yang, C.; Cheng, K. A 4.5 μW 2.4 GHz wake-up receiver based on complementary current-reuse RF detector. In Proceedings of the IEEE International Symposium on Circuits and Systems (ISCAS), Lisbon, Portugal, 24–27 May 2015; pp. 1214–1217.

16. Zhou, Y.; Chia, M. A low-power ultra-wideband CMOS true RMS power detector. *IEEE Trans. Microw. Theory Tech.* **2008**, *56*, 1052–1058. [CrossRef]

17. Prummel, J.; Papamichail, M.; Ancis, M.; Willms, J.; Todi, R.; Aartsen, W.; Kruiskamp, W.; Haanstra, J.; Opbroek, E.; Rievers, S.; et al. A 10 mW bluetooth low-energy transceiver with on-chip matching. *IEEE J. Solid State Circuits* **2015**, *50*, 3077–3088. [CrossRef]

18. Sano, T.; Mizokami, M.; Matsui, H.; Ueda, K.; Shibata, K.; Toyota, K.; Saitou, T.; Sato, H.; Yahagi, K.; Hayashi, Y. A 6.3mW BLE transceiver embedded RX image-rejection filter and TX harmonic-suppression filter reusing on-chip matching network. In Proceedings of the IEEE International Solid-State Circuits Conference—(ISSCC) Digest of Technical Papers, San Francisco, CA, USA, 22–26 February 2015; pp. 240–241.

19. Masuch, J.; Delgado-Restituto, M. A 1.1-mW-RX 81.4 dBm sensitivity CMOS transceiver for bluetooth low energy. *IEEE Trans. Microw. Theory Tech.* **2013**, *61*, 1660–1673. [CrossRef]

2

Resonant Cavity Antennas for 5G Communication Systems

Azita Goudarzi * 🆔, Mohammad Mahdi Honari 🆔 and Rashid Mirzavand 🆔

IWT lab, University of Alberta, Edmonton, AB T6G 2R3, Canada; honarika@ualberta.ca (M.M.H.);
mirzavan@ualberta.ca (R.M.)

* Correspondence: agoudarz@ualberta.ca

Abstract: Resonant cavity antennas (RCAs) are suitable candidates to achieve high-directivity with a low-cost and easy fabrication process. The stable functionality of the RCAs over different frequency bands, as well as, their pattern reconfigurability make them an attractive antenna structure for the next generation wireless communication systems, i.e., fifth generation (5G). The variety of designs and analytical techniques regarding the main radiator and partially reflective surface (PRS) configurations allow dramatic progress and advances in the area of RCAs. Adding different functionalities in a single structure by using additional layers is another appealing feature of the RCA structures, which has opened the various fields of studies toward 5G applications. This paper reviews the recent advances on the RCAs along with the analytical methods, and various capabilities that make them suitable to be used in 5G communication systems. To discuss different capabilities of RCA structures, some applicable fields of studies are followed in different sections of this paper. To indicate different techniques in achieving various capabilities, some recent state-of-the-art designs are demonstrated and investigated. Since wideband high-gain antennas with different functionalities are highly required for the next generation of wireless communication, the main focus of this paper is to discuss primarily the antenna gain and bandwidth. Finally, a brief conclusion is drawn to have a quick overview of the content of this paper.

Keywords: fabry–pérot cavity antenna; fifth generation (5G); millimeter wave spectrum; partially reflective surface; resonant cavity antenna

1. Introduction

Demand of high traffic capacity and speed in wireless communication systems led to the fifth-generation (5G) technologies [1]. The upcoming 5G technologies provide a multitude of advantages including high data rate, high reliability, and low power consumption. More important, it brings new born technologies to have smart cities and factories based on the industry 4.0 [2]. The millimeter-wave (MMW) frequency band has attracted significant attention among academic and industrial sectors, since it has enormous unlicensed bandwidth in comparison with other frequency bands [3]. Thus, MMW band can take an integral role in 5G communication systems. The MMW spectrum brings about compact structures and higher data rate. However, many concerns are remained, which should be addressed in the future communication technologies. One of these concerns is the high cost and complexity of fabrication processes within the MMW band. Another concern is the high energy loss of the MMW spectrum in comparison with the other frequency bands, which can be addressed by increasing the antenna gain.

New research directions have been done to find the effective solutions to address the aforementioned concerns over the MMW frequency band. Different antenna types with a variety of configurations have been proposed to compensate the high loss and propagation issues such as interference of the MMW spectrum. Directional antennas with medium to high gain characteristic are excellent candidates to compensate the high loss of the MMW spectrum compared to the conventional planar antennas. Antenna structures such as reflectors [4] or waveguide horns [5,6] and even array antennas [7] and dielectric lenses [8] are conventional structures, which have potential to achieve high-gain and wideband characteristic. However, they have some issues such as being bulky, heavy and complicated, and having lossy feeding networks, which in turn draws attention to other possible alternative solutions. In fact, designing a proper antenna that meets the requirements of the upcoming 5G communication systems in a simple, efficient, low-profile structure is of prime importance, which has been the subject of many researches.

There are different types of antennas, such as corrugated antennas [9–11], cavity-backed antennas [12,13], and SIW aperture antennas [14,15] that meet the requirement of having a high gain for 5G applications. Recently, much attention has been focused on the design of multi-functional antennas, which have a combination of characteristics in one single structure for different applications. In fact, having a high-gain antenna with features such as beam steering and circular polarization while being multi frequency band, compact, and wideband is desired. Such structures are highly required for the next communication systems, since they lead to lower cost, compact size and even lower power consumption. It is still challenging to obtain multiple functionalities in just one antenna structure.

Recently, resonant cavity antennas have been attracting a growing attention due to their planar configuration, low fabrication difficulty, high-gain characteristic and their capability of integration with other systems. Therefore, the RCAs with the capabilities of reconfigurability, polarization conversion, being wideband or multi-band, while keeping high gain features have been investigated in a variety of studies. This type of antennas, which are a promising candidate to be used in the future 5G communication systems over the MMW frequency band, will be reviewed in this paper. redAn example of their potential is the capability to be used in 5G wireless multiple-input-multiple-output (MIMO) systems due to having a low loss and compact structure with a high-gain characteristic. Besides, RCA structures can achieve steerable radiation patterns with a great radiation performance, which is an essential key for the future 5G base stations and mobile devices. Additionally, the RCAs can offer tilted radiation beam towards a desired direction, which is the demand of mobile communication base stations. Another potential use of the RCAs is in the wireless sensor network (WSN) for future 5G systems due to the requirement of using an efficient antenna with multi-function features to enrich the communication between the nodes [16–20].

1.1. History

RCAs were first introduced in 1956 by Trentini [21], who demonstrated how placing a partially reflective surface (PRS) above a waveguide aperture antenna structure can increase the antenna direactivity, significantly. Since then, further studies have been carried out in this area leading to introducing several PRS configurations with different functionalities in combination with various radiating elements inside the structure. In Reference [22,23], Alexopoulos et showed that using full dielectric PRS layers above the antennas can provide a remarkable directivity improvement. Then, in Reference [24,25], Jackson and Oliner conducted more studies on RCAs with multi-layer dielectric PRS. In Reference [26], James et al. added extra discussion to Terentini study and used a three-layer PRS to increase the gain of an aperture antenna. The conventional RC structures have utilized thick full dielectric or multi-layer PRS structures. By emerging electromagnetic band gap (EBG), metamaterial (MTM), and frequency selective surface (FSS) structures, the trend of studies changed to the design of metallo-dielectric PRS layers to take advantage of fewer layers, thinner layers, more degrees of freedom, and layers with flexible properties.

In the literature, RCAs have associated with different terminologies such as electromagnetic band gap (EBG) [27,28], RC structures [29], 2-D leaky-wave (LW) structures [30,31], PRS [32], and Fabry–Pérot Cavity (FPC) structures [33]. RCA structures consist of a PRS in parallel with a perfect electric/magnetic conductor or an impedance surface, which establish a cavity, fed by a main radiating element inside the cavity to excite the entire structure [21]. Open-ended waveguide, patch antenna, stacked antenna, dielectric resonant antenna (DRA), dipole antenna, and crossed bowtie dipole can be used as the main radiating element inside the cavity. The PRS layer might have different configurations as will be discussed later. It can be a full dielectric structure or a periodic structure composed of an array of metallic unit cells. Due to having a cavity with reflective surfaces, multiple reflections of electromagnetic wave happen. A proper cavity thickness (the distance between the PRS and the ground plane) can superimpose in-phase transmitted waves, which enhances the antenna gain significantly. The phase and magnitude of the PRS reflection behaviour have remarkable impact on the performance of the RCAs in terms of gain, bandwidth, beam angle and aperture efficiency. Therefore, designing the PRS structures has an imperative role in the design of the RCAs to achieve the desired performance.

1.2. Analytical Methods

The design of RCA structures for desired radiation performances, requires an appropriate theoretical analysis. Many studies have been focused on how these structures can be analyzed, which led to a variety of analytical methods such as ray tracing, transmission line (TL), LW, EBG, and the principle of reciprocity methods. Among the mentioned analytical methods, the ray tracing, TL, and LW methods are more used in the literature. In this subsection, a brief review of these three methods is prepared, and a short comparison between them is drown.

RCAs are parallel-plate waveguides leading to the leakage of the ray and known as 2-D periodic leaky-wave antennas (LWAs) [34]. Leaky-wave antennas are considered as antennas with a directive beam, scanning the space as a function of frequency. Leaky-wave antennas are a kind of phased array antennas without phase shifters, which leads to a compact structure with a low energy consumption. There are many studies in the literature in which the functionality of the resonant cavity is discussed by LW method [25,30,31,35–38]. Since comparing to other methods, this method is more efficient and accurate for different configurations of the RCAs; recent advanced studies are carried out using LW models, especially those with steering beam functionalities [34,39–41]. The transverse equivalent network model might be used to derive proper formulas in terms of the angle of the beam, gain, beamwidth, leaky-wave phase and attenuation constants of the structure. The propagation constant is dependent on the PRS reflection coeficient and the distance between the ground plane and the PRS structure placed above the main radiating element.

Another analysis technique for the RCA structures is the ray tracing method. The ray tracing method was first introduced by Terentini [21], in which the RCA is analyzed as a resonant microwave cavity structure known as a Fabry–Pérot cavity. The resonance condition, which is dependent on the phase and magnitude of the reflection coefficient of the PRS needs to be satisfied in order to improve the radiation characteristics of the antenna. According to the resonance condition, the distance between the PRS and the ground plane layers are adjusted to create in-phase superposition of waves leaking out from the structure, which leads to high directive radiation pattern. In the ray tracing model, the diffracted rays are not considered since the structure size is assumed infinite. Consequently, this model gives initial design values, which facilitates the designing process; however, it is not as accurate and general as leaky-wave analyses, due to having some approximations assumed. In fact, this method is more applicable when the goal is to increase the antenna gain and bandwidth, and even to change the antenna polarization [29,42].

The transmission-line (TL) method, as the ray tracing method, gives the initial values of antenna design, making it straightforward and time efficient. Many studies have been carried out to demonstrate the functionality of the resonant cavity structure by the TL model [22,23,30,35,38,43,44]. In [22], the TL analysis is used to derive some formulas related to the bandwidth, gain, and beamwidth of the RCAs. For this purpose, the entire RCA structure is modeled by TLs with different characteristics, and some lumped elements. As a result, the thickness of different parts and the properties of the PRS in terms of refection coefficient are calculated in order to improve radiation performance of the RCA structure. Using this method resulted in better evaluation of the directivity of RCAs [44].

2. Research Directions

The demand for compact high-gain antennas with simple feeding networks has highly increased for the next generation of communication systems. The RCAs have attracted the interest of antenna designers for a variety of study directions. The latest studies on the resonant cavity structures combined with other newly PRSs and main radiating structures demonstrate the flexibility of RCAs to be designed over MMW spectrum. In this section, the study directions of the RCAs with the recent investigations and applicable examples, specially over the MMW frequency band, are reviewed.

2.1. 3-dB Gain Bandwidth Improvement

RCAs are regarded as resonant structures with the drawback of narrow 3-dB gain bandwidth [36,37,45,46]. In Reference [45], the inverse proportionality between the maximum gain and 3-dB gain bandwidth of the RCAs is discussed and proved theoretically by the ray tracing analytical method. Higher gain contributes to narrower bandwidth, which has been considered as one of the concerns that has got the attention of researchers for many years. Consequently, many studies with different methods have been introduced to provide a solution to tackle the mentioned shortcomings of the RCAs. This section deals with presenting different methods carried out to increase the bandwidth of the RCAs.

2.1.1. Positive Reflection Phase Gradient

Inverting the reflection phase gradient of the PRS unit cell to achieve a positive slope is among the most famous and applicable methods in order to obtain wider 3-dB gain bandwidth in the design of the RCAs. References [47,48] are among the first studies conducted to demonstrate the possibility of achieving a positive reflection phase gradient to increase the 3-dB gain bandwidth. Changing the phase gradient behaviour can be achieved by using multi-layer PRS structures [49–56], thick full-dielectric PRSs [57], and thin one-layer metallo-dielectric PRSs [29,48,58–62]. One reason behind using multi-layer PRS structures is creating multiple resonances at different frequencies, which can satisfy the multi resonance conditions over a desired bandwidth. Permitivity and thicknesses of the dielectric slabs, the distances between layers, and other parameters have impact on creating multiple resonant frequencies. It is worth noting that using multi-layer PRSs make the RCAs thicker which might be one of the concerns in some applications.

Several studies have been focused on achieving wider 3-dB gain bandwidth using PRS structures with positive phase slope over millimeter wave spectrum [40,63–65]. As an example, in Reference [65], a wideband high-gain MMW FPCA with the operating frequency of 60 GHz is introduced. A printed ridge-gap waveguide (PRGW) technology is used as the slot antenna feed, because it is a proper candidate to suppress the surface waves with a good functionality over the MMW spectrum. The PRS is composed of gridded square patch (GSP) and square slot-loaded patch (SSLP) structures etched on two different dielectric layers. The configuration of the PRS unit cell, PRGW, and FPCA structures are demonstrated in Figure 1. The wideband characteristic is achieved by using a double-layer PRS unit cell with positive reflection phase gradient as demonstrated in Figure 2a. The "a" is used as a scale factor to control the phase of the unit cell. A maximum gain of 16.8 dBi and 3-dB gain bandwidth of 12.5% are achieved as shown in Figure 2b.

2.1.2. PRS Unit Cell with Sharp Resonance

In the studies reviewed in the previous subsection, it was indicated that having a resonant frequency at the middle of a desired frequency band with a positive phase gradient leads to a wider 3-dB gain bandwidth. In some studies, it is proved that having a sharp resonance at the centre of a frequency band results in 3-dB gain improvement [56,66–68]. The phase variation can get a 180 degree jump at a resonant frequency to achieve a wide 3-dB gain bandwidth.

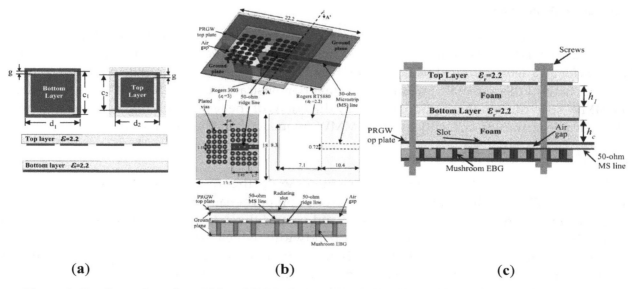

(a)   (b)   (c)

Figure 1. Configuration of a wideband RCA along with a PRS with positive reflection phase: (**a**) Unit cell of PRS; (**b**) Main radiating element (PRGW); (**c**) Entire structure [65].

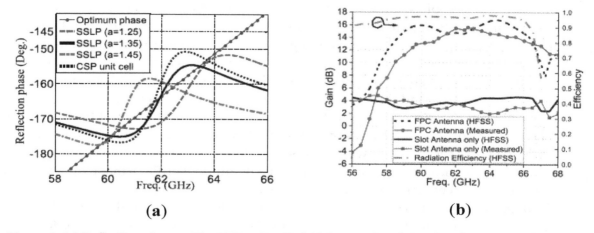

(a)   (b)

Figure 2. (**a**) Reflection phase of the PRS unit cell; (**b**) Measured and simulated gain and efficiency of the RCA in [65].

In Reference [67], as demonstrated in Figure 3, a dual-layer full-dielectric PRS structure with a high permitivity laminate substrate is proposed to provide a sharp resonant frequency. A crossed dipole is used as the main radiator inside the RCA structure which can result in a wide impedance bandwidth and suitable CP characteristic. Figure 4a demonstrates the behaviour of reflection coefficient of the proposed PRS with a sharp resonant frequency. As can be seen, the sharp resonance creates extra frequencies (except the center frequency) to satisfy the resonance conditions which is the reason to make the antenna wideband. The incorporation of the PRS and crossed dipole contributes to a 3-dB gain bandwidth of 50.9% with a maximum gain of 15 dBic, as can be seen in Figure 4b. It should be noted that the optimum results are provided by optimizing the distance between the PRS layers along with their size, and the distance between the ground plane and the PRS layers. The proposed antenna possess simple and compact geometry while providing a high gain and circular polarization, which makes it a suitable candidate to be used in base stations.

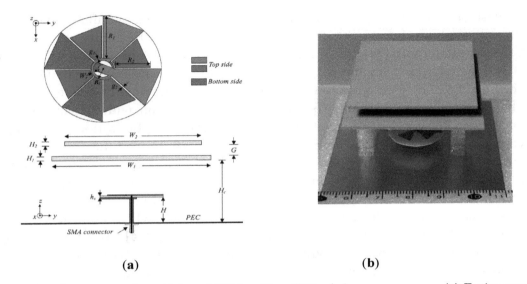

(a)   **(b)**

Figure 3. Configuration of a wideband RCA with a PRS of sharp resonance: (**a**) Entire geometry; (**b**) Fabricated structure [67].

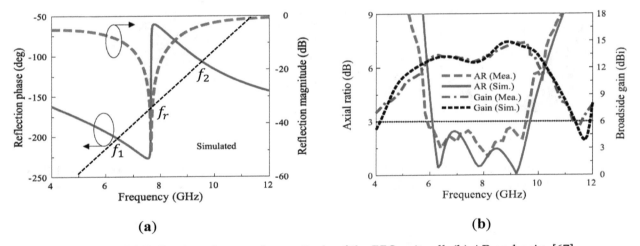

(a)   **(b)**

Figure 4. (**a**) Reflection phase and magnitude of the PRS unit cell; (**b**) AR and gain. [67].

2.1.3. New Configuration of PRS Structures: Nonuniform PRS Structures

Recent non-uniform configurations of superstrates mostly presented by Baba and Hashmi et al., have been designed to provide a significant increment in 3-dB gain bandwidth of the RCAs [28,69–72]. The proposed PRS structures have taken advantage of the integration of different dielectric substrate slabs with different either thickness or permittivity. Basically, such structures are mainly used to compensate the non-uniform phase distribution of the RCA aperture as will be discussed later in the next section. In Reference [69], a planar PRS layer consisted of dielectric slabs with different permitivities is proposed. The proposed PRS named as transverse permitivity gradient (TPG) is a single-layer planar structure with the capability of the aperture phase correction using different sections with different permittivities. Next, Hashmi et al. demonstrated the possibility of a PRS structure composed of multiple dielectric slabs with different permitivity and thickness in order to improve the 3-dB gain bandwidth of a RCA [28,70]. They also investigated the PRS structures with stepped configurations and indicated how these stepped configurations can increase the antenna gain over a wide bandwidth. In Reference [71], the PRS structure is a stepped configuration with laminate substrates of different permitivity and thickness, whereas in [72], the PRS is the similar stepped configuration with just different thicknesses. A slot radiator fed by a waveguide is used as the main radiator for these structures.

Recent investigations with stepped configurations have been done in millimetre wave frequency band to increase the antenna bandwidth [32,73,74], which shows the flexibility of RCAs for different frequency bands. In Reference [74], the application of two different non-uniform PRS structures to enhance the 3-dB gain bandwidth and maximum gain of the RCA are investigated over the MMW spectrum. The first PRS is composed of a four concentric full dielectric rings of different permitivity and thickness, whereas the second PRS is made of a single laminate substrate with the same permitivity and different thicknesses. Both of the PRS structures have an stepped configuration, and an open-ended WR-15 waveguide is used as the main radiator inside the cavity to feed the antenna. The antenna structure without PRSs and PRS prototypes are demonstrated in Figure 5. The antenna with the second PRS has a maximum measured gain of 19.5 dBi with a proper matching from 55.2 GHz to 65 GHz. The simulated and measured results are displayed in Figure 6. These kinds of non-uniform PRS structures achieve remarkable gain-bandwidth product (GBP), which is a true merit used in the comparison between different antennas. The proposed antenna in [74] has a simple structure with a high gain and low cross-polarization, which makes it beneficial for base stations, point-to-point communication systems, autonomous radars, remote sensing satellites, and Internet of things (IoT).

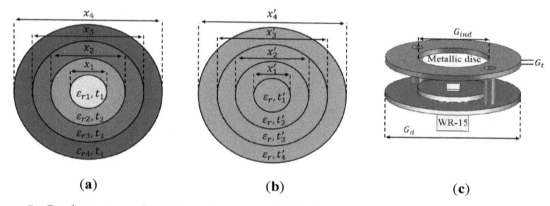

Figure 5. Configuration of a RCA with stepped PRS: (**a**) First PRS configuration; (**b**) Second PRS configuration; (**c**) Entire structure [74].

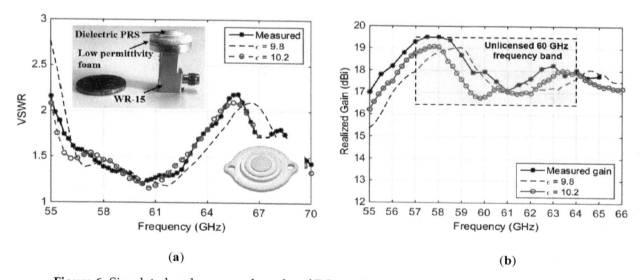

Figure 6. Simulated and measured results of RCA with stepped PRS: (**a**) VSWR; (**b**) Gain [74].

2.1.4. Shape Manipulation of the Conventional RCA Configuration

The demand for wideband high-gain antennas have led researchers to seek different and novel methods to efficiently enhance the antenna bandwidth without sacrificing the antenna performance. It is proved that the performance of FPCAs can be improved by curving the ground plane or PRS architecture. In Reference [75–77], manipulating the configuration of the ground plane and PRS structures so that the distance between the PRS layer and the ground plane gradually gets unequal values for different parts of the RCA structures are considered by different methods. These manipulated structures are capable to compensate and correct the phase and magnitude distribution far from the center of the PRS structure that results in a broader 3-dB gain bandwidth. In Reference [75], a shaped ground plane with a semi-spherical configuration is used to make the 3-dB gain bandwidth wider. The configuration of the proposed RCA structure is shown in Figure 7a. The antenna performance is compared with the performance of a conventional RCA with a flat ground plane, and the results are shown in Figure 7b. The RCA structure reported in [75] provides a measured 3-dB gain bandwidth of 25% with a maximum gain of 17.7 dBi as demonstrated in Figure 8. Similar works have been done for MMW spectrum by using unconventional RCA structures [78,79].

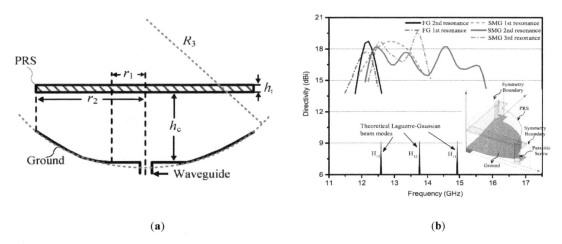

(a)   (b)

Figure 7. (a) RCA with a modified ground plane; (b) Directivity [75].

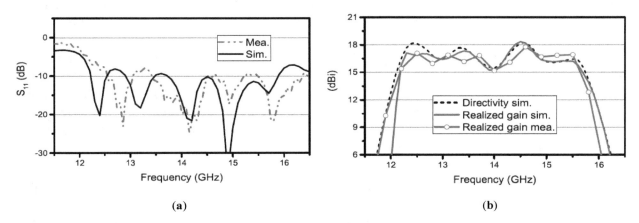

(a)   (b)

Figure 8. Performance of the RCA with modified ground plane: (a) Reflection coefficient; (b) Realized gain [75].

2.1.5. Array Feed

Using array antennas instead of single main radiator inside the cavity structure is another conventional method to increase the 3-dB gain bandwidth. This idea has been investigated in many studies [27,33,43,80], which mostly used complicated feeding networks. In Reference [43], an array of patch elements are used as the radiators. Besides, by adding two PRS layers above the array antennas, the radiation performance of the antenna is improved. The antenna configuration and the results of the antenna gain are demonstrated in Figure 9. As can be seen, the maximum gain of a 2 × 2 array patch antenna without any PRS layer is almost the same as a RCA with a single patch as the main radiator. Similarly, the results are same for a 4×4 patch array without PRS and a RCA with a 2×2 patch array as the main radiator.

Although using array source inside the cavity results in reasonable improvement of the gain and 3-dB gain bandwidth, it is not a good idea for millimeter wave spectrum. Difficult fabrication process and high loss caused by feed networks are two main issues that reduce the tendency to use array structures at those frequencies.

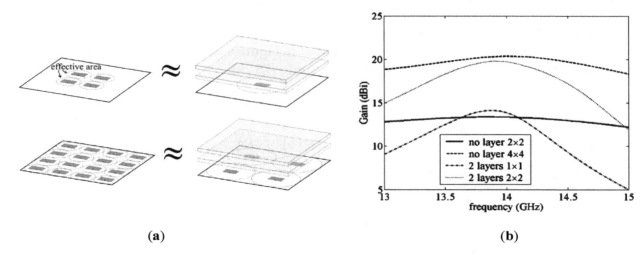

(a)   (b)

Figure 9. RCA with an array source radiator: (**a**) Schematic configuration; (**b**) Gain of the antenna [43].

2.2. Directivity Enhancement: Aperture Efficiency Improvement

There is another group of studies which have been conducted to compensate the non-uniform aperture magnitude and phase distribution of the conventional RCA structure [66,81–88] and as a result to enhance the antenna directivity. Conventional and classical structures of the RCAs have non-uniform aperture phase and magnitude distribution, which result in the reduction of the antenna radiation performance. Compensating the non-uniform field distribution of a RCA aperture is an effective attempt to increase the antenna directivity and decrease side lobe level. Some of the studies have focused on the phase compensation [66,81–85], whereas the others have investigated the aperture magnitude compensation [86,87] for enhancing the antenna performance. In a few studies such as [88], both phase and magnitude are compensated to create remarkable aperture efficiency. Using a non-uniform PRS structure is the most common method to compensate the non-uniform magnitude and phase distribution of electric filed over the antenna aperture. Using phase compensation method is more applicable, because it leads to greater improvement in comparison with the magnitude compensation.

In Reference [82], an stepped superstrate is used to compensate the non-uniform aperture phase distribution of a RCA, which leads to a higher directivity and aperture efficiency. The structure is a phase rectifying transparent superstrate (PRTS), which was fabricated by 3-D printing technology and the main radiating element is a microstrip patch antenna with a full-dielectric PRS structure illustrated in Figure 10. Figure 11 shows the phase distribution and the directivity of the antenna, before and after applying PRTS structure. As can be seen, the phase distribution becomes almost uniform all over the aperture, since the PRTS placed above the PRS compensates the phase delay. According to the Figure 11c, the directivity gets a significant improvement when that the PRTS is placed above the entire RCA structure.

Additionally, a few other methods, such as using different geometries and sizes for the PRS structures located at different distances from the center of the aperture, have been used in the literature to improve the gain and aperture efficiency of the RCAs [79,89,90].

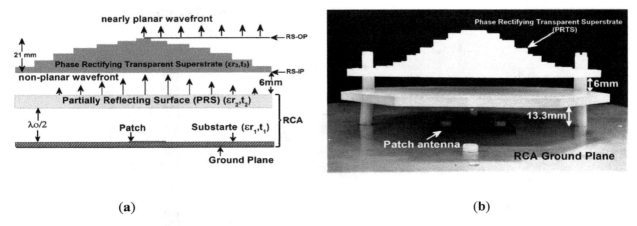

(a)   (b)

Figure 10. Configuration of a RCA with an aperture phase distribution compensation: (**a**) Schematic; (**b**) Fabricated prototype [82].

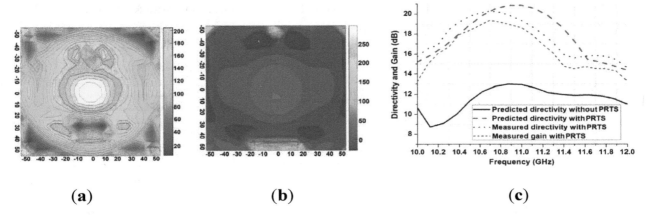

(a)   (b)   (c)

Figure 11. Results of the RCA with the aperture phase correction: (**a**) Aperture phase distribution after correction; (**b**) Aperture phase distribution before correction; (**c**) Antenna directivity [82].

2.3. Circular Polarization

High-gain wideband circularly-polarized (CP) antennas are required in many applications such as radar, satellite, and mobile communication systems. In comparison with LP, CP antennas take advantage of lower the interference in multipath environments. Besides, CP antennas are independent of polarization between transmitters and receivers. Having RCAs with high-gain, wideband, and circular polarization has been a challenge which has been addressed in many studies [67,68,91–104].

Usually two different methods are used to achieve a high-gain CP RCA. The first one is to use a CP main radiator element and the radiation improvement of the entire structure is obtained through utilizing a PRS structure whose behaviour is independent of the polarization [67,68,91–97].

The other method is known as self-polarizing RCA. In this method, the linear-to-circular polarization conversion is the subject that is studied. A linearly polarized (LP) antenna is used as the main radiating feed and its polarization is converted to circular through using a proper PRS placed above the RCA [98–104]. This method is preferred, since it does not need any feeding networks to make circular polarization; however it might not obtain a wide bandwidth.

Many studies have been reported for designing a high-gain CP RCA over the MMW spectrum in the literature [42,105–107]. Some of them present a practical application of the CP RCAs for the 5G communication systems. In Reference [42], a CP high-gain RCA is designed for 5G multiple-input multiple output (MIMO) applications over 26 GHz to 31 GHz. A single-layer full dielectric PRS structure with a sharp resonant frequency is used to enhance the antenna radiation performance over a wide frequency band. A CP truncated patch antenna with a proper slot is placed inside the cavity to illuminate the entire structure as demonstrated in Figure 12. Figure 13 exhibits a 3-dB AR bandwidth of 17.5% with a maximum gain of 14.1 dBic achieved by the structure.

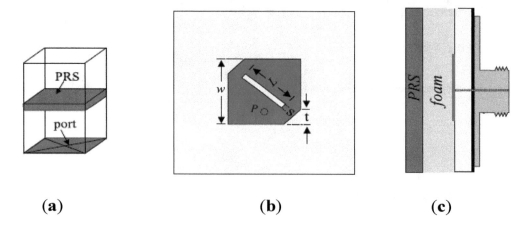

(a)   **(b)**   **(c)**

Figure 12. Configuration of a CP RCA: (**a**) Unit cell; (**b**) Main radiator; (**c**) Entire RCA structure [42].

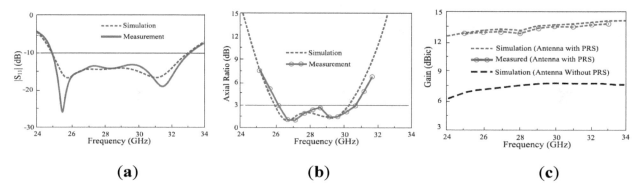

(a)   **(b)**   **(c)**

Figure 13. Measured and simulated results of the CP RCA reported in [42]: (**a**) Reflection coefficient; (**b**) AR; (**c**) Gain.

2.4. Reconfigurable RCAs

Reconfigurability of the RCAs is the capability of their structures to alter their radiation features by electrical elements or mechanical mechanism. A reconfigurable antenna might use different methods to alter operation frequency, radiation pattern, polarization, beamwidth, or even a combination of these variables.

Multi functional antennas are proper solution for newborn 5G millimeter wave frequency band, where the large size of the systems and the extra equipment to provide other functions are considered as potential problems. Reconfigurable antennas can offer other functionalities over multiple bands and can be efficient in variable environments in case they face a limitation or a new situation. Consequently, reconfigurable RCAs, due to their simple feed network and high-gain characteristic, can be considered as an efficient and low-cost option to mitigate the significant challenges in 5G applications. There are many studies in the literature concentrated on reconfigurable RCA structures. These studies investigate frequency [108–112], beamwidth [113–115], and polarization [116–127], and pattern reconfigurability [128–139].

Polarization reconfigurability can be achieved by using either a reconfigurable main radiating element [116–122] or a dependent polarization PRS [123–127]. In Reference [140], a four-polarization-reconfigurable aperture-coupled patch antenna is designed as the main radiator of a RCA structure, which works with two different linear polarizations. The PRS structure is designed with adjustable reflection phase consisting of four PIN diodes to steer the RCA beam. The entire RCA configuration is shown in Figure 14. By a proper arrangement of the main radiating elements and PRS structure, a dual-polarized 2-D beam-steering functionality is achieved as shown in Figure 15.

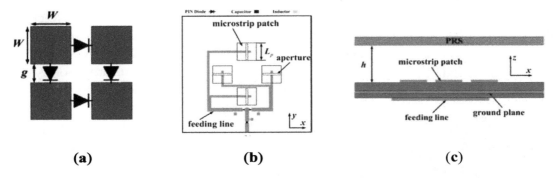

(a)   **(b)**   **(c)**

Figure 14. Configuration of a RCA with dual-reconfigurability feature: (**a**) Unit cell; (**b**) Main radiator; (**c**) Entire RCA structure [140].

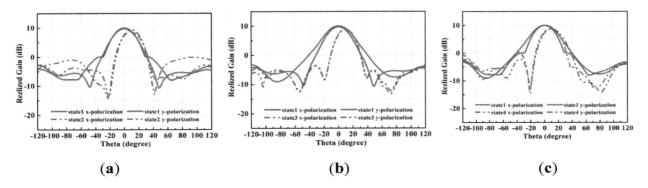

(a)   **(b)**   **(c)**

Figure 15. Radiation patterns of the RCA structure reported in [140] at three different planes: (**a**) 0°; (**b**) 45°; (**c**) 90°.

Beam steering property is one of the demands of the new generation of wireless communications that enables the wide coverage. Using phased array antennas is the conventional solution for the communication coverage, in which the beam steering is realized by carefully adjusting the phase difference between the array elements. However, conventional phased array antennas suffer from high-cost fabrication process and complicated feeding networks with significant loss.

A variety of investigations have been carried out to demonstrate the capability of the RCAs to achieve reconfigured pattern in the millimeter wave frequency band. Many studies proposed that the main radiator placed inside the RCA structure takes the responsibility of the pattern reconfiurability [141–143] and the antenna radiation performance can be improved by using PRS structures. Other studies have proposed suitable PRS structures to manipulate the RCA radiation pattern instead of using a fed antenna with complicated feed networks [128–139]. PRS structures provide more degree of freedom to control the beam of the main radiator with other functionalities simultaneously without applying extra equipment. Therefore, many RCA structures have been proposed with either both pattern and polarization reconfigurability or pattern and beamwidth reconfigurability [115,140,144–146].

In References [139,146], the possibility of using RCA with reconfigurable pattern characteristic for 5G is illustrated. In Reference [139], the beam tilting of a RCA is investigated through four different techniques at 60 GHz. The techniques used for tilting the beam include wedge-shaped dielectric lens (WSDL), discrete multilevel grating dielectric (DMGD), printed gradient surface (PGS), and perforated dielectric gradient surface (PDGS), which are based on the phase gradient surface method. Among the mentioned techniques, PGS and PDGS have better performance inorder to provide reconfigurability feature. Figure 16a shows a printed ridge gap waveguide (PRGW) used as the main radiating element in [139]. Six different structures of the PGS and PDGS were proposed and their functionality is investigated while they are placed above the main radiator. Three designs #1, #2, #3 for PGS and PDGS structures were done for different tilt angles. A maximum gain of around 22 dBi is achieved when these structures are used to tilt the beam of the RCA. The simulated and measured results achieved by using these six superstrates are illustrated in Figure 17. As shown, the tilted beam angles of $\theta = 14^0$, $\theta = 27^0$ and $\theta = 44^0$ are achieved by placing every design of PGS or PDGS above the main radiator. The proposed antenna in [139] is a potential candidate for narrow-band communication systems. It can be used in mobile devices due to its compact configuration with reconfigurable radiation pattern.

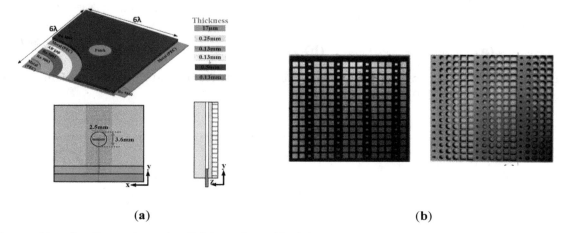

(a)   (b)

Figure 16. Configuration of a RCA with a tilted beam: (a) Main radiating element (PRGW); (b) Fabricated prototype of PGS and PDGS [139].

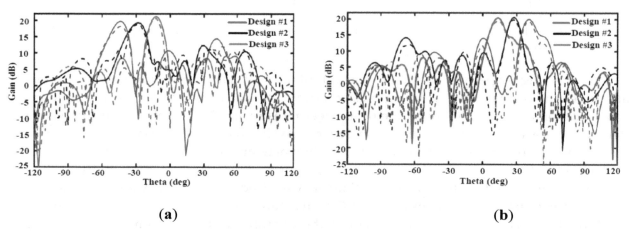

(a)   (b)

Figure 17. Simulated and measured results of the radiation patterns of the RCA in [139]: (a) PGS; (b) PDGSs configuration.

2.5. Further Fields of Study: Low Profile and Multi-Band RCAs

Reducing the total size of the RCAs, especially, the cavity height makes them more appealing for the future communication systems. Many works have been done to reduce the profile of RCAs [52,100,101,147–157]. Minimizing the cavity height so that the antenna performance does not have any degradation has been challenging. Based on the analytical models, the height of RCA structure is dependent on the sum of the reflection phase of the ground plane and PRS layer, which is approximately half of the wavelength. Consequently, a ground plane with zero reflection phase, which can be realized by employing high impedance surfaces (HIS) called artificial magnetic conductors (AMC) can minimize the cavity thickness. A compact multi-band high-gain RCA is presented in [156] using a FSS structure as ground plane with zero reflection phase. In some cases, using a ground plane with an associate reflection phase opposite to the PRS reflection phase sign shrinks the thickness of the cavity a lot [150]. In Reference [153], a miniaturized-element frequency selective surface (MEFSS) cover consisted of a PRS and HIS is proposed to improve the radiation gain of the antenna in chip while having a small cavity height. The MEFSS cover is designed to be placed on the chip to increase the antenna radiation performance in RFICs. As shown in Figure 18, the antenna radiation gain can be increased by 9 dB by placing the MEFSS cover on top of the antenna, while the cover is designed to have a very small cavity height by designing the PRS and HIS layers. A maximum gain of 14 dBi with a cavity height of $\lambda/30$ is achieved through using the proposed MEFSS. This antenna is applicable for radar and communication systems, as well as wireless sensor networks for the future 5G systems.

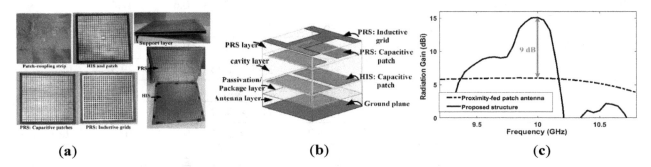

Figure 18. Configuration of the RCA in [153]: (**a**) Fabricated MEFSS cover; (**b**) MEFSS unit cell; (**c**) Radiation gain of the main radiator with and without MEFSS cover.

Multi-band high-gain antennas are highly required for the next communication systems. The capability of the RCA structures to possess the multi-band characteristic has been extensively investigated in the literature [156,158–163]. Generally, creating different resonances that satisfy the resonance conditions is the way to design multi-band RCAs. In Reference [159], a dual-band high-gain RCA with two vertical and horizontal polarizations is proposed. A dual-feed microstrip patch antenna is used as the main radiator, which has dual-band and dual-polarization characteristics. A PRS composed of double-layer orthogonal dipole arrays to create V-pol and H-pol is placed above the main radiator. The RPS unit cell is designed so that the resonance condition is satisfied over two different frequency bands with orthogonal polarizations in order to improve the radiation characteristic of the RCA. The configuration of the main radiator and PRS are shown in Figure 19. The lower band and upper band have a maximum gain of 19.6 dBi and 18 dBi at 10 GHz and 11.6 GHz, respectively, as shown in Figure 20. The polarization of the RCA is vertical and horizontal at lower and upper frequency bands, respectively.

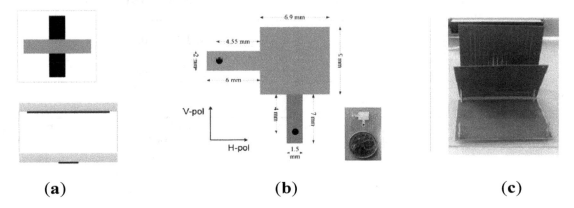

Figure 19. Configuration of the design in [159]: (**a**) Unit cell; (**b**) Main radiator; (**c**) Entire RCA structure.

Finaly, Table 1 lists some of the RCAs discussed in this paper to comprehensively summarize different works. The performance of the antennas in terms of maximum gain, overlapped bandwidth, operating frequency, polarization, reconfigurability, entire size, the height of cavity, and the number of PRSs used is demonstrated.

Table 1. Comparison between different RCAs.

Ant	Max-gain (dB/dBic)	Pol.	Reconfigurability	f_{min} (GHz)	Size $(\lambda_{min})^2$	H (λ_{min})	BW (%)	PRS Layer (#)
[58]	17.2	LP	-	8.9	2.37 × 2.37	0.52	25.5	1
[61]	12.5	CP	-	6.5	1.56 × 1.56	0.44	35.5	1
[67]	15	CP	-	6	1.2 × 1.2	0.6	47.4	2
[68]	13.5	CP	-	6	2 × 2	0.7	73.7	3
[93]	15.5	CP	-	6.3	1.7 × 1.7	0.46	25.6	1
[96]	11.45	CP	-	12.4	1.45 × 1.45	0.4	29.3	1
[72]	19.3	LP	-	10	2.5 × 2.5	0.4	51.8	stair-case
[65]	16.8	LP	-	58.6	4.33 × 3.52	0.5	12.5	2
[74]	19.5	LP	-	55.2	3.11 × 3.11	0.55	16.3	stair-case
[75]	17.7	LP	-	12.2	4.32 × 4.32	1	25	1
[82]	19	LP	-	10.3	3.95 × 3.95	0.5	9.4	one flat and one stair-case
[42]	14.1	CP	-	26	1.58 × 1.58	0.7	18.5	1
[140]	9.7	LP	• pattern • polarization	5.4	1.8 × 1.8	0.5	3.7	1
[109]	10	LP	• frequency	5.2	4.16 × 4.16	0.4	13.4	1
[111]	14.1	LP	• frequency	9.05	2.74 × 2.74	0.53	9.97	1
[114]	- - -	LP	• beamwidth	1.985	3.3 × 3.3	0.5	5	1
[125]	14.1	LP/CP	• polarization	9.6	2.83 × 2.83	1.06	6.7	3
[139]	22	LP	• pattern	59	5.9 × 5.9	0.5	3.4	2
[115]	14.7	LP	• pattern • beamwidth	1.97	2.26 × 2.26	0.51	3	1
[145]	8	LP	• pattern • beamwidth	4.9	0.98 × 1.1	0.1	4	1
[153]	14	LP	- - -	9.94	3.2 × 3.2	0.03	2.97	1
[159]	19.6 / 18	LP(V) / LP(H)	- - -	9.8 / 11.4	5 × 5 / 5.85 × 5.85	0.5 / 0.6	3 / 2.56	2

f_{min} is the minimum operation frequency. 'H' is the height of the cavity which is the distance between the ground plane and the first layer of superstrates. BW is overlapped bandwidths of 3-dB AR, 3-dB gain, and $|S_{11}|$.

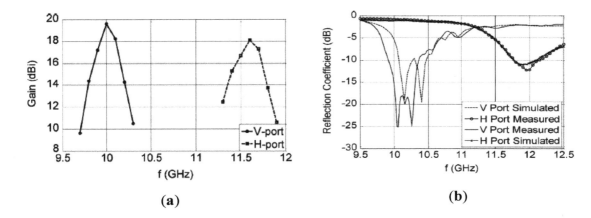

Figure 20. Results of the RCA in [159]: (**a**) Gain; (**b**) Reflection coefficient.

3. Conclusions

Since there is a high interest both in academia and industry for the high-gain antennas, this paper provides a comprehensive discussion on the RCAs in terms of their challenges, applications, and research trend. The RCAs owing to their advantages such as simple feed structure, planar configuration, high-gain, low-cost fabrication, and ease of integration with other systems are promising candidates to be used in the next communication systems.

The mechanism of the RCAs has been presented by different techniques; among them are ray tracing, TL, and LW models. The research fields of the RCAs were briefly discussed through demonstrating the practical examples and recent studies covering their low-profile, high aperture efficiency, wide 3-dB bandwidth, beam-steering and CP features and multi-beam and reconfigurability capabilities, especially in the millimeter-wave frequencies for future developments.

Having wider 3-dB gain bandwidth is still challenging in the implementation of the RCAs. Efficient designs and techniques to make wider 3-dB gain bandwidth including using PRS unit cells with positive reflection phase gradient, sharp resonance, and non-uniform configurations were reviewed. Besides, the possibility to generate a wider 3-dB gain bandwidth by manipulating the configuration of the ground plane and PRS structures was briefly discussed.

Multi functional antennas are the desired solution for the 5G millimeter wave frequencies to avoid no extra equipment, which lead to bulky structures. Reconfigurability of RCAs offer altering operation frequency, polarization, radiation pattern, and beamwidth to address the potential problems of the next communication systems. Beam-steering and CP characteristics add more flexibility to the antennas to tolerate the environment issues and to have stable performance, as explained in this paper.

Multi-band and low-profile RCAs are attractive topics covered in this review paper. In order to decrease the cavity height, decreasing the sum value of reflection phase of the PRS unit cell and ground plane is required, which can be obtained by using AMC or HIS structures as the ground plane. It is explained that to design a multi-band RCA, it is required to create different resonances that satisfy the resonance condition, simultaneously.

In summary, it is the stable functionality and stunning performance of the RCAs over different frequency bands that make them attractive for the next generation wireless communication systems, i.e., fifth generation (5G).

Author Contributions: A.G. and M.M.H. prepared the first draft of the paper. A.G., M.M.H. and R.M. discussed the presentation development of the manuscript. A.G., M.M.H. and R.M. reviewed the manuscript. All authors have read and agreed to the published version of the manuscript

Abbreviations

The following abbreviations are used in this manuscript:

5G	Fifth-Generation
MMW	Millimeter-Wave
RC	Resonant Cavity
PRS	Partially Reflective Surface
FPCA	Fabry–Pérot Cavity Antenna
EBG	Electromagnetic-Band-gap
LW	Leaky- Wave
MTM	Metamaterial
FSS	Frequency Selective Surface
TPG	Transverse Permitivity Gradient
GBP	Gain-Bandwidth Product
FZP	Fresnel Zone Plate
CP	Circularly-Polarized
HIS	High Impedance Surfaces
AMC	Artificial Magnetic Conductors
FSS	Frequency Selective Surface
PRGW	Printed Ridge-Gap Waveguide
GSP	Gridded Square Patch
SSLP	Square Slot-Loaded Patch
WSDL	Wedge-Shaped Dielectric Lens
DMGD	Discrete Multilevel Grating Dielectric
PGS	Printed Gradient Surface
PDGS	Perforated Dielectric Gradient Surface
CM	Characteristic Mode
MEFSS	Miniaturized-Element Frequency Selective Surface

References

1. Thompson, J.; Ge, X.; Wu, H.C.; Irmer, R.; Jiang, H.; Fettweis, G.; Alamouti, S. 5G wireless communication systems: Prospects and challenges [Guest Editorial]. *IEEE Commun. Mag.* **2014**, *52*, 62–64. [CrossRef]
2. Rappaport, T.S.; Sun, S.; Mayzus, R.; Zhao, H.; Azar, Y.; Wang, K.; Wong, G.N.; Schulz, J.K.; Samimi, M.; Gutierrez, F. Millimeter wave mobile communications for 5G cellular: It will work! *IEEE Access* **2013**, *1*, 335–349. [CrossRef]
3. Honari, M.M.; Mirzavand, R.; Melzer, J.; Mousavi, P. A new aperture antenna using substrate integrated waveguide corrugated structures for 5G applications. *IEEE Antennas Wirel. Propag. Lett.* **2016**, *16*, 254–257. [CrossRef]
4. Mahajan, M.; Jyoti, R.; Sood, K.; Sharma, S.B. A Method of Generating Simultaneous Contoured and Pencil Beams From Single Shaped Reflector Antenna. *IEEE Trans. Antennas Propag.* **2013**, *61*, 5297–5301. [CrossRef]
5. Deguchi, H.; Tsuji, M.; Shigesawa, H. Compact low-cross-polarization horn antennas with serpentine-shaped taper. *IEEE Trans. Antennas Propag.* **2004**, *52*, 2510–2516. [CrossRef]
6. Honari, M.M.; Abdipour, A.; Moradi, G. Aperture-coupled multi-layer broadband ring-patch antenna array. *IEICE Electron. Expr.* **2012**, *9*, 250–255. [CrossRef]
7. Cheng, Y.J.; Guo, Y.X.; Liu, Z.G. W-Band Large-Scale High-Gain Planar Integrated Antenna Array. *IEEE Trans. Antennas Propag.* **2014**, *62*, 3370–3373. [CrossRef]
8. Fan, C.; Yang, W.; Che, W.; He, S.; Xue, Q. A Wideband and Low-Profile Discrete Dielectric Lens Using 3-D Printing Technology. *IEEE Trans. Antennas Propag.* **2018**, *66*, 5160–5169. [CrossRef]
9. Honari, M.M.; Mirzavand, R.; Mousavi, P. A high-gain planar surface plasmon wave antenna based on substrate integrated waveguide technology with size reduction. *IEEE Trans. Antennas Propag.* **2018**, *66*, 2605–2609. [CrossRef]
10. Honari, M.M.; Sarabandi, K.; Mousavi, P. Dual-Band High-Gain Planar Corrugated Antennas with Integrated Feeding Structure. *IEEE Access* **2020**, *8*, 67075–67084. [CrossRef]

11. Honari, M.M.; Sarabandi, K.; Mousavi, P. Design and Analysis of Corrugated Antennas Based on Surface Susceptance of a Single Cell of Corrugation. *IEEE Trans. Antennas Propag.* **2020**. [CrossRef]
12. Amjadi, S.M.; Sarabandi, K. A low-profile, high-gain, and full-band subarray of cavity-backed slot antenna. *IEEE Trans. Antennas Propag.* **2017**, *65*, 3456–3464. [CrossRef]
13. Guan, D.F.; Ding, C.; Qian, Z.P.; Zhang, Y.S.; Guo, Y.J.; Gong, K. Broadband high-gain SIW cavity-backed circular-polarized array antenna. *IEEE Trans. Antennas Propag.* **2016**, *64*, 1493–1497. [CrossRef]
14. Yang, D.; Cao, F.; Pan, J. A single-layer dual-frequency shared-aperture SIW slot antenna array with a small frequency ratio. *IEEE Antennas Wirel. Propag. Lett.* **2018**, *17*, 1048–1051. [CrossRef]
15. Honari, M.M.; Mirzavand, R.; Saghlatoon, H.; Mousavi, P. A dual-band low-profile aperture antenna with substrate-integrated waveguide grooves. *IEEE Trans. Antennas Propag.* **2016**, *64*, 1561–1566. [CrossRef]
16. Mirzavand, R.; Honari, M.M.; Mousavi, P. Direct-Conversion Sensor for Wireless Sensing Networks. *IEEE Trans. Ind. Electron.* **2017**, *64*, 9675–9682. [CrossRef]
17. Saghlatoon, H.; Mirzavand, R.; Honari, M.M.; Mousavi, P. Sensor Antenna Transmitter System for Material Detection in Wireless-Sensor-Node Applications. *IEEE Sens. J.* **2018**, *18*, 8812–8819. [CrossRef]
18. Khalid, N.; Mirzavand, R.; Saghlatoon, H.; Honari, M.M.; Mousavi, P. A Three-Port Zero-Power RFID Sensor Architecture for IoT Applications. *IEEE Access* **2020**, *8*, 66888–66897. [CrossRef]
19. Mirzavand, R.; Honari, M.M.; Laribi, B.; Khorshidi, B.; Sadrzadeh, M.; Mousavi, P. An unpowered sensor node for real-time water quality assessment (humic acid detection). *Electronics* **2018**, *7*, 231. [CrossRef]
20. Mirzavand, R.; Honari, M.M.; Mousavi, P. High-resolution dielectric sensor based on injection-locked oscillators. *IEEE Sens. J.* **2017**, *18*, 141–148. [CrossRef]
21. Trentini, G.V. Partially reflecting sheet arrays. *IRE Trans. Antennas Propag.* **1956**, *4*, 666–671. [CrossRef]
22. Jackson, D.; Alexopoulos, N. Gain enhancement methods for printed circuit antennas. *IEEE Trans. Antennas Propag.* **1985**, *33*, 976–987. [CrossRef]
23. Yang, H.; Alexopoulos, N. Gain enhancement methods for printed circuit antennas through multiple superstrates. *IEEE Trans. Antennas Propag.* **1987**, *35*, 860–863. [CrossRef]
24. Jackson, D.R.; Oliner, A.A. A leaky-wave analysis of the high-gain printed antenna configuration. *IEEE Trans. Antennas Propag.* **1988**, *36*, 905–910. [CrossRef]
25. Jackson, D.; Oliner, A.; Ip, A. Leaky-wave propagation and radiation for a narrow-beam multiple-layer dielectric structure. *IEEE Trans. Antennas Propag.* **1993**, *41*, 344–348. [CrossRef]
26. James, J.; Kinany, S.; Peel, P.; Andrasic, G. Leaky-wave multiple dichroic beamformers. *Electron. Lett.* **1989**, *25*, 1209–1211. [CrossRef]
27. Leger, L.; Monediere, T.; Jecko, B. Enhancement of gain and radiation bandwidth for a planar 1-D EBG antenna. *IEEE Microw. Wirel. Compon. Lett.* **2005**, *15*, 573–575. [CrossRef]
28. Hashmi, R.M.; Esselle, K.P. A wideband EBG resonator antenna with an extremely small footprint area. *Microw. Opt. Technol. Lett.* **2015**, *57*, 1531–1535. [CrossRef]
29. Meng, F.; Sharma, S.K. A Wideband Resonant Cavity Antenna with Compact Partially Reflective Surface. *IEEE Trans. Antennas Propag.* **2020**, *68*, 1155–1160. [CrossRef]
30. Zhao, T.; Jackson, D.R.; Williams, J.T.; Yang, H.Y.; Oliner, A.A. 2-D periodic leaky-wave antennas-part I: Metal patch design. *IEEE Trans. Antennas Propag.* **2005**, *53*, 3505–3514. [CrossRef]
31. Zhao, T.; Jackson, D.R.; Williams, J.T. 2-D periodic leaky-wave Antennas-part II: Slot design. *IEEE Trans. Antennas Propag.* **2005**, *53*, 3515–3524. [CrossRef]
32. Baba, A.A.; Hashmi, R.M.; Esselle, K.P.; Marin, J.G.; Hesselbarth, J. Broadband Partially Reflecting Superstrate-Based Antenna for 60 GHz Applications. *IEEE Trans. Antennas Propag.* **2019**, *67*, 4854–4859. [CrossRef]
33. Vaidya, A.R.; Gupta, R.K.; Mishra, S.K.; Mukherjee, J. High-gain low side lobe level Fabry Perot cavity antenna with feed patch array. *Prog. Electromagn. Res. C* **2012**, *28*, 223–238. [CrossRef]
34. Sengupta, S.; Jackson, D.R.; Almutawa, A.T.; Kazemi, H.; Capolino, F.; Long, S.A. A Cross-Shaped 2D Periodic Leaky-Wave Antenna. *IEEE Trans. Antennas Propag.* **2019**. [CrossRef]
35. Zhao, T.; Jackson, D.R.; Williams, J.T.; Oliner, A.A. General formulas for 2-D leaky-wave antennas. *IEEE Trans. Antennas Propag.* **2005**, *53*, 3525–3533. [CrossRef]
36. Lovat, G.; Burghignoli, P.; Capolino, F.; Jackson, D. Highly-directive planar leaky-wave antennas: A comparison between metamaterial-based and conventional designs. *Proc. EuMA* **2006**, *2*, 12–21.

37. Lovat, G.; Burghignoli, P.; Jackson, D.R. Fundamental properties and optimization of broadside radiation from uniform leaky-wave antennas. *IEEE Trans. Antennas Propag.* **2006**, *54*, 1442–1452. [CrossRef]

38. Foroozesh, A.; Shafai, L. Investigation into the effects of the patch-type FSS superstrate on the high-gain cavity resonance antenna design. *IEEE Trans. Antennas Propag.* **2009**, *58*, 258–270. [CrossRef]

39. Sengupta, S.; Jackson, D.; Long, S. Modal Analysis and Propagation Characteristics of Leaky Waves on a 2-D Periodic Leaky-Wave Antenna. *IEEE Trans. Microw. Theory Tech.* **2018**, 1–11. [CrossRef]

40. Almutawa, A.T.; Hosseini, A.; Jackson, D.R.; Capolino, F. Leaky-Wave Analysis of Wideband Planar Fabry–Pérot Cavity Antennas Formed by a Thick PRS. *IEEE Trans. Antennas Propag.* **2019**, *67*, 5163–5175. [CrossRef]

41. Almutawa, A.T.; Capolino, F.; Jackson, D.R. Overview of Wideband Fabry-Perot Cavity Antennas with Thick Partially Reflective Surface. In Proceedings of the 2019 International Conference on Electromagnetics in Advanced Applications (ICEAA), Granada, Spain, 9–13 September 2019; pp. 1308–1310.

42. Hussain, N.; Jeong, M.; Park, J.; Kim, N. A Broadband Circularly Polarized Fabry-Perot Resonant Antenna Using A Single-Layered PRS for 5G MIMO Applications. *IEEE Access* **2019**, *7*, 42897–42907. [CrossRef]

43. Gardelli, R.; Albani, M.; Capolino, F. Array thinning by using antennas in a Fabry–Perot cavity for gain enhancement. *IEEE Trans. Antennas Propag.* **2006**, *54*, 1979–1990. [CrossRef]

44. Boutayeb, H.; Denidni, T.A. Internally excited Fabry-Perot type cavity: Power normalization and directivity evaluation. *IEEE Antennas Wirel. Propag. Lett.* **2006**, *5*, 159–162. [CrossRef]

45. Liu, Z.G.; Guo, Y.X. Effect of primary source location on fabry-perot resonator antenna with PEC or PMC ground plate. *J. Infrared Millim. Terahertz Waves* **2010**, *31*, 1022–1031. [CrossRef]

46. Hashmi, R.M.; Esselle, K.P. Enhancing the performance of EBG resonator antennas by individually truncating the superstructure layers. *IET Microw. Antennas Propag.* **2016**, *10*, 1048–1055. [CrossRef]

47. Feresidis, A.P.; Vardaxoglou, J. High gain planar antenna using optimised partially reflective surfaces. *IEE Proc. Microw. Anten. Propag.* **2001**, *148*, 345–350. [CrossRef]

48. Ge, Y.; Esselle, K.P.; Bird, T.S. The use of simple thin partially reflective surfaces with positive reflection phase gradients to design wideband, low-profile EBG resonator antennas. *IEEE Trans. Antennas Propag.* **2011**, *60*, 743–750. [CrossRef]

49. Mateo-Segura, C.; Feresidis, A.P.; Goussetis, G. Bandwidth enhancement of 2-D leaky-wave antennas with double-layer periodic surfaces. *IEEE Trans. Antennas Propag.* **2013**, *62*, 586–593. [CrossRef]

50. Al-Tarifi, M.A.; Anagnostou, D.E.; Amert, A.K.; Whites, K.W. Bandwidth enhancement of the resonant cavity antenna by using two dielectric superstrates. *IEEE Trans. Antennas Propag.* **2013**, *61*, 1898–1908. [CrossRef]

51. Wang, N.; Li, J.; Wei, G.; Talbi, L.; Zeng, Q.; Xu, J. Wideband Fabry–Perot resonator antenna with two layers of dielectric superstrates. *IEEE Antennas Wirel. Propag. Lett.* **2014**, *14*, 229–232. [CrossRef]

52. Konstantinidis, K.; Feresidis, A.P.; Hall, P.S. Broadband sub-wavelength profile high-gain antennas based on multi-layer metasurfaces. *IEEE Trans. Antennas Propag.* **2014**, *63*, 423–427. [CrossRef]

53. Konstantinidis, K.; Feresidis, A.P.; Hall, P.S. Multilayer partially reflective surfaces for broadband Fabry-Perot cavity antennas. *IEEE Trans. Antennas Propag.* **2014**, *62*, 3474–3481. [CrossRef]

54. Hashmi, R.M.; Zeb, B.A.; Esselle, K.P. Wideband high-gain EBG resonator antennas with small footprints and all-dielectric superstructures. *IEEE Trans. Antennas Propag.* **2014**, *62*, 2970–2977. [CrossRef]

55. Wang, N.; Talbi, L.; Zeng, Q.; Xu, J. Wideband Fabry-Perot resonator antenna with electrically thin dielectric superstrates. *IEEE Access* **2018**, *6*, 14966–14973. [CrossRef]

56. Nguyen-Trong, N.; Tran, H.H.; Nguyen, T.K.; Abbosh, A.M. Wideband Fabry–Perot antennas employing multilayer of closely spaced thin dielectric slabs. *IEEE Antennas Wirel. Propag. Lett.* **2018**, *17*, 1354–1358. [CrossRef]

57. Zeb, B.; Hashmi, R.; Esselle, K. Wideband gain enhancement of slot antenna using one unprinted dielectric superstrate. *Electron. Lett.* **2015**, *51*, 1146–1148. [CrossRef]

58. Liu, Z.; Liu, S.; Bornemann, J.; Zhao, X.; Kong, X.; Huang, Z.; Bian, B.; Wang, D. A Low-RCS, High-GBP Fabry–Perot Antenna With Embedded Chessboard Polarization Conversion Metasurface. *IEEE Access* **2020**, *8*, 80183–80194. [CrossRef]

59. Wang, N.; Liu, Q.; Wu, C.; Talbi, L.; Zeng, Q.; Xu, J. Wideband Fabry-Perot resonator antenna with two complementary FSS layers. *IEEE Trans. Antennas Propag.* **2014**, *62*, 2463–2471.

60. Ge, Y.; Sun, Z.; Chen, Z.; Chen, Y.Y. A high-gain wideband low-profile Fabry–Pérot resonator antenna with a conical short horn. *IEEE Antennas Wirel. Propag. Lett.* **2016**, *15*, 1889–1892. [CrossRef]

61.   Goudarzi, A.; Movahhedi, M.; Honari, M.M.; Saghlatoon, H.; Mirzavand, R.; Mousavi, P. Wideband High-Gain Circularly Polarized Resonant Cavity Antenna with a Thin Complementary Partially Reflective Surface. *IEEE Trans. Antennas Propag.* **2020**, 1–6, [CrossRef]

62.   Goudarzi, A.; Movahhedi, M.; Honari, M.M.; Mousavi, P. A Wideband CP Resonant Cavity Antenna with a Self-Complimentary Partially Reflective Surface. In Proceedings of the 2020 IEEE AP-S Symposium on Antennas and Propagatio, Montréal, QC, Canada, 5–10 July 2020.

63.   Ge, Y.; Wang, C. A millimeter-wave wideband high-gain antenna based on the fabry-perot resonator antenna concept. *PIER C* **2014**, *50*, 103–111. [CrossRef]

64.   Han, W.; Yang, F.; Ouyang, J.; Yang, P. Low-cost wideband and high-gain slotted cavity antenna using high-order modes for millimeter-wave application. *IEEE Trans. Antennas Propag.* **2015**, *63*, 4624–4631. [CrossRef]

65.   Attia, H.; Abdelghani, M.L.; Denidni, T.A. Wideband and high-gain millimeter-wave antenna based on FSS Fabry–Perot cavity. *IEEE Trans. Antennas Propag.* **2017**, *65*, 5589–5594. [CrossRef]

66.   Singh, A.K.; Abegaonkar, M.P.; Koul, S.K. High-gain and high-aperture-efficiency cavity resonator antenna using metamaterial superstrate. *IEEE Antennas Wirel. Propag. Lett.* **2017**, *16*, 2388–2391. [CrossRef]

67.   Nguyen-Trong, N.; Tran, H.H.; Nguyen, T.K.; Abbosh, A.M. A compact wideband circular polarized Fabry-Perot antenna using resonance structure of thin dielectric slabs. *IEEE Access* **2018**, *6*, 56333–56339. [CrossRef]

68.   Tran, H.H.; Le, T.T.; Bui, C.D.; Nguyen, T.K. Broadband circularly polarized Fabry-Perot antenna utilizing Archimedean spiral radiator and multi-layer partially reflecting surface. *Int. J. RF Microw. Comput.-Aided Eng.* **2019**, *29*, e21647. [CrossRef]

69.   Hashmi, R.M.; Esselle, K.P. A class of extremely wideband resonant cavity antennas with large directivity-bandwidth products. *IEEE Trans. Antennas Propag.* **2015**, *64*, 830–835. [CrossRef]

70.   Baba, A.; Hashmi, R.; Esselle, K. Wideband gain enhancement of slot antenna using superstructure with optimised axial permittivity variation. *Electron. Lett.* **2016**, *52*, 266–268. [CrossRef]

71.   Baba, A.A.; Hashmi, R.M.; Esselle, K.P. Achieving a large gain-bandwidth product from a compact antenna. *IEEE Trans. Antennas Propag.* **2017**, *65*, 3437–3446. [CrossRef]

72.   Baba, A.A.; Hashmi, R.M.; Esselle, K.P.; Weily, A.R. Compact high-gain antenna with simple all-dielectric partially reflecting surface. *IEEE Trans. Antennas Propag.* **2018**, *66*, 4343–4348. [CrossRef]

73.   Baba, A.A.; Hashmi, R.M.; Asadnia, M.; Matekovits, L.; Esselle, K.P. A Stripline-Based Planar Wideband Feed for High-Gain Antennas with Partially Reflecting Superstructure. *Micromachines* **2019**, *10*, 308. [CrossRef] [PubMed]

74.   Baba, A.A.; Hashmi, R.M.; Esselle, K.P.; Ahmad, Z.; Hesselbarth, J. Millimeter-Wave Broadband Antennas with Low Profile Dielectric Covers. *IEEE Access* **2019**. [CrossRef]

75.   Wu, F.; Luk, K.M. Wideband high-gain open resonator antenna using a spherically modified, second-order cavity. *IEEE Trans. Antennas Propag.* **2017**, *65*, 2112–2116. [CrossRef]

76.   Chen, Q.; Chen, X.; Xu, K. 3-D printed Fabry–Perot resonator antenna with paraboloid-shape superstrate for wide gain bandwidth. *Appl. Sci.* **2017**, *7*, 1134. [CrossRef]

77.   Ji, L.Y.; Qin, P.Y.; Guo, Y.J. Wideband Fabry-Perot cavity antenna with a shaped ground plane. *IEEE Access* **2017**, *6*, 2291–2297. [CrossRef]

78.   Qing-Yi, G.; Hang, W. A Fabry-Pérot Cavity Antenna for Millimeter Wave Application. In Proceedings of the 2019 Cross Strait Quad-Regional Radio Science and Wireless Technology Conference (CSQRWC), Taiyuan, China, 18–21 July 2019; pp. 1–2.

79.   Guo, Q.Y.; Wong, H. A Millimeter-Wave Fabry–Pérot Cavity Antenna Using Fresnel Zone Plate Integrated PRS. *IEEE Trans. Antennas Propag.* **2019**, *68*, 564–568. [CrossRef]

80.   Weily, A.R.; Esselle, K.; Bird, T.S.; Sanders, B.C. Dual resonator 1-D EBG antenna with slot array feed for improved radiation bandwidth. *IET Microw. Antennas Propag.* **2007**, *1*, 198–203. [CrossRef]

81.   Afzal, M.U.; Esselle, K.P.; Lalbakhsh, A. A methodology to design a low-profile composite-dielectric phase-correcting structure. *IEEE Antennas Wirel. Propag. Lett.* **2018**, *17*, 1223–1227. [CrossRef]

82.   Hayat, T.; Afzal, M.U.; Lalbakhsh, A.; Esselle, K.P. 3-D-printed phase-rectifying transparent superstrate for resonant-cavity antenna. *IEEE Antennas Wirel. Propag. Lett.* **2019**, *18*, 1400–1404. [CrossRef]

83. Hayat, T.; Afzal, M.U.; Lalbakhsh, A.; Esselle, K.P. Additively Manufactured Perforated Superstrate to Improve Directive Radiation Characteristics of Electromagnetic Source. *IEEE Access* **2019**, *7*, 153445–153452. [CrossRef]
84. Xie, P.; Wang, G.; Li, H.; Gao, X. A Novel Methodology for Gain Enhancement of the Fabry-Pérot Antenna. *IEEE Access* **2019**, *7*, 176170–176176. [CrossRef]
85. Zhou, L.; Duan, X.; Luo, Z.; Zhou, Y.; Chen, X. High Directivity Fabry-Perot Antenna with a Nonuniform Partially Reflective Surface and a Phase Correcting Structure. *IEEE Trans. Antennas Propag.* **2020**. [CrossRef]
86. Zhou, L.; Chen, X.; Duan, X. Fabry–Pérot resonator antenna with high aperture efficiency using a double-layer nonuniform superstrate. *IEEE Trans. Antennas Propag.* **2018**, *66*, 2061–2066. [CrossRef]
87. Lalbakhsh, A.; Afzal, M.U.; Esselle, K.P.; Smith, S.L.; Zeb, B.A. Single-dielectric wideband partially reflecting surface with variable reflection components for realization of a compact high-gain resonant cavity antenna. *IEEE Trans. Antennas Propag.* **2019**, *67*, 1916–1921. [CrossRef]
88. Lalbakhsh, A.; Afzal, M.U.; Esselle, K.P.; Smith, S.L. Wideband near-field correction of a Fabry-Perot resonator antenna. *IEEE Trans. Antennas Propag.* **2019**, *67*, 1975–1980. [CrossRef]
89. Ge, Y.; Esselle, K.P.; Hao, Y. Design of low-profile high-gain EBG resonator antennas using a genetic algorithm. *IEEE Antennas Wirel. Propag. Lett.* **2007**, *6*, 480–483. [CrossRef]
90. Zheng, Y.; Gao, J.; Zhou, Y.; Cao, X.; Yang, H.; Li, S.; Li, T. Wideband gain enhancement and RCS reduction of Fabry–Perot resonator antenna with chessboard arranged metamaterial superstrate. *IEEE Trans. Antennas Propag.* **2017**, *66*, 590–599. [CrossRef]
91. Qin, F.; Gao, S.; Wei, G.; Luo, Q.; Mao, C.; Gu, C.; Xu, J.; Li, J. Wideband circularly polarized Fabry-Perot antenna [antenna applications corner]. *IEEE Antennas Propag. Mag.* **2015**, *57*, 127–135. [CrossRef]
92. Ratni, B.; De Lustrac, A.; Villers, S.; Nawaz Burokur, S. Low-profile circularly polarized fabry-perot cavity antenna. *Microw. Opt. Tech. Lett.* **2016**, *58*, 2957–2960. [CrossRef]
93. Tran, H.H.; Park, I. Compact wideband circularly polarised resonant cavity antenna using a single dielectric superstrate. *IET Microw. Antenna Propag.* **2016**, *10*, 729–736. [CrossRef]
94. Ta, S.; Nguyen, T. AR bandwidth and gain enhancements of patch antenna using single dielectric superstrate. *Electron. Lett.* **2017**, *53*, 1015–1017. [CrossRef]
95. Ta, S.X.; Nguyen, T.H.Y.; Nguyen, K.K.; Dao-Ngoc, C. Bandwidth-enhancement of circularly-polarized fabry-perot antenna using single-layer partially reflective surface. *Int. J. RF Microw. Comput. Aided Eng.* **2019**, e21774. [CrossRef]
96. Cao, W.; Lv, X.; Wang, Q.; Zhao, Y.; Yang, X. Wideband circularly polarized Fabry–Pérot resonator antenna in Ku-band. *IEEE Antennas Wirel. Propag. Lett.* **2019**, *18*, 586–590. [CrossRef]
97. Cheng, Y.; Dong, Y. Bandwidth Enhanced Circularly Polarized Fabry-Perot Cavity Antenna Using Metal Strips. *IEEE Access* **2020**, 60189–60198. [CrossRef]
98. Diblanc, M.; Rodes, E.; Arnaud, E.; Thevenot, M.; Monediere, T.; Jecko, B. Circularly polarized metallic EBG antenna. *IEEE Microw. Wirel. Compon. Lett.* **2005**, *15*, 638–640. [CrossRef]
99. Ju, J.; Kim, D.; Lee, W.; Choi, J. Design Method of a Circularly-Polarized Antenna Using Fabry-Pérot Cavity Structure. *ETRI J.* **2011**, *33*, 163–168. [CrossRef]
100. Orr, R.; Goussetis, G.; Fusco, V. Design method for circularly polarized Fabry–Perot cavity antennas. *IEEE Trans. Antennas Propag.* **2014**, *62*, 19–26. [CrossRef]
101. Liu, Z.G.; Lu, W.B. Low-profile design of broadband high gain circularly polarized Fabry-Perot resonator antenna and its array with linearly polarized feed. *IEEE Access* **2017**, *5*, 7164–7172. [CrossRef]
102. Srivastava, K.; Kumar, A.; Chaudhary, P.; Kanaujia, B.K.; Dwari, S.; Verma, A.K.; Esselle, K.P.; Mittra, R. Wideband and high-gain circularly polarised microstrip antenna design using sandwiched metasurfaces and partially reflecting surface. *IET Microw. Antenna Propag.* **2018**, *13*, 305–312. [CrossRef]
103. Muhammad, S.A.; Sauleau, R.; Valerio, G.; Le Coq, L.; Legay, H. Self-polarizing Fabry–Perot antennas based on polarization twisting element. *IEEE Trans. Antennas Propag.* **2012**, *61*, 1032–1040. [CrossRef]
104. Arnaud, E.; Chantalat, R.; Monédière, T.; Rodes, E.; Thevenot, M. Performance enhancement of self-polarizing metallic EBG antennas. *IEEE Antennas Wirel. Propag. Lett.* **2010**, *9*, 538–541. [CrossRef]
105. Hussain, N.; Jeong, M.J.; Abbas, A.; Kim, T.J.; Kim, N. A Metasurface-Based Low-Profile Wideband Circularly Polarized Patch Antenna for 5G Millimeter-Wave Systems. *IEEE Access* **2020**, *8*, 22127–22135. [CrossRef]
106. Lin, X.; Seet, B.C.; Joseph, F.; Li, E. Flexible fractal electromagnetic bandgap for millimeter-wave wearable antennas. *IEEE Antennas Wirel. Propag. Lett.* **2018**, *17*, 1281–1285. [CrossRef]

107. Tang, M.C.; Shi, T.; Ziolkowski, R.W. A Study of 28 GHz, Planar, multilayered, electrically small, broadside radiating, huygens source antennas. *IEEE Trans. Antennas Propag.* **2017**, *65*, 6345–6354. [CrossRef]
108. Ourir, A.; Burokur, S.; de Lustrac, A. Electronically reconfigurable metamaterial for compact directive cavity antennas. *Electron. Lett.* **2007**, *43*, 698–700. [CrossRef]
109. Weily, A.R.; Bird, T.S.; Guo, Y.J. A reconfigurable high-gain partially reflecting surface antenna. *IEEE Trans. Antennas Propag.* **2008**, *56*, 3382–3390. [CrossRef]
110. Burokur, S.; Daniel, J.P.; Ratajczak, P.; De Lustrac, A. Tunable bilayered metasurface for frequency reconfigurable directive emissions. *Appl. Phys. Lett.* **2010**, *97*, 064101. [CrossRef]
111. Huang, C.; Pan, W.; Ma, X.; Luo, X. A frequency reconfigurable directive antenna with wideband low-RCS property. *IEEE Trans. Antennas Propag.* **2016**, *64*, 1173–1178. [CrossRef]
112. Xie, P.; Wang, G.M. Design of a frequency reconfigurable Fabry-Pérot cavity antenna with single layer partially reflecting surface. *Prog. Electromagn. Res.* **2017**, *70*, 115–121. [CrossRef]
113. Edalati, A.; Denidni, T.A. Reconfigurable beamwidth antenna based on active partially reflective surfaces. *IEEE Antennas Wirel. Propag. Lett.* **2009**, *8*, 1087–1090. [CrossRef]
114. Debogovic, T.; Perruisseau-Carrier, J.; Bartolic, J. Partially reflective surface antenna with dynamic beamwidth control. *IEEE Antennas Wirel. Propag. Lett.* **2010**, *9*, 1157–1160. [CrossRef]
115. Debogović, T.; Perruisseau-Carrier, J. Array-fed partially reflective surface antenna with independent scanning and beamwidth dynamic control. *IEEE Trans. Antennas Propag.* **2013**, *62*, 446–449. [CrossRef]
116. Vaidya, A.R.; Gupta, R.K.; Mishra, S.K.; Mukherjee, J. Right-hand/left-hand circularly polarized high-gain antennas using partially reflective surfaces. *IEEE Antennas Wirel. Propag. Lett.* **2014**, *13*, 431–434. [CrossRef]
117. Tan, G.N.; Yang, X.; Xue, H.G.; Lu, Z. A dual-polarized Fabry-Perot cavity antenna at Ka band with broadband and high gain. *Prog. Electromagnet. Res.* **2015**, *60*, 179–186. [CrossRef]
118. Lian, R.; Tang, Z.; Yin, Y. Design of a broadband polarization-reconfigurable Fabry–Perot resonator antenna. *IEEE Antennas Wirel. Propag. Lett.* **2017**, *17*, 122–125. [CrossRef]
119. Qin, P.Y.; Ji, L.Y.; Chen, S.L.; Guo, Y.J. Dual-polarized wideband Fabry–Perot antenna with quad-layer partially reflective surface. *IEEE Antennas Wirel. Propag. Lett.* **2018**, *17*, 551–554. [CrossRef]
120. Zhu, H.; Qiu, Y.; Wei, G. A Broadband Dual-Polarized Antenna with Low Profile Using Nonuniform Metasurface. *IEEE Antennas Wirel. Propag. Lett.* **2019**, *18*, 1134–1138. [CrossRef]
121. Wu, Z.; Liu, H.; Li, L. Metasurface-inspired low profile polarization reconfigurable antenna with simple DC controlling circuit. *IEEE Access* **2019**, *7*, 45073–45079. [CrossRef]
122. Liu, P.; Jiang, W.; Sun, S.; Xi, Y.; Gong, S. Broadband and Low-Profile Penta-Polarization Reconfigurable Metamaterial Antenna. *IEEE Access* **2020**, *8*, 21823–21831. [CrossRef]
123. Comite, D.; Baccarelli, P.; Burghignoli, P.; Galli, A. Omnidirectional 2-D leaky-wave antennas with reconfigurable polarization. *IEEE Antennas Wirel. Propag. Lett.* **2017**, *16*, 2354–2357. [CrossRef]
124. Chen, C.; Liu, Z.G.; Wang, H.; Guo, Y. Metamaterial-inspired self-polarizing dual-band dual-orthogonal circularly polarized Fabry–Pérot resonator antennas. *IEEE Trans. Antennas Propag.* **2018**, *67*, 1329–1334. [CrossRef]
125. Wu, F.; Luk, K.M. Circular Polarization and Reconfigurability of Fabry–Pérot Resonator Antenna Through Metamaterial-Loaded Cavity. *IEEE Trans. Antennas Propag.* **2019**, *67*, 2196–2208. [CrossRef]
126. Niaz, M.W.; Yin, Y.; Zheng, S.; Zhao, Z. Dual-polarized low sidelobe Fabry-Perot antenna using tapered partially reflective surface. *Int. J. RF Microw. Comput.-Aided Eng.* **2019**, e22070. [CrossRef]
127. Swain, R.; Chatterjee, A.; Nanda, S.; Mishra, R.K. A Linear-to-Circular Polarization Conversion Metasurface Based Wideband Aperture Coupled Antenna. *J. Electr. Eng. Technol.* **2020**, 1–7. [CrossRef]
128. Ourir, A.; Burokur, S.; de Lustrac, A. Passive and active reconfigurable resonant metamaterial cavity for beam deflection. In Proceedings of the 2007 IEEE Antennas and Propagation Society International Symposium, Honolulu, HI, USA, 9–15 June 2007; pp. 4973–4976.
129. Ourir, A.; Burokur, S.N.; Yahiaoui, R.; de Lustrac, A. Directive metamaterial-based subwavelength resonant cavity antennas–Applications for beam steering. *C.R. Phys.* **2009**, *10*, 414–422. [CrossRef]
130. Edalati, A.; Denidni, T.A. High-gain reconfigurable sectoral antenna using an active cylindrical FSS structure. *IEEE Trans. Antennas Propag.* **2011**, *59*, 2464–2472. [CrossRef]
131. Guo, Y.J.; Gómez-Tornero, J.L. Reconfigurable Fabry-Perot leaky-wave antennas. In Proceedings of the 2013 International Workshop on Antenna Technology (iWAT), Karlsruhe, Germany, 4–6 March 2013; pp. 390–393.

132. Ji, L.Y.; Guo, Y.J.; Qin, P.Y.; Gong, S.X.; Mittra, R. A reconfigurable partially reflective surface (PRS) antenna for beam steering. *IEEE Trans. Antennas Propag.* **2015**, *63*, 2387–2395. [CrossRef]

133. Afzal, M.U.; Esselle, K.P. Steering the Beam of Medium-to-High Gain Antennas Using Near-Field Phase Transformation. *IEEE Trans. Antennas Propag.* **2017**, *65*, 1680–1690. [CrossRef]

134. Singh, K.; Afzal, M.U.; Kovaleva, M.; Esselle, K.P. Controlling the Most Significant Grating Lobes in Two-Dimensional Beam-Steering Systems with Phase-Gradient Metasurfaces. *IEEE Trans. Antennas Propag.* **2019**, *68*, 1389–1401.

135. Ji, L.Y.; Qin, P.Y.; Li, J.Y.; Zhang, L.X. 1-D Electronic Beam-Steering Partially Reflective Surface Antenna. *IEEE Access* **2019**, *7*, 115959–115965. [CrossRef]

136. Ji, L.Y.; Zhang, Z.Y.; Liu, N.W. A two-dimensional beam-steering partially reflective surface (PRS) antenna using a reconfigurable FSS structure. *IEEE Antennas Wirel. Propag. Lett.* **2019**, *18*, 1076–1080. [CrossRef]

137. Ji, L.Y.; Fu, S.; Zhang, L.X.; Li, J.Y. One-dimensional beam-steering Fabry–Perot cavity (FPC) antenna with a reconfigurable superstrate. *Int. J. Microw. Wirel. Technol.* **2020**, *12*, 233–239. [CrossRef]

138. Das, P.; Mandal, K.; Lalbakhsh, A. Single-layer polarization-insensitive frequency selective surface for beam reconfigurability of monopole antennas. *J. Electromagnet. Waves Appl.* **2020**, *34*, 86–102. [CrossRef]

139. Akbari, M.; Farahani, M.; Ghayekhloo, A.; Zarbakhsh, S.; Sebak, A.R.; Denidni, T. Phase Gradient Surface Approaches for 60 GHz Beam Tilting Antenna. *IEEE Trans. Antennas Propag.* **2020**, *68*, 4372–4385.

140. Xie, P.; Wang, G.; Li, H.; Liang, J. A dual-polarized two-dimensional beam-steering Fabry–Pérot cavity antenna with a reconfigurable partially reflecting surface. *IEEE Antennas Wirel. Propag. Lett.* **2017**, *16*, 2370–2374. [CrossRef]

141. Ji, L.; Fu, G.; Gong, S.X. Array-fed beam-scanning partially reflective surface (PRS) antenna. *Prog. Electromagnet. Res.* **2016**, *58*, 73–79. [CrossRef]

142. Guo, Q.Y.; Wong, H. Wideband and high-gain Fabry–Pérot cavity antenna with switched beams for millimeter-wave applications. *IEEE Trans. Antennas Propag.* **2019**, *67*, 4339–4347. [CrossRef]

143. Tran, H.H.; Le, T.T. A metasurface based low-profile reconfigurable antenna with pattern diversity. *AEU Int. J. Electron. Commun.* **2020**, *115*, 153037. [CrossRef]

144. Guzmán-Quirós, R.; Weily, A.; Gómez-Tornero, J.; Guo, Y. A Fabry–Pérot antenna with two-dimensional electronic beam scanning. *IEEE Trans. Antennas Propag.* **2016**, *64*, 1536–1541. [CrossRef]

145. Towfiq, M.A.; Bahceci, I.; Blanch, S.; Romeu, J.; Jofre, L.; Cetiner, B.A. A reconfigurable antenna with beam steering and beamwidth variability for wireless communications. *IEEE Trans. Antennas Propag.* **2018**, *66*, 5052–5063. [CrossRef]

146. Mantash, M.; Denidni, T.A. CP Antenna Array With Switching-Beam Capability Using Electromagnetic Periodic Structures for 5G Applications. *IEEE Access* **2019**, *7*, 26192–26199. [CrossRef]

147. Wang, S.; Feresidis, A.; Goussetis, G.; Vardaxoglou, J. Low-profile resonant cavity antenna with artificial magnetic conductor ground plane. *Electron. Lett.* **2004**, *40*, 405–406. [CrossRef]

148. Feresidis, A.P.; Goussetis, G.; Wang, S.; Vardaxoglou, J.C. Artificial magnetic conductor surfaces and their application to low-profile high-gain planar antennas. *IEEE Trans. Antennas Propag.* **2005**, *53*, 209–215. [CrossRef]

149. Zhou, L.; Li, H.; Qin, Y.; Wei, Z.; Chan, C. Directive emissions from subwavelength metamaterial-based cavities. *Appl. Phys. Lett.* **2005**, *86*, 101101. [CrossRef]

150. Ourir, A.; de Lustrac, A.; Lourtioz, J.M. All-metamaterial-based subwavelength cavities ($\lambda/60$) for ultrathin directive antennas. *Appl. Phys. Lett.* **2006**, *88*, 084103. [CrossRef]

151. Kelly, J.R.; Kokkinos, T.; Feresidis, A.P. Analysis and design of sub-wavelength resonant cavity type 2-D leaky-wave antennas. *IEEE Trans. Antennas Propag.* **2008**, *56*, 2817–2825. [CrossRef]

152. Mateo-Segura, C.; Goussetis, G.; Feresidis, A.P. Sub-wavelength profile 2-D leaky-wave antennas with two periodic layers. *IEEE Trans. Antennas Propag.* **2010**, *59*, 416–424. [CrossRef]

153. Honari, M.M.; Mousavi, P.; Sarabandi, K. Miniaturized-Element Frequency Selective Surface Metamaterials: A Solution to Enhance Radiation Off of RFICs. *IEEE Trans. Antennas Propag.* **2019**, *68*, 1962–1972.

154. Konstantinidis, K.; Feresidis, A.P.; Hall, P.S. Dual subwavelength Fabry–Perot cavities for broadband highly directive antennas. *IEEE Antennas Wirel. Propag. Lett.* **2014**, *13*, 1184–1186. [CrossRef]

155. Lin, F.H.; Chen, Z.N. Low-profile wideband metasurface antennas using characteristic mode analysis. *IEEE Trans. Antennas Propag.* **2017**, *65*, 1706–1713. [CrossRef]

156. Qin, F.; Gao, S.; Luo, Q.; Wei, G.; Xu, J.; Li, J.; Wu, C.; Gu, C.; Mao, C. A triband low-profile high-gain planar antenna using Fabry–Perot cavity. *IEEE Trans. Antennas Propag.* **2017**, *65*, 2683–2688. [CrossRef]

157. Deng, F.; Qi, J. Shrinking Profile of Fabry-Perot Cavity Antennas with Stratified Metasurfaces: Accurate Equivalent Circuit Design and Broadband High-Gain Performance. *IEEE Antennas Wirel. Propag. Lett.* **2019**, *19*, 208–212.

158. Ge, Y.; Esselle, K.P.; Bird, T.S. A method to design dual-band, high-directivity EBG resonator antennas using single-resonant, single-layer partially reflective surfaces. *Prog. Electromagnet. Res.* **2010**, *13*, 245–257. [CrossRef]

159. Moghadas, H.; Daneshmand, M.; Mousavi, P. A dual-band high-gain resonant cavity antenna with orthogonal polarizations. *IEEE Antennas Wirel. Propag. Lett.* **2011**, *10*, 1220–1223. [CrossRef]

160. Zeb, B.A.; Ge, Y.; Esselle, K.P.; Sun, Z.; Tobar, M.E. A simple dual-band electromagnetic band gap resonator antenna based on inverted reflection phase gradient. *IEEE Trans. Antennas Propag.* **2012**, *60*, 4522–4529. [CrossRef]

161. Moghadas, H.; Daneshmand, M.; Mousavi, P. Single-layer partially reflective surface for an orthogonally-polarised dual-band high-gain resonant cavity antenna. *IET Microw. Antennas Propag.* **2013**, *7*, 656–662. [CrossRef]

162. Meng, F.; Sharma, S.K. A dual-band high-gain resonant cavity antenna with a single layer superstrate. *IEEE Trans. Antennas Propag.* **2015**, *63*, 2320–2325. [CrossRef]

163. Lv, Y.H.; Ding, X.; Wang, B.Z. Dual-Wideband High-Gain Fabry-Perot Cavity Antenna. *IEEE Access* **2019**, *68*, 1389–1401.

3

Design of True Time Delay Millimeter Wave Beamformers for 5G Multibeam Phased Arrays

Dimitrios I. Lialios [1], **Nikolaos Ntetsikas** [2], **Konstantinos D. Paschaloudis** [2], **Constantinos L. Zekios** [1,*], **Stavros V. Georgakopoulos** [1] **and George A. Kyriacou** [2,*]

[1] College of Engineering & Computing, Florida International University, Miami, FL 33174, USA; dlial001@fiu.edu (D.I.L.); georgako@fiu.edu (S.V.G.)

[2] Department of Electrical & Computer Engineering, Democritus University of Thrace, 67100 Xanthi, Greece; nikontet@ee.duth.gr (N.N.); kopascha@ee.duth.gr (K.D.P.)

[*] Correspondence: kzekios@fiu.edu (C.L.Z.); gkyriac@ee.duth.gr (G.A.K.)

Abstract: Millimeter wave (mm-Wave) technology is likely the key enabler of 5G and early 6G wireless systems. The high throughput, high capacity, and low latency that can be achieved, when mm-Waves are utilized, makes them the most promising backhaul as well as fronthaul solutions for the communication between small cells and base stations or between base stations and the gateway. Depending on the channel properties different communication systems (e.g., beamforming and MIMO) can accordingly offer the best solution. In this work, our goal is to design millimeter wave beamformers for switched beam phased arrays as hybrid beamforming stages. Specifically, three different analog beamforming techniques for the frequency range of 27–33 GHz are presented. First, a novel compact multilayer Blass matrix is proposed. Second, a modified dummy-ports free, highly efficient Rotman lens is introduced. Finally, a three-layer true-time-delay tree topology inspired by microwave photonics is presented.

Keywords: 5G; early 6G; hybrid beamforming networks; millimeter waves; switched beam phased arrays; Blass matrix; Rotman lens; tree beamformer

1. Introduction

The evolution of fifth generation (5G) wireless systems toward high microwave and millimeter wave (mm-Wave) frequencies provides significant advantages because of the wide available spectrum (i.e., high throughput, high capacity, and low latency) and improved features resulting from the smaller wavelength [1–12]. Of course, higher frequencies are characterized by challenges of higher propagation losses, higher phase noise in local oscillators, and higher insertion loss in the RF-front-end. This is accompanied by possible shadowing even by trees, limited propagation range due to rain attenuation, atmospheric and molecular absorption, and lower or impossible penetration to buildings [13–16]. The latter can be compensated by ensuring line-of-sight links, thereby requiring dense base stations even down to distances of 150–200 m [17]. Therefore, mm-Wave communications are mainly used for indoor environments and small cell access. This densification is anyways needed to ensure the excessively high communication rates of 5G. Of course, densely deployed small cells demand high cost to connect 5G base stations (BSs) to other BSs and to the network by fiber-based backhaul [18]. In contrast, wireless backhaul in mm-Wave bands that offers high speed, wide bandwidth, several *Gbps* data rates is by far more cost-effective, flexible and much easier to deploy. As shown in Figure 1, a wireless mm-Wave communication system is an extremely promising backhaul solution for small cells that can support the desired high speed transmission between the small cell base stations (BSs) or between BSs and the gateway. However, high antenna gain (or narrow beamwidth) is needed to compensate for high propagation losses; therefore, beamforming is required.

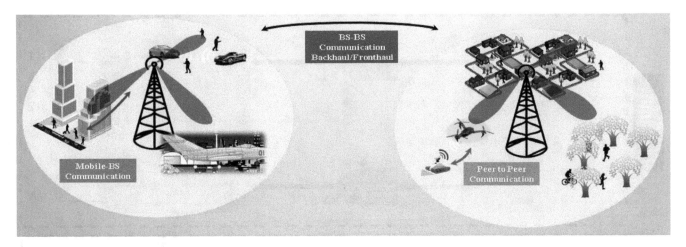

Figure 1. Illustration of different 5G communication cells.

Besides beamforming, antenna diversity (two or more antennas at different positions) has been used at low microwave frequencies (i.e., below 6 GHz) to compensate the effects of multipath propagation. Even though multipath effects are negligible at millimeter waves or above 7 GHz (*FR3* band) as they require line–of–sight paths, antenna diversity is still beneficial at these frequencies. Specifically, it has been shown that the capacity of communication systems at these frequencies can be increased by employing multiple antennas (i.e., MIMO) to receive signals.

A MIMO configuration can be deployed in single-user systems, where it offers the best throughput, or for multi-user systems to increase the overall cell capacity. On the other hand, beamforming needs antenna elements with the same polarization. A beam can be made directive even with two antenna elements, but with very poor results. Instead, four antennas offer acceptable directivity, while eight antennas provide the best compromise [e.g., especially for the 7–24 GHz (*FR3* range)]. Additionally, it has been shown that beamforming is more efficient than MIMO in hard–to–cover areas (cell edges or through buildings) where the SNR is low [19]. On the contrary, MIMO is more efficient than beamforming in areas with strong SNR (best for $SNR > 17$ dB), i.e., near the center of a cell, where multiple layers of MIMO are used.

Therefore, the most practically feasible configuration of a passive antenna at frequencies below 6 GHz, is one that implements both beamforming and MIMO operations. The established configuration, which incorporates both beamforming and MIMO, is known as "Hybrid Beamforming" and it is generally accepted as the best option for the upper microwave (7–33 GHz) and millimeter wave bands [20–23].

Hybrid beamforming has been adopted since 1970. In addition, various hybrid beamforming systems have been presented in recent papers (e.g., [22,24]). Specifically, such systems are comprised of the following components, as shown in Figure 2: (i) the antenna array; (ii) the "RF precoding", i.e., the analog RF beamformer that is driven by a number of transmitting RF chains; and (iii) a number of digital–to–analog converters (DACs) that combine the "digital baseband precoding" (which plays the role of the software defined radio (SDR) for the transmitter) into the analog RF beamforming networks. The receiver follows the same logic and appears at the bottom of Figure 2. Notably, hybrid beamforming combines the low cost, simplicity, and ease of implementation of analog beamformers with powerful abilities of digital beamformers, such as comprehensive interference rejection and beamsteering towards any desired position (at least theoretically) [25]. Therefore, by combining the analog and digital beamformers, novel communication systems can be developed that meet the needs of 5G [26]. Our paper is focused on analogue beamformers that represent an important component of hybrid beamforming systems.

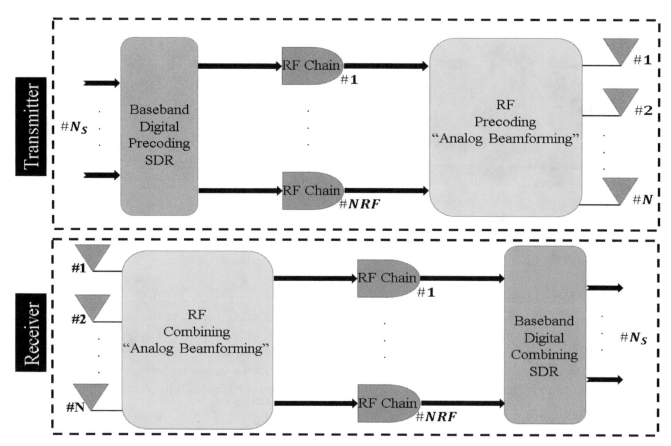

Figure 2. Hybrid beamforming architecture.

Many different analog RF beamforming implementations have been developed during the last 60 years following the topologies proposed by Butler [27], Blass [28], and Nolen [29] matrices, as well as Rotman lenses [30]. The most common analog beamforming network is based on the Butler matrix, which is known for its high efficiency. However, it can only produce uniform amplitude distributions that are known for their high side-lobe level (*SLL*) values [31], and high number of crossovers that cause many undesired effects, such as increased insertion loss, mismatched junctions, etc. [32]. In the literature, limited works have been proposed for the development of Butler matrices with non-uniform amplitude distributions [33] and the reduction of crossovers [34]. Additionally, Butler matrices exhibit significantly narrower bandwidths than the most wideband antenna arrays currently available. A design with bandwidth of 3:1 (based on a $10 - dB$ return loss), $1 - dB$ insertion loss variation, and $7°$ phase deviation was recently reported by Chen et al. [35]. Even though Butler matrices have been extensively used at the frequency range of 1–4 GHz, there are limited works at higher frequencies and mm-Waves. Nedil et al. [36] introduced a 4×4 two-layer Butler matrix at 5.8 GHz using coplanar wave-guide technology. Cao et al. [37], introduced a compact substrate integrated waveguide (SIW) multi-folded 4×8 Butler matrix combined with a 8×10 SIW slot-coupled patch array at 38 GHz accomplishing a size reduction of 53.5% in its longitudinal direction. Tornielli di Crestvolant et al. [38] introduced a new class of Butler matrices with inherent bandpass filter transfer functions. Specifically, they presented the synthesis and design of a 2×2, 180 hybrid coupler at 10 GHz and a 4×4 Butler matrix with an equal-ripple four-pole Chebyshev bandpass characteristic centered at 12.5 GHz. More recently Dyad et al. [39] presented a dually-polarized Butler matrix for base stations with polarization diversity operating at 60 GHz. Finally, Tamayo-Dominguez et al. [40] utilized additive manufacturing techniques to introduce a 3D-printed modified butler matrix for a monopulse radar based on gap waveguides at W-Band.

Nolen matrices outperform Butler matrices as they can produce an arbitrary number of beams. However, similar to Butler matrices, they are usually limited to uniform amplitude tapering and only few reported attempts have attempted to fix this problem [41]. In addition, Nolen networks operate at relatively narrowband frequency range, compared to wideband antenna arrays. Djerafi et al. [42], introduced a relatively broadband Nolen matrix using substrate integrated waveguide (SIW) technology, which provided constant phase over a 11.7% frequency bandwidth centered at 77 GHz. More recently, a Nolen matrix with slightly wider bandwidth of 1.4:1, amplitude imbalance less than 0.75 dB, phase variation 6°, and return loss better than 16 dB was proposed by Ren et al. [43].

The Blass matrix is the ordinary case of the Nolen matrix. In addition, the Blass matrix can provide non-uniform amplitudes (thereby overcoming Nolen matrix limitations for uniform amplitudes) but exhibits higher losses due to the use of matched loads. Moreover, the Blass matrix can be designed as a true time-delay network, thereby resolving (at least in theory) the bandwidth limitations of both Butler and Nolen matrices. A true time-delay Blass matrix with an instantaneous bandwidth of 5:1 was first reported by Chu et al. [44]. However, this was an active network design that used amplifiers as directional couplers.

Finally, Rotman lenses are passive topologies that have been used to implement true time-delay beamforming networks. The main advantage of Rotman lenses is that they require low number of switching elements, compared to Blass matrices. Rotman lenses have been extensively used with relatively wideband performance, such as 3.4:1 by Lambrecht et al. [45] and 10:1 by Merola et al. [46]. However, Rotman lenses suffer from relatively low efficiency (e.g., 20–50% efficiency was reported by Merola [46]).

In this work, three novel analog beamforming networks are proposed as 5G mm-wave candidates. Each network separately tackles and solves specific limitations and drawbacks of its prior counterparts. First, a multilayer Blass matrix design is presented. To our knowledge, this is the first multilayer, circular Blass topology. Typical Blass networks are big in size and suffer from excessive losses. Aiming towards compact and highly efficient mm-Wave beamforming networks, we first transform the classical Blass matrix topology [28] into a circular design [47], and then implement it in a two-layer topology. The challenge of this attempt is the appropriate design of the couplers. Typical approaches use bondwires and bridges [48], two factors that decrease the efficiency of beamformers and increase the design complexity. In this work, to minimize the losses and significantly reduce the design complexity, a dual-layer directional coupler is designed and used. As a proof of concept, a 3 × 4 Blass matrix is designed, which shows excellent performance. Compared to the typical single-layer Blass matrix the proposed dual-layer network achieves 60% size reduction.

The second topology, which is presented here, is a modified Rotman lens topology. Rotman lenses, similar to the Blass matrix topologies have large footprints, while their efficiency is limited due to the existence of dummy ports. Aiming towards a compact, low cost and highly efficient design, we minimize the number of dummy ports by using field absorbers. Specifically, by imitating perfect matched layers (PMLs), we engineer the substrate by doping it with conductive material. This key modification essentially increases the loss tangent about ten times, yielding an almost perfect absorber and thereby eliminating the undesired reflections. As a proof of concept, a simple 3 × 3 structure, which shows very good results, is presented here.

Our last topology is a tree topology inspired by microwave photonics. Specifically, this implementation provides ultra-wideband operation; eliminates the problems of crossovers faced by Butler, Nolen, and Blass matrices; and is capable of providing significantly higher efficiencies than ordinary Rotman lenses. The primary novelty of this topology is the minimal number of delay lines. By hierarchically repeating the same feature, using successive power divisions (equal and un-equal) and by doubling the delay as we move diagonally from the first antenna port to its last port and towards the network input, we achieve the desired amplitude and phase difference. As a proof of concept, a three-layer 1 × 4 network, which shows excellent results, is designed.

2. Beamforming Networks Design

2.1. Blass Matrix Architecture

In this task, we pursue the development of a novel dual-layer Blass matrix. Typical Blass matrices are large in size and suffer from excessive losses. Aiming to compact and highly efficient mm-Wave beamforming networks in this work, the classical Blass matrix topology is first transformed into a circular design, and then implemented in a two-layer topology. To our knowledge, this is the first multi-layer Blass design. Compared to the typical single-layer Blass matrix, the proposed dual-layer network achieves 60% size reduction. An additional key point of our implementation are the couplers. Typical approaches use bondwires and bridges [48], two factors that decrease the efficiency of the beamformer and increase the design complexity in regular Blass matrices. Here, to minimize the losses and significantly reduce the design complexity, a dual-layer directional coupler is designed and used.

The Blass matrix was first introduced in 1960 [28], and its network is shown in Figure 3. In its original form, the Blass matrix consists of N traveling wave feed lines (rows) cross-connected to a set of M transmission lines (columns), each one feeding an antenna element of the array. M and N are independent and they can be chosen arbitrarily, but always the number of beams M should be less (or equal) to the number of antenna elements. The other end of the line is terminated in a matched load (see Figure 3). This beamforming network provides M beams with an array of N elements. The interconnections are implemented with directional couplers of unequal power divisions as shown in Figure 3. For a uniform amplitude excitation of the array, the first (out of N) row of couplers divides the power to a $1/N$ portion that flows towards the antenna element and a $(N-1)/N$ portion that flows along the column towards the rest of the network. Likewise, the second row of couplers divides the power into ratios of $1/(N-1)$ and $(N-2)/(N-1)$ and so on. In addition, as long as we are moving farther away from the feeding ports and towards the radiating elements (element 1, element 2, $\cdots$, element N) additional electrical length is introduced causing a true time delay (TTD) difference between the elements, which is responsible for the beam-steering of our array. By appropriately choosing the lengths of these paths, we can steer the beam at the desired direction. However, as long as we are moving towards higher port number (port 1, port 2, $\cdots$, port M), the signal may travel to a specific radiating element along different paths (Figure 3). Take for example port 2 when it excites the radiating element 2. As shown in Figure 3, the desired signal is the one that can follow the path denoted with green line. However, there is also an additional path that the signal can take, which is the one denoted with red line. This second path is undesired and is referred to as spurious path, which is responsible for some performance degradation of the Blass matrix.

Aiming to a compact mm-Wave beamforming network, a dual-layer semi-circular Blass matrix adopted herein is introduced [47]. Figure 4 shows a graphical representation of the circular topology. The semi-circular Blass matrix is designed so that beams 2 and 3 are identical to beam 1, but smartly rotated so that the transmission line between them introduces the desired time delays. Therefore, for the first beam, the time delay for the nth element is $n\tau_{vertical}$, for the second beam is $n(\tau_{vertical} + \tau_{horizontal})$, while for the third beam is $n(\tau_{vertical} + 2\tau_{horizontal})$. By determining $\tau_{vertical}$ and $\tau_{horizontal}$, each beam can be accurately steered in the desired direction.

A critical component of the Blass matrix is, as expected, the coupler. The couplers used in this work are directional, but, unlike traditional couplers [48], they have their through port on the opposite side of the coupled port. Traditional couplers can exhibit this behavior by using either bondwires and bridges, such as the Lange directional coupler (see Figure 5b) or branch-line directional couplers (see Figure 5a). However, the first approach increases significantly the losses as well as the fabrication complexity especially at mm-Wave frequencies, while the second approach leads into a significantly larger design. To avoid both the bondwires' complexity and minimize the undesired losses, a dual layer directional coupler is used [49], as shown in Figure 6. This coupler is comprised of two rectangular patches (Figure 6) printed on the top and bottom surface of a two-sided substrate with a common ground plane. The two patches are coupled through a rectangular slot etched on the common ground plane. The dimensions of the patches and the slot are selected to achieve the desired coupling coefficient. To design the coupler, a quasi-static approach is initially used, utilizing the design formulas given by Wong [50]. Then, the design is optimized by running full-wave simulations.

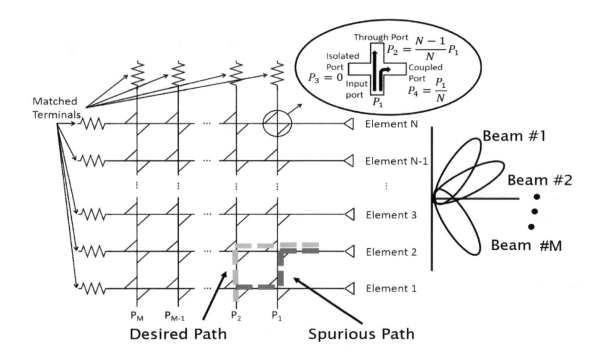

Figure 3. Blass matrix schematic.

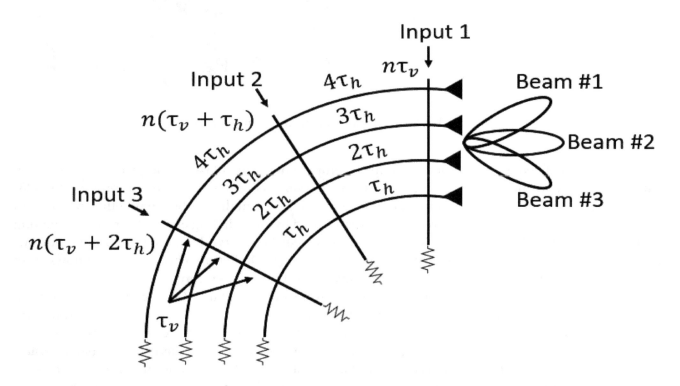

Figure 4. Circular Blass matrix topology.

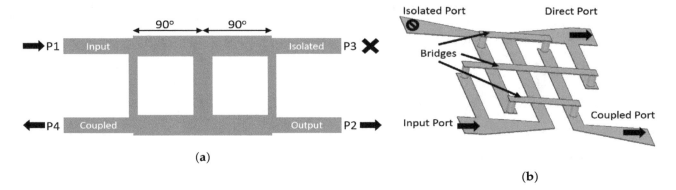

Figure 5. Traditional implementations of directional cross couplers: (**a**) multi-section branch line cross-coupler; and (**b**) Lange directional coupler.

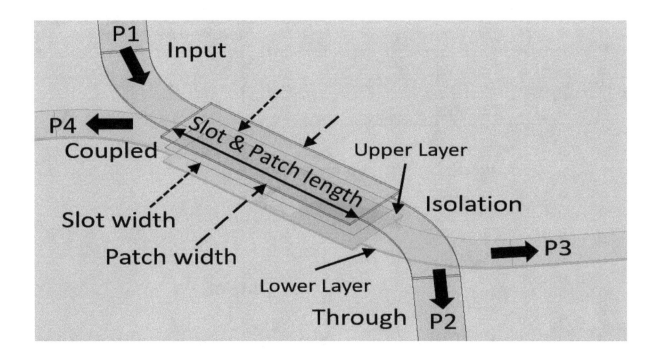

Figure 6. Layout of the dual-layer directional coupler.

As mentioned above, one of the most critical components of the Blass topology are the directional couplers. Let us take the case when all couplers have the same coupling coefficient as proposed in [28]. In this case, the higher are the coupling values (see Figure 3), the higher is the circuit efficiency we can achieve. This is expected as more power will be distributed to the radiating elements, instead of getting dissipated at the lines' terminations. Moreover, it can be easily observed that, for every non-spurious path case, the input signal realizes only one coupling value (see in Figure 3 how the signal flows from input port 2 to output port 2) before it reaches any output port. Therefore, an effect of the order of C [e.g., $O(C)$] is introduced on this signal, both in amplitude and phase. On the other hand, for the case of a spurious path, the signal has to travel through at least three "coupled ports", introducing an effect of at least $O(C^3)$. Thus, the lower the coupling coefficient of the coupler is, the lower the effect of the spurious path is. Therefore, there is a trade-off between the efficiency of the network and the errors introduced on both the amplitude and phase of the excitations. For our case, we choose as a good balance coupling values of $C = -11.5$ dB [see Figure 7a]. In the next subsection, the design methodology of the proposed $N \times M$ Blass matrix beamforming network is presented.

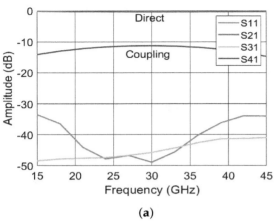
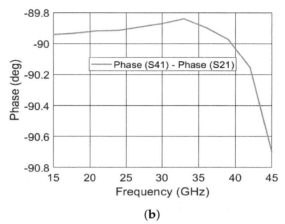

(a)   (b)

Figure 7. Frequency response of the double layer directional coupler: (a) amplitude response; and (b) phase difference between the output ports of the double layer directional coupler.

Blass Matrix Design Methodology

As mentioned above, a Blass matrix consists of M traveling wave feed lines (columns) cross-connected to a set of N transmission lines (rows), each one feeding an antenna element of the array. The first step of the design process is to choose the number of inputs M and outputs N, respectively (in this work, $M = 3$ and $N = 4$). Aiming to steer the beams of our antenna array in a specific direction, the desired phase differences $\Delta\phi_{(n+1,n)}^{(m)}$ between the output ports for each m beam are computed, utilizing array theory [51]. Recall that, for a linear phased array with interelement distance d, phase difference $\Delta\phi$, or group delay τ, to steer the beam toward an angle θ_0 from broadside is (e.g., [51]):

$$\Delta\phi = \frac{\omega d \sin\theta_0}{c} \quad or \quad \tau = \frac{d \sin\theta_0}{c} \tag{1}$$

where c is the speed of light in vacuum and ω is the angular frequency. Although a true-time-delay beamformer is sought, its design is based on the corresponding phase differences. These are first estimated at the center frequency based on standard antenna array synthesis. In turn, phase differences are transformed to the corresponding group delays to implement the desired wideband true-time-delay antenna array. Note that, since we are proposing a Blass matrix implemented in a semi-circular dual-layer topology, which to our knowledge has never been reported before, significant modifications have to be done on the design approach compared to the classical Blass matrix [28]. Namely, the directional couplers as well as the feed-lines' lengths between the couplers have to be appropriately chosen and designed.

The Blass matrix design methodology is a serial process, which is performed column-wise. Namely, we start with the design of column 1 adding the appropriate couplers and lengths of lines at each row $1 \cdots M$, respectively. A dual-layer directional coupler in Figure 6 is used herein to accommodate our novel implementation. Since this Blass matrix has $N = 4$ output ports, four directional couplers are needed at each column. Between the couplers, a transmission line with a group delay τ_v is added. The purpose of this line is to introduce some physical separation between the couplers for fabrication purposes and is usually chosen to be around $8.3ps$ at the frequency band we have chosen to operate. In addition, the couplers have to be symmetric and form a "cross", as shown in the schematic of Figure 3. To achieve this geometrical configuration, the top and bottom patches of the coupler are tilted at $45°$, while in addition $45°$ arcs are introduced at the ends of the patches, as shown in Figure 6. Note that the radius of the arcs has to be chosen wisely as firstly the characteristic impedance of the corresponding lines have to remain unaffected, and secondly the coupler has to remain compact. The length as well as the width of the slot and the patch are designed following the methodology in [49,50]. The total phase difference between the coupler's input-port and its through-port is denoted as $\Delta\phi_c$ and the corresponding delay τ_c. In addition, the coupling value is decided by the patch and slot width, and are always $90°$ in terms of electrical length. Thus, the phase introduced due to the coupling value is negligible. Therefore by looking at the Blass matrix schematic (see Figure 8) the total phase introduced at the nth output-port, when input port $m = 1$ is used as a reference, is:

$$\Delta\phi_{n,1}^{(1)} = n\Delta\phi_c - (N-n)\Delta\phi_0 \quad \leftrightarrow \quad \tau_{n,1}^{(1)} = n\tau_c - (N-n)\tau_0 \tag{2}$$

where $\Delta\phi_0$ is the phase introduced by line τ_0. From Equation (2), it can be seen that the phase difference between two adjacent output-ports $(n, n+1)$ is:

$$\Delta\phi_{n+1,n}^{(1)} = \Delta\phi_c - \Delta\phi_0 \quad \leftrightarrow \quad \tau_{n+1,n}^{(1)} = \tau_c - \tau_0 = \tau \tag{3}$$

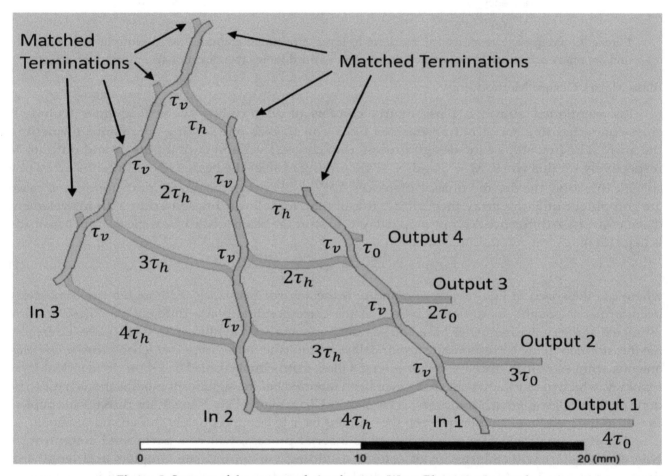

Figure 8. Layout of the proposed circular mm-Wave Blass matrix topology.

Using Equation (3), the electrical length of line τ_0 can be specified. Since τ is the desired group delay estimated from the antenna array in order for the beam maximum to point toward the pre-specified direction θ_0, it can similarly be seen that for column 2 the phase differences between adjacent output ports are:

$$\Delta\phi_{n+1,n}^{(2)} = \Delta\phi_c - \Delta\phi_0 - \Delta\phi_h \quad \leftrightarrow \quad \tau_{n+1,n}^{(2)} = \tau_c - \tau_0 - \tau_h \tag{4}$$

where $\Delta\phi_h$ is the phase introduced by line τ_h. In a similar manner, the rest of the matrix columns are designed. The last step of the design methodology is to choose the coupling coefficient of the couplers. As mentioned above, all beams except from the first one suffer from spurious excitations. These excitations degrade the performance of the Blass matrix by introducing perturbations in amplitude and phase, as shown in [28] (Equation (6)). It can be seen that the higher the coupling value is, the stronger the effect of these perturbations is, ultimately degrading SLL. On the other hand, if a low coupling value is chosen, only a small portion of the power will be coupled to the output ports, and the rest will be consumed by the terminations (see Figure 3). Therefore, the coupling value is determined by the application and how tolerant this application is to high SLL values.

2.2. Rotman Lens Architecture

In our second approach, a novel true time delay Rotman lens is used. Rotman lenses, similar to classical Blass topologies, are known for two main limitations: their large footprint and their limited efficiencies (typically 50–60%) due to the dummy ports. Aiming at high efficiency, we propose a novel dummy-ports-free Rotman lens utilizing the idea of field absorbers. Specifically, imitating perfect matching layers (PMLs), we engineer the substrate by doping it with conductive material. This key modification essentially increases the loss tangent about ten times, yielding an almost perfect absorber and eliminating the undesired reflections.

As is well known, the original Rotman lens is a metallic plate waveguide, loaded with dielectric for miniaturization purposes, where its input ports lie upon the beam contour and the output ports lie upon the array contour. As its grounded substrate is extended beyond the waveguide boundaries for practical purposes, it resembles a printed microstrip structure, although the area between the two metallic planes remains a parallel plate waveguide operating at its TEM mode.

As shown in Figure 9, the Rotman lens is considered as a $M \times N$ beamforming network. Its functionality is: when an input port is excited, the input signal is divided into N output ports with different amplitudes and phase sequences, which create a radiated beam with its maximum directed along a specific direction. When the same input signal is fed to another input port, a beam oriented to a different scanning angle is created. The energy fed to each input port emerges as rays directed towards all possible directions inside the parallel plate region. Some of them collimate, feeding the antennas through the output ports. However, there are also rays that propagate towards the top and bottom sides of the waveguide (in Figure 9, the top and bottom sides are considered the sides where the dummy ports are placed). These rays may be reflected to arrive at the output ports (even at the input ports) interfering with the directed rays at arbitrary phase, thus "contaminating" the desired signals (or causing mismatching at the inputs). Hence, the energy of these side-propagating rays must be absorbed and dissipated when they arrive at the side edges of the lens. To reduce the reflections in the parallel plate region, the dummy ports are used, as shown in Figure 9. Essentially, the dummy ports serve as absorbers for the spillovers of the lens, reducing the multiple reflections and standing waves, which undermine the lens performance [52].

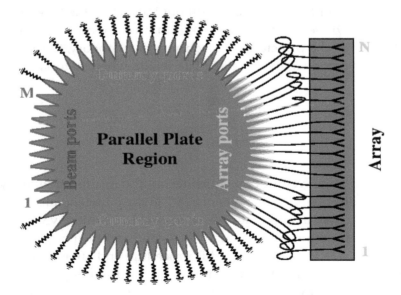

Figure 9. Principle Rotman lens architecture.

Rotman Lens Design Methodology

In this work we follow the classical Rotman lens design approach, as depicted in Figure 10 given by [52].

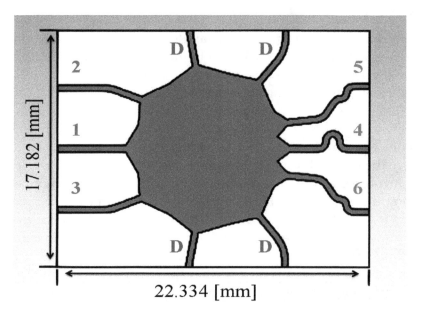

Figure 10. Proposed Rotman lens architecture with three inputs (red), three outputs (green), and four dummy (D) ports (blue) in the side walls. The scale is 10:3.06 mm.

Aiming at a compact, low cost, and highly efficient design, our goal is to minimize the number of dummy ports. However, when a small number of dummy ports is used, the Rotman lens performance degrades radically, as the inputs are no longer well matched, and the outputs' phase response fluctuates leading to a non-constant group delay. This is a well-recognized problem and numerous attempts have been made by researchers to confront it. Indicatively, Sorrentino et al. [53] utilized field absorbers at the side edges built up on a metamaterial approach. Here, we take a different path, imitating the perfect matching layers (PML). For this purpose, we appropriately engineer the substrate by doping it with conductive material, (e.g., carbon), on the area directly beyond the sides of the upper metal. This is considered to retain about the same dielectric constant but increasing the loss tangent about ten times. Thus, we can efficiently absorb the incident waves and in turn attenuate the standing waves of the parallel plate region. Following the principles of electromagnetic theory, the attenuation constant for the case of a TEM propagating mode in the parallel plate region when material losses are introduced is given by [48]:

$$a_d = \frac{k \tan \delta}{2} \tag{5}$$

where $k = \omega\sqrt{\epsilon\mu}$ stands for the wavenumber. Considering the desired frequency of operation and keeping in mind that the mode attenuates at a rate of $e^{-a_d x}$ in the parallel plate region, we can evaluate the distance the wave has to cover to attenuate $1/e$ in respect to its initial value. Based on this initial estimation the Rotman lens' absorbing area is initially designed and then in turn is optimized using an EM simulator. In our proposed design, following this procedure and after the necessary optimization, the length of the structure increases almost 6% with respect to the dummy-ports-based design, ensuring sufficient absorption of the incident waves in the side walls. Figure 11 presents how the dummy-port based design has been modified when absorbers are introduced in the side walls.

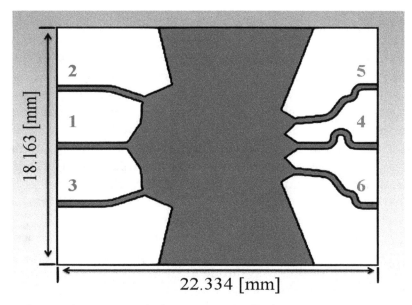

Figure 11. Rotman lens architecture with three inputs (red), three outputs (green), and two absorbers in the side walls. The scale is 10:3.06 mm.

This approach is proven to yield an almost perfect absorber, eliminating reflections along with their deleterious effects on the input matching and output phase fluctuations. The challenge here is to invent practical ways for the implementation of the assumed "conductive doping". A simple structure with only three input and three output ports is considered herein as a proof of concept, and it is proven to work very well.

2.3. Tree Type Beamforming Network Architecture

Our last design is focused on the development of a novel tree-based beamforming network, which mimics topologies that have been employed in microwave photonics [54]. We expect that this design will enable the development of new tunable three-dimensional (3D) beamformers for next-generation hybrid phased array systems providing game-changing capabilities compared to other beamformers. Specifically, the proposed implementation provides ultra-wideband operation, eliminates the problems of crossovers faced by Butler, Nolen, and Blass matrices, and is capable of exhibiting significantly higher efficiencies than Rotman lenses. Figure 12 shows our proposed tree topology. As can be seen, the building blocks of this topology are transmission lines and power dividers, making its topology relatively simple. The primary novelty of this topology is the minimal number of the involved delay lines. This idea stems from microwave photonics, and it is exploited in microwave and millimeter wave regimes. The key feature to observe refers to the hierarchical usage of delay lines with multiples of the basic delay τ from the antenna ports level towards the transceiver port. The idea is to hierarchically repeat the same feature by doubling the delay while moving diagonally from the first antenna port towards its last port and towards the network input. Specifically, the signal entering the input port undergoes successive power divisions by reaching the output ports with both the desired amplitude and phase. Even though our proposed design looks similar to corporate-type beamforming networks [55], the time delays in the form of transmission line paths are introduced in the common branches and not before the output ports of the network. This modification is crucial as it offers two main advantages. First, in the tree topology, the total delay (or phase shift) is introduced through the common branch, instead of the multiple small branches used in the corporate network that introduce the desired time delays in parallel. This key modification makes the tree beamforming network significantly more compact compared to the corporate network. Secondly, and most importantly, the transmission lines playing the role of phase shifters are true time delay elements, which eliminates the beam squint phenomenon. As is expected, a tree-based network can introduce only a specific time delay, steering the beam towards only one specific angle. However, this limitation can be overcome using multi-layer topologies with a different tree network on each layer, where each layer is devoted to steer the beam towards a specific angle. Figure 13 shows in detail our proposed multilayer design. To connect the different layers a single pole multiple throw (SPNT) type switch is used at the input and at each antenna port to effectively activate the corresponding layer (tree-topology) and provide the desired radiated beam (thereby achieving beamsteering). The challenge here is to appropriately design the interconnecting lines so that they are compatible with the monolithic microwave integrated circuits (MMICs) of SPNT switches. To address this challenge, all the interconnecting lines are designed as striplines, an approach that has never been implemented before on these networks. This design approach not only provides the needed multilayer beamformers but also ensures low losses at mm-wave frequencies.

As is expected, a critical component of the tree beamformer is the power divider. The power divider has to be appropriately designed to support the required bandwidth while it simultaneously introduces the desired amplitude distributions at the radiating elements of the network.

At this stage of our research, we are only focused on the second aim, leaving the wideband behavior as a future task. Our goal is to design a beamforming network that can introduce Chebyshev distribution at the antenna ports. To achieve the Chebyshev distribution, asymmetric power dividers have to be designed by appropriately distributing the power at their ports and throughout the network.

Conventional unequal power division Wilkinson power dividers can achieve the unequal power split by introducing an asymmetry in the characteristic impedance of their two branches. This way excellent performance is achieved in relatively small power division ratios. As the power division ratio increases, however, the characteristic impedances of the branches acquire very large or very small values, which are unrealizable in microstrip or stripline technology. Large characteristic impedance lines tend to be very thin and sometimes outside of the manufacturing tolerances, while small characteristic impedance lines are very wide, introducing large parasitic capacitances, thereby making them inappropriate for our application. Therefore, unequal power division Wilkinson power dividers are not suitable for our design due to the different power ratios that we need to achieve.

Specifically, Figure 14 shows the power divider used in this work. This unequal power divider is based on the design introduced in [56], which is capable of utilizing 50 Ω lines and introducing the desired asymmetry in the electrical length of its corresponding transmission lines. In our case, a 21 dB SLL Chebyshev amplitude distribution is excited among the elements. Therefore, the ratio of power arriving at the elements located at the edge of the array over the ones in the middle is 5.15 dB.

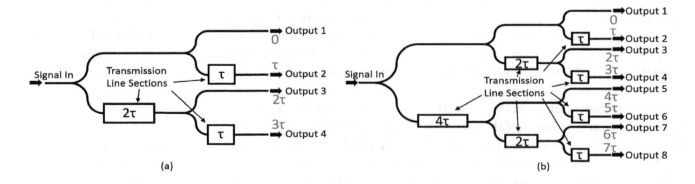

Figure 12. Schematic of the hierarchical true time delay tree beamformer: (**a**) 1 to 4; and (**b**) 1 to 8.

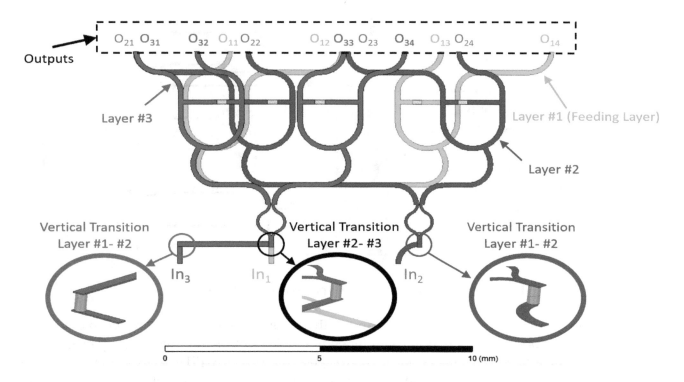

Figure 13. Multilayer tree topology layout.

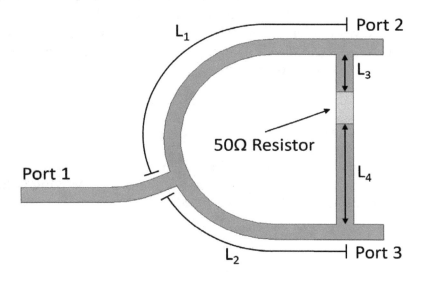

Figure 14. Layout of unequal power divider.

Tree Topology Design Methodology

A tree beamforming network is comprised of delay lines and power dividers. Similar to any other beamforming network design methodology, we first start by defining the number of desired output elements N and beams M. Since the tree network is capable of introducing only a specific time delay, steering the beam towards only one specific angle, for each beam direction $m \in [1, M]$, a different tree network has to be designed. In our proposed design, as we are implementing a multi-layer topology, each beam is produced by a different layer of our network. N by default can only be a power of two $N = 2^{\nu}$, with ν denoting the number of stages used to implement the tree network. Figure 15 shows one layer of a general schematic for the tree type network implemented with ν stages and N outputs.

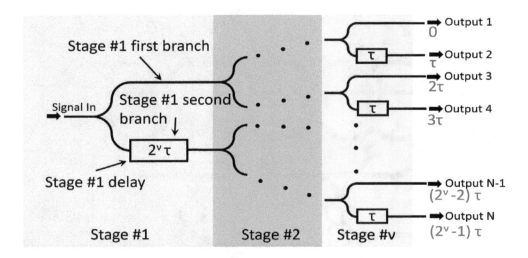

Figure 15. Schematic of the tree type network with N outputs.

Aiming to steer the beam at a specific angle, the phase difference $\Delta\phi$ between the adjacent output-ports, corresponding to a frequency independent delay of τ as in Equation (1). Thus, a $\Delta\phi = \omega_0\tau$ at the center frequency ω_0, is chosen, following the array theory [51]. In addition, as the tree topology offers us the ability to apply any tapering distribution we desire, the appropriate amplitude coefficients at the output ports are chosen. Herein, a Chebyshev array with a predefined $SLL = 21$ dB is employed. These are achieved by appropriately choosing the power division through the network's cascaded power dividers. The whole design philosophy is deployed as we move from stage 1 (left) to stage ν (right), as depicted with different colors in Figure 15. Assuming that P_n is the power arriving at each output port, the first stage power divider (in Figure 15, it is denoted with purple color) is designed to have a power division ratio of:

$$C_1 = \frac{\sum_{i=1}^{N/2} P_i}{\sum_{i=N/2+1}^{N} P_i} \tag{6}$$

Note that, when the amplitude tapering is symmetric, this ratio is equal to one. In addition, the electrical length of the first stage delay line is $\Delta\phi_1 = (N/2)\Delta\phi$. Moving to higher-order stages (e.g., stage 2–ν), the power division ratios are evaluated as:

$$C_\nu^j = \frac{\displaystyle\sum_{i=\text{ ports} \in \text{ first branch of j}^{\text{th}} \text{ divider}} P_i}{\displaystyle\sum_{i=\text{ ports} \in \text{ second branch of j}^{\text{th}} \text{ divider}} P_i} \tag{7}$$

while the electrical lengths and the delays of the corresponding delay lines are:

$$\Delta\phi_\nu = \frac{N}{2^\nu}\Delta\phi = \omega_0\tau_\nu \tag{8}$$

where ω_0 is the center angular frequency. Following this design methodology, any desired amplitude distribution is feasible maintaining the beam at the desired steered angle.

3. Numerical Results

In this section, the results of our proposed beamforming networks are organized into three corresponding subsections. Excellent performance is attained for each case network, thereby proving that they are suitable for our envisioned switched-beam phased arrays.

3.1. Blass Matrix Results

Figure 8 shows our proposed compact double-layer semi-circular Blass matrix that can achieve uniform amplitude distribution and steer its beam between the following three directions: broadside and $\pm 34°$. Notably, the design rules of the proposed Blass architecture are based on the design equations given in the original paper by Blass [28]. The 11.5-dB dual-layer directional coupler used in our design is implemented in a double 0.127-mm-thick duroid 5870 substrate with a dielectric constant of $\epsilon_r = 2.33$, as shown in Figure 16. The slot width is 0.4744 mm and the patch width is 0.5277 mm, which are estimated according to the methods in [49,50] to ensure an 11.5-dB coupling. The analytically estimated dimensions are 0.48 and 0.533 mm for the slot and the patch, respectively. These design parameters are optimized through the ANSYS HFSS simulation software, and the final slot and the patch lengths (see Figure 6) are 1.8 mm. Curved 50 Ω microstrip lines are attached to each side of both patches with a radius of 1.3 mm and a 45° arc angle. As shown in Figure 7a, the 3-dB bandwidth in terms of coupling coefficient (S_{41}) ranges from 15 to 45 GHz, while both the return loss S_{11} and isolation S_{31} are satisfactory. Finally, the phase of the coupling coefficient with respect to the direct path is designed to be $\Delta\phi = phase(S_{41}) - phase(S_{21}) = 90°$ for all desired bands of operation, as shown in Figure 7b.

Both the second and third columns of the Blass matrix are similar to the first but rotated by 31° and 62°, respectively. The required delay lines of multiples of τ_h according to Figure 4 are designed and implemented as microstrip lines, and the results are presented in Figure 17. In addition, τ_0 and its multiples are achieved through 1.987 mm straight microstrip line segments, while τ_h is achieved with curved microstrip lines with a length of 2.01 mm and an arc angle of 31°. Multiples of τ_h have the same arc angle but double the length. The vertical delay lines are chosen to have zero electrical length as, based on our simulation analysis, the coupler inserts a $18.08ps$ delay on its own. Therefore, there is enough physical separation and no extra vertical lines are needed. The desired delays τ_h for an array with an interelement distance $d = 4.5$ cm at the maximum operating frequency of $f_{max} = 33$ GHz are illustrated in Figure 18. Notably, $d = \lambda_{min}/2$ at f_{max} to avoid grating lobes. However, this could be related to higher values due to the restricted angular deflection of the radiated beam from 0° (broadside) to $\theta_{max} = 34°$ degrees as $d = \lambda_{0,min}/(1 + \theta_{max})$.

Figure 16. Blass matrix cross section.

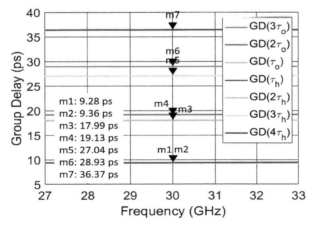

Figure 17. Group delay response of the delay lines of the Blass matrix.

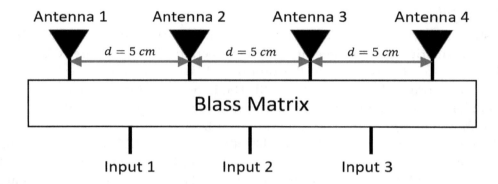

Figure 18. Example of array fed by the Blass matrix

The simulation results of the Blass matrix topology are presented in Figure 19a–f. Specifically, Figure 19a shows that, for the first beam, which aims towards +34°, the amplitudes at the corresponding elements exhibit only a ±0.4 dB or ±9.6% deviation from the desired uniform amplitude distribution. As expected, the time delays of the first beam are flat for the total operational bandwidth (see Figure 19b) since spurious paths are avoided by utilizing directional couplers. As shown in the inset of Figure 19b, the time delay differences between adjacent ports are fairly constant, deviating by no more than ±0.7*ps* or ±8.4%. Regarding the second beam, which aims towards broadside, the amplitudes at the output ports deviate from the uniform distribution by ±0.35 dB or ±8.4%, as shown in Figure 19c. Figure 19d presents the corresponding time delay of the output ports, showing a maximum difference of no more than ±2*ps* between them. Finally, for the third beam that aims towards −34°, which is affected the most from the spurious paths, compared to any other beam, the maximum amplitude deviation is ±0.35 dB ±8.4% (see Figure 19e), and the time delay differences between the output ports differ by ±2*ps* or ±24% at most. It is important to note that, even though the errors seem large at a first glance, they are not systematic and they exhibit a random-like behavior. As a consequence, both the directivity and the beam pointing angle of the resulting array factors are largely unaffected. The only degradation introduced by the spurious paths can be observed at the slightly increased SLL shown in Figure 20a–d as calculated directly from the S-parameters of our network. The maximum SLL of the three beams is 10 dB and occurs for the second beam's entire operational bandwidth. This is due to the spurious excitations, as predicted. The beam crossover level is at 4.5 dB at 33 GHz and goes up to 3 dB at 27 GHz. In summary, the operational bandwidth of the Blass design presented in this section is 27–33 GHz.

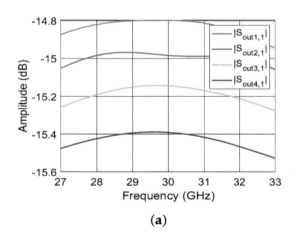

(a)

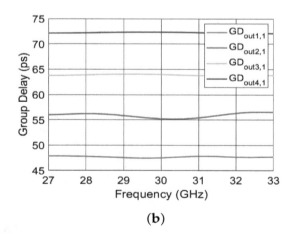

(b)

Figure 19. *Cont.*

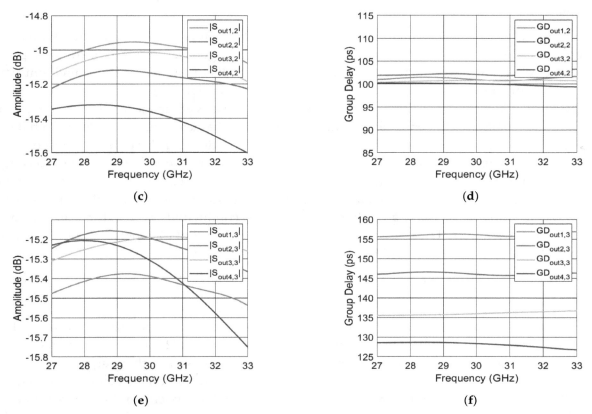

Figure 19. Proposed Blass matrix simulation results: (**a,c,e**) amplitudes $|S_{ij}|$ for the $i = 1$–4 outputs when the input ports $j = 1, 2, 3$ are excited; and (**b,d,f**) group delay for the $i = 1$–4 outputs when input ports $j = 1, 2, 3$ are excited.

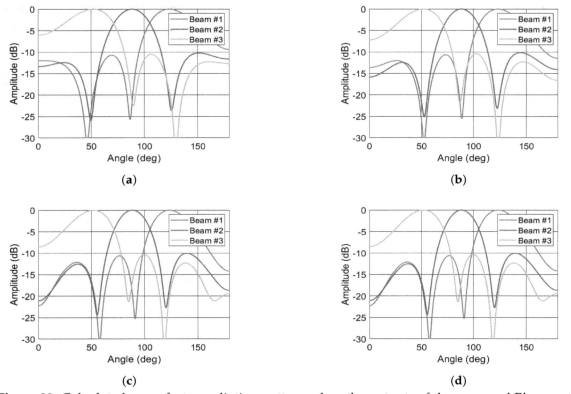

Figure 20. Calculated array factor radiation pattern when the outputs of the proposed Blass matrix (Figure 8) is used to excite the phased array at: (**a**) 27 GHz; (**b**) 29 GHz; (**c**) 31 GHz; and (**d**) 33 GHz.

3.2. Rotman Lens Results

In this work, the classical Rotman lens architecture with dummy ports is compared with a novel design where the dummy ports are replaced by absorbers. In general, the role of dummy ports is to reduce the reflection of the incident waves at the side walls of the lens. These incident waves are responsible for the appearance of standing waves in the parallel plate region and, consequently, the degradation of lens response. By replacing the dummy ports with absorbing layers, the incident waves on the side walls are efficiently absorbed, improving the characteristics of the lens. Both designs consist of three input ports, and three output ports. Notably, the design rules of the proposed Rotman lens architectures are based on the design equations of Rotman and Turner [30]. Aiming at a compact design, a substrate Rogers RT6006 with $\epsilon_r = 6.15$, and $tan\delta = 0.0027$ is chosen. All simulations were performed in the range of 27–33 GHz. The absorber material is assumed as the same substrate, but doped with conductive media (e.g., carbon) to increase its conductivity and, in turn, its loss tangent as: $\epsilon_r = 6.15$ and $tan\delta = 0.7$. The presented topology serves as a proof of concept and more practical designs with higher port numbers will be investigated in the future.

3.2.1. Rotman Lens with Four Dummy Ports

The classical Rotman lens topology studied herein consists of three input, three output, and four dummy ports, as depicted in Figure 10, with dimensions $22.334 \times 17.182 \times 0.3$ mm^3. The reduction of dummy ports number results in a reduced efficiency. Figure 21a–f shows the magnitude and phase distributions of the designed lens. Although the device seems to work satisfactorily in the frequency range 27–30 GHz, it has a prohibitive response in the second half of spectrum that ranges from 30 to 33 GHz. This unwanted behavior is due to the ineffective absorption of the incident waves on the side walls of the lens. These waves do not get absorbed, thereby exciting standing waves in the parallel plate region. Therefore, this deficiency affects the array factor of the feeding antenna array. Figure 22 reveals this distortion showing the occurring beam squint (i.e., undesired shifts of the beam's pointing angle) at 27 GHz (dotted lines) and at 30 GHz (continuous lines). The magnitudes of the input reflection coefficients (return loss) are depicted in Figure 23 and as expected they present poor matching.

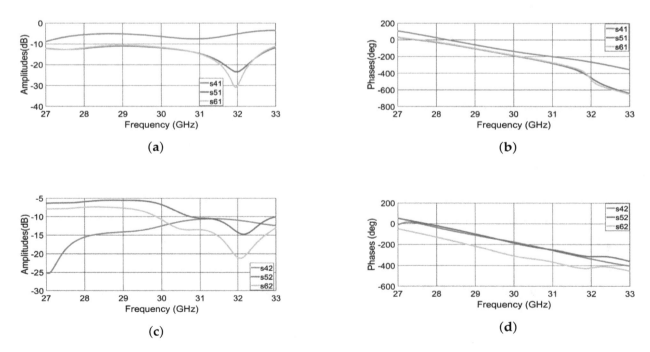

Figure 21. *Cont.*

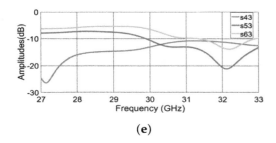

(e)

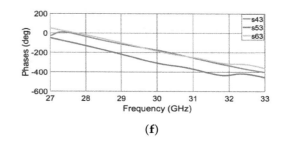

(f)

Figure 21. Ordinary Rotman lens with 4 dummy ports simulation results: (**a,c,e**) amplitudes $|S_{ij}|$ for the $i = 1$–3 outputs when the input ports $j = 1, 2, 3$ are excited; and (**b,d,f**) group delay for the $i = 1$–3 outputs when input ports $j = 1, 2, 3$ are excited.

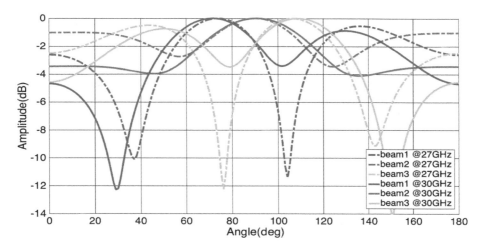

Figure 22. Calculated array factor radiation patterns at 27 GHz and 30 GHz when the outputs of the proposed Rotman lens (Figure 10) are used to excite the phased array.

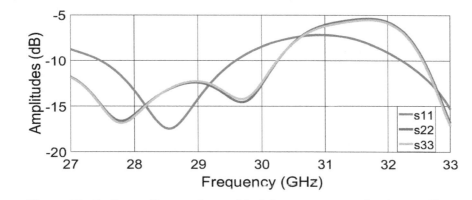

Figure 23. Ordinary Rotman lens with 4 dummy ports reflection coefficients.

3.2.2. Rotman Lens with Absorbers

Aiming at reducing the unwanted reflections of incident waves in the side walls and enhancing the total efficiency of the lens, the dummy ports are replaced by two absorbing layers covered with metallic shields on each side wall, as shown in Figure 11. The dimensions of this structure are defined as $22.334 \times 18.163 \times 0.3$ mm^3. The height of the absorbers is equal to the height of the substrate, and their material characteristics are chosen with an $\epsilon_r = 6.15$ and a $tan\delta = 0.7$. These material values have been chosen theoretically, by adjusting the value between the initial material and the material is then doped with the conductive molecules as described above. By engineering the substrate, we can efficiently absorb the incident waves and thus attenuate the standing waves of the parallel plate region.

Besides, the ultimate scope of this design is to prove that there is an easier to design and implement approach to absorb the incident waves in the side walls than the dummy ports.

Figure 24a–f reveals the improved Rotman lens response in the whole frequency range of 27–33 GHz. This improvement is also justified from the array factor distribution in Figure 25, which is valid in the entire frequency range.

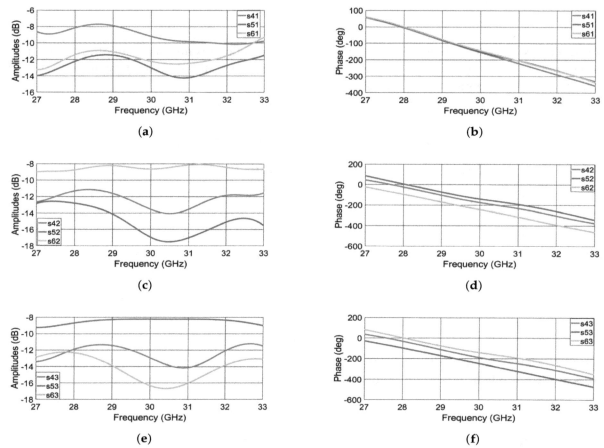

Figure 24. Proposed Rotman lens with absorbers in the side walls simulation results: (**a,c,e**) amplitudes $|S_{ij}|$ for the $i = 1$–3 outputs when the input ports $j = 1, 2, 3$ are excited; and (**b,d,f**) group delay for the $i = 1$–3 outputs when input ports $j = 1, 2, 3$ are excited.

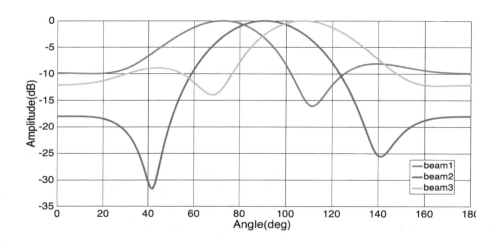

Figure 25. Calculated array factor radiation pattern when the outputs of the proposed Rotman lens (Figure 11) are used to excite the phased array.

3.2.3. Comparison of the Two Rotman Lens Topologies

The return loss (input reflection coefficient) of the two designed Rotman lenses are depicted in Figures 23 and 26. An improvement by at least 4 dB is clearly observed in the topology with the absorbers, while an acceptable matching over the entire band is offered at the same time. The corresponding coefficients are presented in Figures 21a–f and 24a–f when input ports 1, 2, 3 are activated. A reduction in the amplitude variation is first observed. However, the direct path transmission coefficient S_{41} is higher. Although a theoretically uniform distribution is sought, higher signal levels at the middle output ports are observed (Figures 21a–f and 24a–f), which yield shaped antenna array aperture excitation and thus lower sidelobes. Furthermore, the observed large amplitude variations in transmission coefficients are degrading the antenna array excitation, as shown in Figure 26. This may be due to the small number of output ports and we expect an improvement in a design with more output ports.

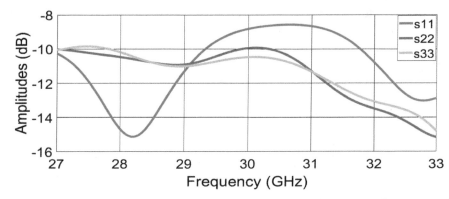

Figure 26. Proposed Rotman lens with absorbers in the side walls reflection coefficients.

Comparing Figures 21b,d,f and 24b,d,f, the output phase versus frequency linearity is impressively improved when the proposed design of the absorbers is used. Thus, the Rotman design with absorbers provides group delays with smaller variation versus frequency and thus more accurate beamsteering.

3.3. Tree Topology Results

Figure 13 shows the proposed three-layer tree beamforming network. The implementation of each layer is in stripline technology and the substrate is RT duroid 5870 with $\epsilon_r = 2.33$ and $tan\delta = 0.0012$. The thickness of each board is 0.127 mm and the signal line of each tree network is sandwiched between two single plated boards, as shown in Figure 27.

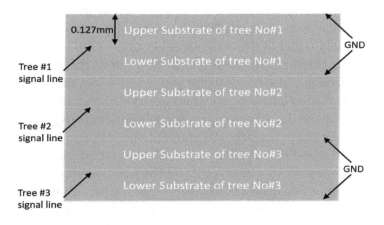

Figure 27. Cross section of the multilayer tree beamforming network.

All the input ports are located at the bottom layer of the network (see detail in Figure 13). Microstrip to stripline transitions are used for each layer separately. The first tree beamformer is located at the feeding layer and is designed to excite beam 1, as shown in Figure 13. The second tree beamformer at the middle of our multi-layer topology steers the beam to the left and is excited through a via that connects layer 1 with layer 2, as shown in detail in Figure 13. Finally, the third tree beamformer, at the top of our topology, is responsible for the broadside beam and is excited by utilizing two cascaded vertical transitions, as shown in detail in Figure 13.

As mentioned above, the power dividers are of high importance in the proposed topology. Two types of power dividers are incorporated in the network; an equal split Wilkinson power divider [57] and an unequal split power divider as the one reported in [56]. They are both implemented using stripline technology. Figure 28a shows the layout of the equal power divider. As shown in Figure 28b, the reflection coefficient is below 20 dB and the power is equally split through the desired bandwidth. Regarding the unequal power divider (shown in Figure 14), based on the design specifications of our tree topology, a 5-dB power division ratio has to be achieved. To achieve this power division ratio, the divider is designed to have a lower arc of an electrical length of $L_1^e = 151.2°$ and an upper arc of $L_2^e = 106.21°$, as shown in Figure 29a. In addition, the lines connecting the 50 Ω resistor to the branches of the divider are designed with electrical lengths of $L_3^e = 14.85°$ and $L_4^e = 59.95°$, respectively.

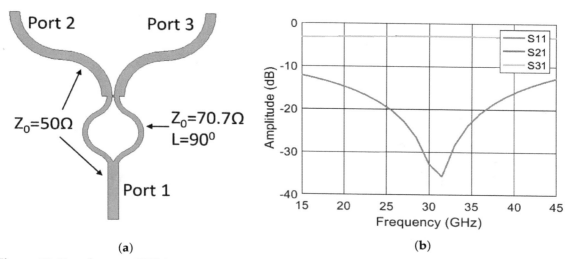

(a)   (b)

Figure 28. Equal power Wilkinson power divider: (a) layout of the equal power split Wilkinson power divider; and (b) amplitude response of the equal power split Wilkinson power divider.

Figure 29a shows the amplitude response of the unequal power division power divider. It can be seen that return loss is kept low for a large bandwidth, whereas the power division ratio limits the bandwidth of this design. As shown in Figure 29b, the phase difference between the two output ports of the divider is approximately equal at the center frequency.

In the proposed design (Figure 13), the physical lengths of the transmission lines connecting the output ports with the unequal power division power dividers on layer 1 are chosen to be 2.36 and 4.05 mm for the left and the right branches, respectively, and they consist of two right angle curved bends. The lengths of the lines connecting the Wilkinson power divider of the first stage to the power dividers of the second stage are 6.24 mm for the left and 2.86 mm for the right branch. In both branches, the arcs attached to the unequal division power divider have a radius of 0.372 mm and an angle of 130°. The tree network of layer 2 is a the same as that of layer 1 but mirrored over the vertical axis. Finally, for the tree network in layer 3, the transmission lines connecting the output ports with the unequal power division power dividers again consist of two right angle curved bends, whose combined physical length is 2.36 mm, while the interconnecting lines between the two stages have a total length of 3.02 mm.

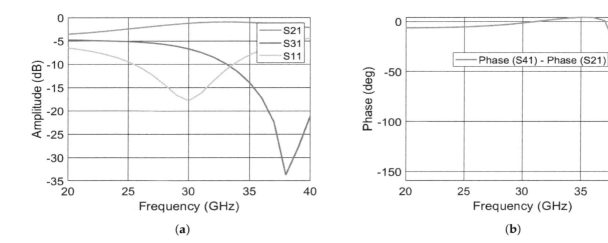

Figure 29. Response of the unequal split power divider: (**a**) amplitudes of the unequal split power divider output ports; and (**b**) phase difference between the unequal split power divider output ports.

Figure 30a–f shows the simulated results of the proposed multilayer tree topology. As shown in Figure 30a, the two middle output ports have similar amplitude responses and are approximately 5 dB higher than the other two ports at the center frequency of 29 GHz. Figure 30b shows the phase differences between the adjacent ports, which are approximately 90° with a minor deviation of ±3° at the center frequency. The second tree exhibits similar behavior. Figure 30c shows its amplitude responses where the difference of the two middle ports compared to the two outer ports is approximately 5 dB. The phase differences, similar to the first tree, maintain the required 90° between the adjacent ports with a slight deviation of ±3°. Lastly, Figure 30e shows the amplitude distribution for the ports of the third tree topology. As is shown, the amplitudes follow the same trend with cases 1 and 2, while the phase differences, as shown in Figure 30f, are at 0° with a minor deviation of ±3°. Figure 31 shows the array factor that was calculated using the S-matrix values of our network. It can be seen that the 19 dB achieved SLL is close to the desired 21 dB. The beam crossover level is at 3 dB and the beams point to broadside, +30°, and −30° with negligible deviations.

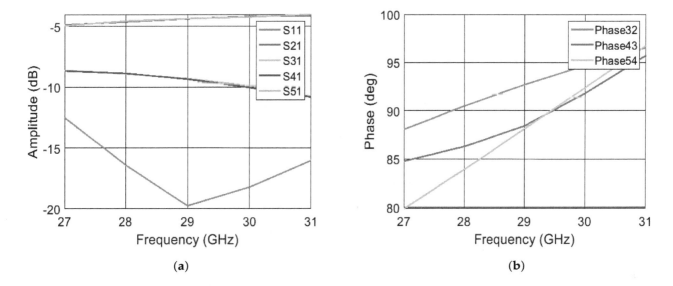

Figure 30. *Cont.*

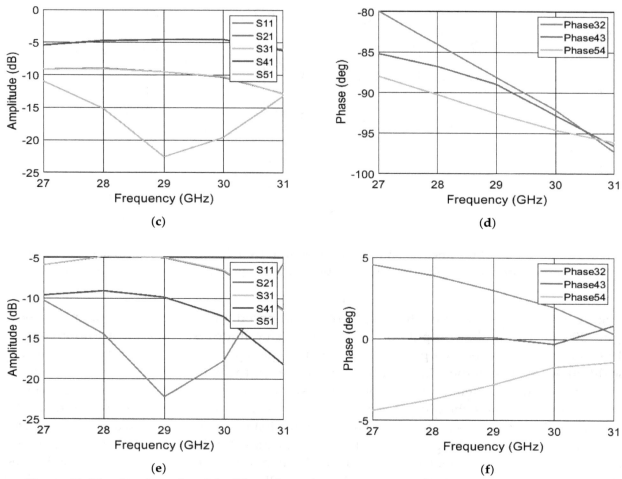

Figure 30. Simulated results of the 3D tree beamformer: (**a,c,e**) amplitude of the s-parameters for the bottom top and middle network respectively; and (**d,e,f**) the phase by the s-parameters for the bottom, top and middle network, respectively.

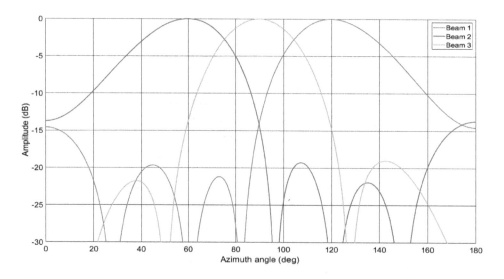

Figure 31. Calculated array factor of the tree topology.

4. Discussion

Wireless backhaul in mm-Wave bands that offers high speed, wide bandwidth, and several *Gbps* data rates will be a key component of 5G and early 6G communication systems. As discussed in the Introduction, depending on the channel properties, different communication systems (e.g., beamforming and MIMO) can be accordingly utilized to meet the system requirements, offering the optimal performance. In this work, we develop and show three different beamforming topologies, each with its own advantages and disadvantages, that can be appropriately used on switched beam phased arrays as hybrid beamforming stages. These are designed mainly as proofs of concept with small number of input- and output- ports. Our future work will aim to develop designs that fulfill the practical needs of 5G systems.

First, a novel semi-circular multilayer Blass matrix is proposed (Figure 8). By appropriately choosing the paths of the Blass network, we design a true-time-delay beamforming network aiming to a uniform amplitude distribution. To avoid the high number of crossovers that the typical couplers introduce at its original configuration (e.g., [28]), a double-layer directional coupler [49], is utilized, showing excellent performance throughout the desired band of operation. Notably, a size reduction is achieved compared to a regular Blass design (Figure 3). As shown in Section 3, both the amplitudes and the group delays observed at the output ports are fairly constant for the entire desired frequency band. This response offers an acceptable array factor radiation pattern with a side lobe of at most -10 dB fulfilling the purpose of the current work. For our second true-time-delay beamforming network, a 3×3 Rotman lens design is utilized with four dummy ports. To improve the typical poor performance of Rotman lenses with a small number of dummy ports, an innovative technique of field absorbers at the position of its dummy ports is introduced. Specifically, the dielectric substrate that surrounds the area of the dummy ports is doped with a conductive material. By engineering the substrate in this fashion, the dielectric constant remains the same, but its loss tangent increases approximately 10 times, which eliminates in turn the reflections that deteriorate the lens' performance. Following this design approach, we were able to build a pivotal implementation of a Rotman lens with only four dummy ports, a number that is significantly smaller compared to the typical number of dummy ports used in related works found in the literature [46]. As our last true-time-delay beamforming network, a multi-layer tree topology is introduced for the first time herein (Figure 13). The primary advantage of this topology is the extremely small number of delay lines used along with the corresponding power dividers. Aiming at a Chebyshev amplitude distribution at the excited elements, power dividers of unequal power division are utilized as a proof of concept. As shown in our results for all three beams, there is only a slight phase deviation of $\pm 3°$ at the center frequency, while the amplitude responses of the outer ports do not deviate more than ± 0.2 dB in respect to the amplitude responses of the center ports, thereby producing a radiation pattern of 19 dB SLL. Notably, even though the tree network is inherently ultra- wideband, since its transmission lines are true time delay elements, the unequal power dividers we use in the proposed design are narrowband. Specifically, the unequal power dividers are limited to operate in a narrow frequency band around the center frequency of 29 GHz thereby making the response of the proposed tree network narrowband as well. However, this is not a big concern, as the specific power divider is only used as a proof of concept and it can be replaced by ultra-wideband unequal power dividers published in the literature (e.g., [58]).

5. Conclusions

Millimeter wave communications have become one of the most promising candidates for the future wireless networks. In this paper, we address the needs of 5G and early 6G wireless communication systems and we explain why hybrid beamforming networks have to be used to address the challenges of these networks and the corresponding channel environments. Specifically, we mention that millimeter wave beamformers are appropriate for switched beam phased arrays as hybrid beamforming stages. Three different true-time-delay analog beamforming networks are developed: a multi-layer Blass matrix, a Rotman lens, and a multi-layer tree topology. For each beamforming network, we develop separately its design methodology as well as analyze its electromagnetic performance, thereby showing their suitability for future communication systems.

Author Contributions: All authors equally contributed to the manuscript. All authors have read and agreed to the published version of the manuscript.

References

1. Parvez, I.; Rahmati, A.; Guvenc, I.; Sarwat, A.I.; Dai, H. A Survey on Low Latency Towards 5G: RAN, Core Network and Caching Solutions. *IEEE Commun. Surv. Tutor.* **2018**, *20*, 3098–3130. [CrossRef]
2. David, K.; Berndt, H. 6G Vision and Requirements: Is There Any Need for Beyond 5G? *IEEE Veh. Technol. Mag.* **2018**, *13*, 72–80. [CrossRef]
3. Li, R. Towards a New Internet for the Year 2030 and Beyond. In Proceedings of the 3rd Annual ITU IMT-2020/5G Workshop and Demo Day, Geneva, Switzerland, 18 July 2018.
4. Strinati, E.C.; Barbarossa, S.; Gonzalez-Jimenez, J.L.; Ktenas, D.; Cassiau, N.; Maret, L.; Dehos, C. 6G: The Next Frontier: From Holographic Messaging to Artificial Intelligence Using Subterahertz and Visible Light Communication. *IEEE Veh. Technol. Mag.* **2019**, *14*, 42–50. [CrossRef]
5. Haerick, W.; Gupta, M. 5G and the Factories of the Future. Available online: https://5g-ppp.eu/wp-content/uploads/2014/02/5G-PPP-White-Paper-on-Factories-of-the-Future-Vertical-Sector.pdf (accessed on 13 August 2020).
6. Huawei. 5G: A Technology Vision. Available online: https://www.huawei.com/ilink/en/download/HW_314849 (accessed on 13 August 2020).
7. Hossain, E.; Rasti, M.; Tabassum, H.; Abdelnasser, A. Evolution toward 5G multi-tier cellular wireless networks: An interference management perspective. *IEEE Wirel. Commun.* **2014**, *21*, 118–127. [CrossRef]
8. Schulz, P.; Matthe, M.; Klessig, H.; Simsek, M.; Fettweis, G.; Ansari, J.; Ashraf, S.A.; Almeroth, B.; Voigt, J.; Riedel, I.; et al. Latency critical IoT applications in 5G: Perspective on the design of radio interface and network architecture. *IEEE Commun. Mag.* **2017**, *55*, 70–78. [CrossRef]
9. Johansson, N.A.; Wang, Y.P.E.; Eriksson, E.; Hessler, M. Radio access for ultra-reliable and low-latency 5G communications. In Proceedings of the 2015 IEEE International Conference on Communication Workshop (ICCW), London, UK, 8–12 June 2015; pp. 1184–1189.
10. Lema, M.A.; Antonakoglou, K.; Sardis, F.; Sornkarn, N.; Condoluci, M.; Mahmoodi, T.; Dohler, M. 5G case study of Internet of skills: Slicing the human senses. In Proceedings of the 2017 European Conference on Networks and Communications (EuCNC), Oulu, Finland, 12–15 June 2017; pp. 1–6.
11. Ghosh, A. 5G mmWave Revolution and New Radio. 2017. Available Online https://futurenetworks.ieee.org/images/files/pdf/5GmmWave_Webinar_IEEE_Nokia_09_20_2017_final.pdf (accessed on 13 August 2020).
12. Ghosh, A.; Maeder, A.; Baker, M.; Chandramouli, D. 5G Evolution: A View on 5G Cellular Technology Beyond 3GPP Release 15. *IEEE Access* **2019**, 127639–127651. [CrossRef]
13. Niu, Y.; Li, Y.; Jin, D.; Su, L.; Vasilakos, A.V. A survey of millimeter wave communications (mmWave) for 5G: opportunities and challenges. *Wirel. Netw.* **2015**, *21*, 2657–2676. [CrossRef]
14. Zhao, Q.; Li, L. Rain attenuation in millimeter wave ranges. In Proceedings of the 2006 7th International Symposium on Antennas, Propagation & EM Theory, Guilin, China, 26–29 October 2006; pp. 1–4.
15. Humpleman, R.J.; Watson, P.A. Investigation of attenuation by rainfall at 60 GHz. *Proc. Inst. Electr. Eng.* **1978**, *125*, 85–91. [CrossRef]
16. E-Band Technology. E-band Communications. Available Online: http://www.e-band.com/index.php?id=86 (accessed on 13 August 2020).
17. Rappaport, T.S.; Sun, S.; Mayzus, R.; Zhao, H.; Azar, Y.; Wang, K.; Wong, G.N.; Schulz, J.K.; Samimi, M.; Gutierrez, F. Millimeter wave mobile communications for 5G cellular: It will work! *IEEE Access* **2013**, *1*, 335–349. [CrossRef]
18. Pleros, N.; Tsagkaris, K.; Tselikas, N.D. A moving extended cell concept for seamless communication in 60 GHz radio-over-fiber networks. *IEEE Commun. Lett.* **2008**, *12*, 852–854. [CrossRef]
19. Mandyan, A. 4G and 5G Capacity Solution, Comparative Study. Available Online: https://telecoms.com/intelligence/4g-and-5g-capacity-solutions-comparative-study/ (accessed on 13 August 2020).
20. Sun, S.; Rappaport, T.S.; Shafi, M.; Tang, P.; Zhang, J.; Smith, P.J. Propagation models and performance evolution for 5G millimeter-wave bands. *IEEE Trans. Vehicular Tech.* **2018**, *67*, 8422–8437. [CrossRef]

21. Molisch, A.F.; Ratnam, V.V.; Han, S.; Li, Z.; Nguyen, S.L.H.; Li, L.; Haneda, K. Hybrid Beamforming for Massive MIMO: A Survey. *IEEE Commun. Mag.* **2017**, *55*, 134–141. [CrossRef]
22. Rappaport, T.S.; Xing, Y.; MacCartney, G.R.; Molisch, A.F.; Mellios, E.; Zhang, J. Overview of millimeter wave communications for fifth generation (5G) wireless networks-with a focus on propagation models. *IEEE Trans. Antennas Propag.* **2017**, *65*, 6213–6230. [CrossRef]
23. Roh, W.; Seol, J.Y.; Park, J.; Lee, B.; Lee, J.; Kim, Y.; Cho, J.; Cheun, K.; Aryanfar, F. Millimeter-Wave beamforming as an enabling technology for 5G cellular communications: The oretical feasibility and prototype results. *IEEE Commun. Mag.* **2017**, *52*, 106–113. [CrossRef]
24. Cameron, T. RF Technology for 5G Mmwave Radios. Available online: /https://www.analog.com/media/en/technical-documentation/white-papers/RF-Technology-for-the-5G-Millimeter-Wave-Radio.pdf (accessed on 13 August 2020).
25. Ali, E.; Ismail, M.; Nordin, R.; Abdullah, N.F. Beamforming techniques for massive MIMO systems in 5G: overview, classification, and trends for future research. *Front. Inf. Technol. Electron. Eng.* **2017**, *18*, 753–772. [CrossRef]
26. Ahmed, I.; Khammari, H.; Shahid, A.; Musa, A.; Kim, K.S.; De Poorter, E.; Moerman, I. A Survey on Hybrid Beamforming Techniques in 5G: Architecture and System Model Perspectives. *IEEE Commun. Surv. Tutor.* **2018**, *20*, 3060–3097. [CrossRef]
27. Butler, J.; Lowe, R. Beam forming matrix simplifiers design of electrically scanned antennas. *Electron. Des.* **1961**, *9*, 170–173.
28. Blass, J. Multidirectional antenna—A new approach to stacked beams. In Proceedings of the 1958 IRE International Convention Record, New York, NY, USA, 21–25 March 1966; pp. 48–50. [CrossRef]
29. Nolen, J. Synthesis of Multiple Beam Networks for Arbitrary Illuminations. Ph.D. Thesis, Radio Division, Bendix Corporation, Baltimore, MD, USA, 1965.
30. Rotman, W.; Turner, R. Wide-angle microwave lens for line source applications. *IEEE Trans. Antennas Propag.* **1963**, *11*, 623–632. [CrossRef]
31. Wincza, K.; Staszek, K.; Gruszczynski, S. Broadband Multibeam Antenna Arrays Fed by Frequency-Dependent Butler Matrices. *IEEE Trans. Antennas Propag.* **2017**, *65*, 4539–4547. [CrossRef]
32. Djerafi, T.; Wu, K. A Low-Cost Wideband 77-GHz Planar Butler Matrix in SIW Technology. *IEEE Trans. Antennas Propag.* **2012**, *60*, 4549–4954. [CrossRef]
33. Fakoukakis, F.E.; Kaifas, T.; Vafiadis, E.E.; Kyriacou, G.A. Design and implementation of Butler matrix-based beam-forming networks for low sidelobe level electronically scanned arrays. *Int. J. Microw. Wirel. Technol.* **2015**, *7*, 69–79. [CrossRef]
34. Fakoukakis, F.E.; Kyriacou, G.A. Novel Nolen Matrix Based Beamforming Networks for Series-Fed Low SLL Multibeam Antennas. *Prog. Electromagn. Res. B* **2013**, *51*, 33–64. [CrossRef]
35. Chen, Q.P.; Zheng, S.Y.; Long, Y.; Ho, D. Design of a Compact Wideband Butler Matrix Using Vertically Installed Planar Structure. *IEEE Trans. Components Packag. Manuf. Technol.* **2018**, *8*, 1420–1430. [CrossRef]
36. Nedil, M.; Denidni, T.A.; Talbi, L. Novel butler matrix using CPW multilayer technology. *IEEE Trans. Microw. Theory Tech.* **2006**, *54*, 499–507. [CrossRef]
37. Cao, Y.; Chin, K.; Che, W.; Yang, W.; Li, E.S. A Compact 38 GHz Multibeam Antenna Array With Multifolded Butler Matrix for 5G Applications. *IEEE Antennas Wirel. Propag. Lett.* **2017**, *16*, 2996–2999. [CrossRef]
38. Tornielli di Crestvolant, V.; Martin Iglesias, P.; Lancaster, M.J. Advanced Butler Matrices With Integrated Bandpass Filter Functions. *IEEE Trans. Microw. Theory Tech.* **2015**, *63*, 3433–3444. [CrossRef]
39. Dyab, W.M.; Sakr, A.A.; Wu, K. Dually-Polarized Butler Matrix for Base Stations With Polarization Diversity. *IEEE Trans. Microw. Theory Tech.* **2018**, *66*, 5543–5553. [CrossRef]
40. Tamayo-Domínguez, A.; Fernández-González, J.; Sierra-Castañer, M. 3-D-Printed Modified Butler Matrix Based on Gap Waveguide at W-Band for Monopulse Radar. *IEEE Trans. Microwave Theory Tech.* **2020**, *68*, 926–938. [CrossRef]
41. Fonseca, N.J. Printed S-Band 4 × 4 Nolen Matrix for Multiple Beam Antenna Applications. *IEEE Trans. Antennas Propag.* **2009**, *57*, 1673–1678. [CrossRef]
42. Djerafi, T.; Fonseca, N.J.G.; Wu, K. Broadband Substrate Integrated Waveguide 4 × 4 Nolen Matrix Based on Coupler Delay Compensation. *IEEE Trans. Microw. Theory Tech.* **2011**, *59*, 1740–1745. [CrossRef]

43. Ren, H.; Zhang, H.; Li, P.; Gu, Y.; Arigong, B. A Novel Planar Nolen Matrix Phased Array for MIMO Applications. In Proceedings of the 2019 IEEE International Symposium on Phased Array System & Technology (PAST), Waltham, MA, USA, 15–18 October 2019; pp. 1–4. [CrossRef]

44. Chu, T.-S.; Hossein, H. True time delay based multi-beam arrays. *IEEE Trans. Microw. Theory Tech.* **2013**, *61*, 3072–3081. [CrossRef]

45. Lambrecht, A.; Beer, S.; Zwick, T. True-time-delay beamforming with a Rotman-lens for ultrawideband antenna systems. *IEEE Trans. Antennas Propag.* **2010**, *10*, 3189–3195. [CrossRef]

46. Merola, C.S.; Vouvakis, M.N. Massive MIMO Beamforming on a Chip. In Proceedings of the 2019 IEEE International Symposium on Antennas and Propagation and USNC-URSI Radio Science Meeting, Atlanta, GA, USA, 7–12 July 2019; pp. 1477–1478.

47. Hansen, R.C. *Microwave Scanning Antennas: Array Systems*; Academic Press: Cambridge, MA, USA, 1964.

48. Pozar, D.M. *Microwave Engineering*; Wiley: Hoboken, NJ, USA, 2012.

49. Tanaka, T.; Tsunoda, K.; Aikawa, M. Slot–coupled directional couplers on a both–sided substrate MIC and their applications. *Electron. Comm. Jpn. Part II* **1989**, *72*, 91–99. [CrossRef]

50. Wong, M.F.; Hanna, V.F.; Picon, O.; Baudrand, H. Analysis and design of slot-coupled directional couplers between double-sided substrate microstrip lines. In Proceedings of the 1991 IEEE MTT-S International Microwave Symposium Digest, Boston, MA, USA, 10–14 June 1991; Volume 2, pp. 755–758. [CrossRef]

51. Balanis, C.A. *Antenna Theory: Analysis and Design*, 3rd ed.; John Wiley: Hoboken, NJ, USA, 2005.

52. Vashist, S.; Soni, M.K.; Singhal, P.K. A Review on the Development of Rotman Lens Antenna. *Chin. J. Eng.* **2014**, *2014*, 1–9. [CrossRef]

53. Sbarra, E.; Marcaccioli, L.; Gatti, R.V.; Sorrentino, R. A novel rotman lens in SIW technology. In Proceedings of the 2007 European Radar Conference, Munich, Germany, 10–12 October 2007; pp. 236–239. [CrossRef]

54. Yao, J. Microwave Photonics. *J. Lightwave Technol.* **2009**, *27*, 314–335. [CrossRef]

55. Mailloux, R.J. *Phased Array Antenna Handbook*, 3rd ed.; Artech House, Inc.: Norwood, MA, USA, 2017.

56. Qi, T.; He, S.; Dai, Z.; Shi, W. Novel Unequal Dividing Power Divider With 50 Ω Characteristic Impedance Lines. *IEEE Microw. Wirel. Components Lett.* **2016**, *26*, 180–182. [CrossRef]

57. Wilkinson, E.J. An N-Way Hybrid Power Divider. *Ire Trans. Microw. Theory Tech.* **1960**, *8*, 116–118. [CrossRef]

58. Chen, J.; Xue, Q. Novel 5to1 Unequal Wilkinson Power Divider Using Offset Double-Sided Parallel-Strip Lines. *IEEE Microw. Wirel. Compon. Lett.* **2007**, *17*, 175–177. [CrossRef]

4-Port MIMO Antenna with Defected Ground Structure for 5G Millimeter Wave Applications

Mahnoor Khalid [1], Syeda Iffat Naqvi [1,*], Niamat Hussain [2], MuhibUr Rahman [3,*], Fawad [1], Seyed Sajad Mirjavadi [4], Muhammad Jamil Khan [1] and Yasar Amin [1]

[1] ACTSENA Research Group, Department of Telecommunication Engineering, University of Engineering and Technology, Taxila, Punjab 47050, Pakistan; mahnoor13tc@gmail.com (M.K.); engr.fawad@students.uettaxila.edu.pk (F.); muhammad.jamil@uettaxila.edu.pk (M.J.K.); yasar.amin@uettaxila.edu.pk (Y.A.)

[2] Department of Computer and Communication Engineering, Chungbuk National University, Cheongju 28644, Korea; hussain@osp.chungbuk.ac.kr

[3] Department of Electrical Engineering, Polytechnique Montreal, Montreal, QC H3T 1J4, Canada

[4] Department of Mechanical and Industrial Engineering, College of Engineering, Qatar University, P.O. Box 2713 Doha, Qatar; seyedsajadmirjavadi@gmail.com

* Correspondence: iffat.naqvi@uettaxila.edu.pk (S.I.N.); muhibur.rahman@polymtl.ca (M.R.)

Abstract: We present a 4-port Multiple-Input-Multiple-Output (MIMO) antenna array operating in the mm-wave band for 5G applications. An identical two-element array excited by the feed network based on a T-junction power combiner/divider is introduced in the reported paper. The array elements are rectangular-shaped slotted patch antennas, while the ground plane is made defected with rectangular, circular, and a zigzag-shaped slotted structure to enhance the radiation characteristics of the antenna. To validate the performance, the MIMO structure is fabricated and measured. The simulated and measured results are in good coherence. The proposed structure can operate in a 25.5–29.6 GHz frequency band supporting the impending mm-wave 5G applications. Moreover, the peak gain attained for the operating frequency band is 8.3 dBi. Additionally, to obtain high isolation between antenna elements, the polarization diversity is employed between the adjacent radiators, resulting in a low Envelope Correlation Coefficient (ECC). Other MIMO performance metrics such as the Channel Capacity Loss (CCL), Mean Effective Gain (MEG), and Diversity gain (DG) of the proposed structure are analyzed, and the results indicate the suitability of the design as a potential contender for imminent mm-wave 5G MIMO applications.

Keywords: Multiple-Input-Multiple-Output (MIMO), array; 5G mm-wave; Defected Ground Structure (DGS), ECC; DG; MEG; CCL

1. Introduction

In the modern era, the eminent increase of wireless devices, inadequate bandwidth, and limited channel capacity have substantially promoted efforts to develop advanced standards for communication networks. Subsequently, this has promoted the development of next-generation (5G) communication systems at the mm-wave spectrum featuring much greater channel capacity and higher data rates [1,2]. The forthcoming 5G technology not only provides greatly increased reliability, high data rate requirements, and low power consumption to meet the massive increase in linked devices, but also promises to increase the prospects of emerging technologies such as virtual reality and smart cities [3–5]. However, critical limitations at the mm-wave spectrum, such as signal fading, atmospheric absorptions, and path loss attenuations need to be resolved, which becomes more significant with the usage of the single antenna [6–8]. Multiple-input multiple-output (MIMO) antenna has been determined to be a key enabling technology for current and future wireless systems, demonstrating concurrent operation of multi-antennas, increasing channel capacity along with the benefits of high data rates and throughput of Gigabits/sec [9–11]. The 5th generation MIMO antenna requires high bandwidth for concurrent functioning, while the high gain is required to reduce the atmospheric diminutions and absorptions at mm-wave frequencies, and compactness of structure is needed to facilitate the assimilation in MIMO systems. In addition to this, the challenges associated with MIMO antenna designing are to design closely packed antenna elements with reduced mutual coupling and high isolation, which subsequently improves the antenna performance.

Recently, several antenna solutions operating at mm-wave bands for 5G applications have been reported in the literature [12–35]. The antenna designs operating at the potential mm-wave bands with low gain reference [12–14] are not proficient enough to deal with the high atmospheric and propagation losses at the mm-wave frequency. In order to mitigate these attenuation effects, several high gain and beam-steering antenna array solutions have been presented to have strong signal strength and to offer large spatial coverage [15–23].

However, multi-element antenna arrays exhibit the same capacity as the single antenna because the antenna arrays are likewise fed with a single port. Conversely, MIMO antennas demonstrate multipath propagation with a higher data rate, increased capacity, and link reliability, which are the main features of 5G. A number of MIMO antenna designs for 5G mm-wave applications have been reported recently in the literature [24–35]. A PIFA array with MIMO configuration with 1 GHz operational bandwidth and a peak simulated gain as 12 dBi is reported in reference [24]. An EBG based mm-wave MIMO antenna is reported in reference [25] with a bandwidth of 0.8 GHz. The multi-element antenna design proposed in reference [26] enables the radiation to bend toward an intended inclined direction, which is suitable for 5G communications. A bandwidth of 1.5 GHz ranging from 27.2 GHz to 28.7 GHz is obtained, whereas the maximum gain attained by the antenna geometry is 7.41 dBi at 28 GHz. DRA based antennas for 5G applications are reported in references [27,28], having a limited data rate of approximately 1 GHz. In reference [27], the SIW feeding technique is reported with a peak gain of 7.37. To ensure MIMO performance, ECC is also evaluated in reference [28]. Likewise, a SIW fed slotted MIMO antenna array for mm-wave communication is reported in reference [29]. The proposed antenna covers the 24.25–27.5 GHz, and the 27.5–28.35 GHz bands for 5G, while the gain varies from 8.2 to 9.6 dBi over the operating frequency range. Moreover, a four-element T-shaped MIMO antenna with overall dimensions of $12 \times 50.8 \times 0.8$ mm^3 is presented for 5G applications [30]. The partial ground at the bottom layer consists of iteratively placed symmetrical split-ring slots. The proposed antenna design is covering a wide bandwidth of 25.1–37.5 GHz with a peak gain of 10.6 dBi. However, only ECC as the MIMO performance metric is investigated in this work. In another article, an 8×8 MIMO antenna design with an overall substrate size of $31.2 \times 31.2 \times 1.57$ mm^3 for future 5G devices is demonstrated [31]. The proposed MIMO structure resonates at 25.2 GHz, having a bandwidth of 5.68 GHz at -6 dB reference, while the maximum value of gain attained is 8.732 dB. Additionally, ECC, MEG, and Diversity gain are also examined. Similarly, a two-port MIMO array with overall dimensions of $31.7 \times 53 \times 0.2$ mm^3, and excited by microstrip feedline has been reported for 5G communication systems [32]. The reported antenna system with reflectors based on EBG provides a wide bandwidth and high gain of up to 11.5 dB in the operating band. Analysis of ECC and Diversity gain is exhibited for the proposed MIMO configuration. Likewise, in reference [33], a 5G MIMO antenna with three pairs of integrated metamaterial arrays is demonstrated. The overall substrate size is $30 \times 30.5 \times 0.508$ mm^3, whereas the maximum gain achieved by the antenna is 7.4 dBi at 26GHz. Moreover, the work in reference [34] presents a Fabry Perot high gain antenna with a superstrate for 5G MIMO applications. The proposed structure covers the mm-wave spectrum ranging from 26–29.5 GHz, with a maximum gain value of 14.1 dBi. In addition, ECC is analyzed to measure the MIMO performance of the proposed antenna configuration. A two-element MIMO dielectric resonator antenna (DRA) is suggested in reference [35]. The reported antenna covers the 27.19–28.48 GHz band for 5G applications. A gain value approaching 10 dB is obtained for the exhibited antenna. Furthermore, ECC, DG, channel capacity, and TARC are also inspected.

The design of a 4-element antenna array with MIMO capabilities at mm-wave 5G frequency bands is demonstrated in this paper. The proposed design is a high gain and wideband antenna with good MIMO characteristics for future 5th generation devices, such as smartwatches and mobile WiFi, etc. The proposed MIMO antenna with a compact and simple geometry facilitates its assimilation into 5G smart devices. The good MIMO performance of reported antenna endorses the appropriateness of the design for future 5G wireless communication applications.

2. Proposed Antenna Design

This work proposes a 4 port MIMO antenna system with overall substrate dimensions of $30 \times 35 \times 0.76$ mm^3, as shown in Figure 1. The antenna is modeled and simulated in a commercially available EM simulator CST microwave studio suite.

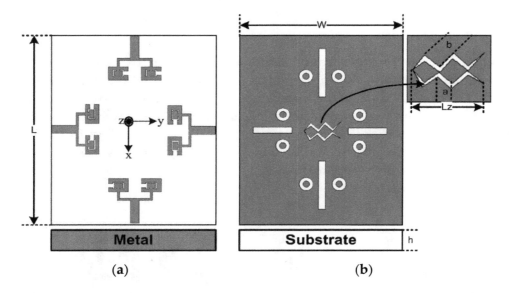

Figure 1. Proposed Multiple-Input-Multiple-Output (MIMO) Antenna module (**a**) Top view (**b**) bottom without a Defected Ground Structure (DGS).

The reported antenna system comprises of four MIMO elements placed on the center of each edge on the top layer, as shown in Figure 1a. Figure 1b depicts that the layer at the bottom side is composed of a Defected Ground Structure (DGS) with rectangular, circular, and zigzag-shaped slots to further enhance the performance of the proposed design. The design is integrated on the Rogers R04350B substrate with a thickness and permittivity (εr) of 0.76 mm and 3.66, respectively. The dimensional details of the design are provided in Table 1, and the progression of the design from a single element to MIMO configuration is comprehensively discussed in the subsequent sections.

2.1. Single Element Antenna

At first, a single element of the patch antenna is designed as shown in Figure 2a. The primary antenna structure resonating at 28 GHz is obtained by following the well-established mathematical equations provided in references [1–4]. The optimized single element antenna consists of an inverted C-shaped patch enclosing a rectangular-shaped slit.

$$W_p = \frac{c}{2f_c \sqrt{\frac{\varepsilon_{relative}+1}{2}}} \tag{1}$$

$$\varepsilon_{eff} = \frac{\varepsilon_{relative} + 1}{2} + \frac{\varepsilon_{relative} - 1}{2} \left(\frac{1}{\sqrt{1 + 12\left(\frac{h}{W_p}\right)}} \right) \tag{2}$$

$$\Delta L = 0.421h \frac{\left(\varepsilon_{eff} + 0.3\right)\left(\frac{W_p}{h} + 0.264\right)}{\left(\varepsilon_{eff} - 0.258\right)\left(\frac{W_p}{h} + 0.8\right)} \tag{3}$$

$$Lp = \frac{c}{2f_o \sqrt{\varepsilon_{eff}}} - 2\Delta L \tag{4}$$

where W_p and L_p are the patch's width and length, h is the height of the substrate, ε_{eff} and $\varepsilon_{relative}$ are the effective permittivity and relative permittivity of substrate respectively. $c, f_c,$, and ΔL are the speed of light, central frequency, and the effective length, respectively.

2.2. Two Element Antenna Array

The design is processed further from a single element to the two-element array, as shown in Figure 2b. The parallel feed network is proposed as the array's excitation mechanism. The main feed is matched at 50 Ω impedance while the impedance of the branched network is matched at 100 Ω. Afterward, a rectangular and two symmetrically placed circular slots are incorporated in the bottom layer in order to further optimize the obtained results, as illustrated in Figure 2c. Consequently, an enhanced bandwidth and gain are achieved by the reported antenna array. Both elements in the array antenna are separated by λ, which is approximately 11 mm at 28 GHz. Hence a compact array structure with proved performance is achieved.

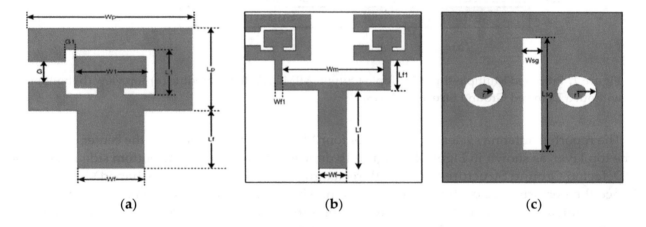

| (a) | (b) | (c) |

Figure 2. Proposed antenna geometry (**a**) Single element (**b**) Top layer of array antenna (**c**) Bottom layer.

2.3. MIMO Configuration

After the attainment of the two-element array, the design is progressed further, and the 4-port MIMO antenna system is obtained. Each MIMO element consists of an antenna array obtained previously in this work and is placed at the center positions of the board sides, as shown in Figure 1a. The overall dimensions of the board are 30×35 mm^2. The MIMO antenna configuration thus obtained exhibits acceptable performance, but to further improve the performance and to reduce mutual coupling among the MIMO antennas, a zigzag-shaped DGS is integrated, as illustrated in Figure 1b. As a result, the isolation for the MIMO configuration is increased.

Table 1. Optimized Design Parameters.

Parameter	Value(mm)	Parameter	Value(mm)	Parameter	Value(mm)
Wp	3	Lp	2	G	0.8
Wf	1.66	Lf	3.4	G1	0.2
W1	1.6	L1	1.35	Wm	3.5
Wf1	0.42	Lf1	1.5	r	0.6
Wsg	1.25	Lsg	5.2	r1	1.3
W	30	L	35	Lz	1.9
a	0.7	b	1.5	h	0.76

3. Simulated Results

3.1. Scattering Parameters

The working principle and radiation characteristics of the reported antenna system are analyzed. Figure 3a shows the analysis of the reflection coefficient curves of the design from a single element to MIMO configuration. It is observed that the single element of the proposed antenna is resonating in the mm-wave frequency spectrum ranging from 26.8–29.6 GHz with a 2.8 GHz bandwidth. The apparent increase in bandwidth is noticed when the antenna is developed from a single element to two-element arrays. The bandwidth obtained for the antenna array is 3.5 GHz covering the 26.2–29.7 GHz frequency band. Moreover, the reflection coefficient curve of the MIMO Ant.1 in Figure 3a demonstrates that the frequency band covered now ranges from 26.1–29.78 GHz with a slight increase in bandwidth to 3.68 GHz. Furthermore, Figure 3b depicts the reflection coefficient curves for the MIMO Ant1-Ant4 with and without DGS. It is observed that the four MIMO antennas cover nearly the same band. In addition, the bandwidth of the MIMO antennas has improved after the incorporation of DGS. The frequency band now covered by the MIMO antenna system is 26.1–30 GHz with a 3.9 GHz bandwidth.

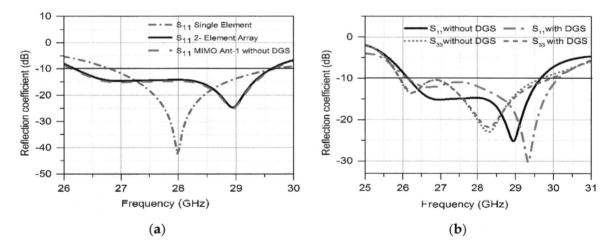

Figure 3. Simulated Reflection Coefficient of (**a**) Single Element, Two Element array, MIMO Ant1 (**b**) Reflection coefficient of MIMO Ant1 and Ant3 with and without DGS.

Figure 4a exhibits the transmission coefficient curves for the four MIMO antennas. It is observed that isolation between antenna 1 and antenna 2 is low. In addition, similar behavior is observed between Ant 3 and Ant 4. Meanwhile, significant isolation is obtained for the antenna pairs 1 and 3, 1and 4, 2 and 3, as well as 2 and 4. A zig-zag shaped DGS is incorporated at the bottom layer of the MIMO antenna structure to reduce mutual coupling effects. Figure 4b validates the isolation enhancement between the MIMO antennas after the assimilation of the DGS. Hence, the minimum isolation obtained for the suggested MIMO antennas is below −10 dB.

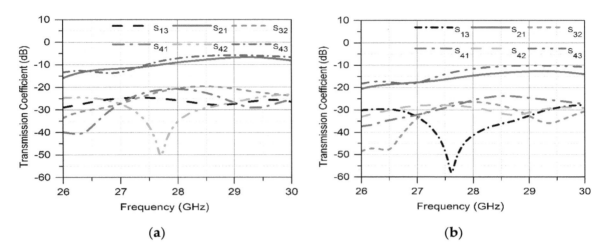

Figure 4. Simulated Transmission Coefficient (**a**) Without DGS (**b**) With DGS.

3.2. Surface Current Distribution

The radiating mechanism of the reported MIMO antenna system was analyzed further by investigating the surface current density. This focused on investigating the antenna parts that are influencing the radiation characteristics and elucidating the amount of coupling between different MIMO antennas. Figure 5 shows the surface current distribution when port 3 is activated at 28 GHz. The current flow is mainly concentrated around the feedline and along the edges of the inverted C-shaped antenna. Moreover, the circular and rectangular slots in the ground exhibit significant current distribution. This determines the contribution of DGS in radiation behavior. In addition, the concentration of the coupling current between MIMO antennas is insignificant due to the DGS, as demonstrated in Figure 5.

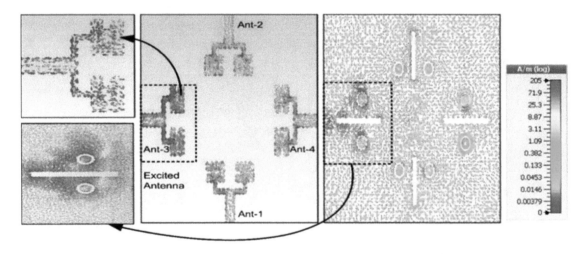

Figure 5. Surface current distribution along with the reported MIMO antenna design.

4. Experimental Results

The reported MIMO antenna system was fabricated on a Rogers RO4350B substrate using the photolithography process, and measurements were performed in order to endorse the antenna capabilities for practical utilization. The fabricated prototype and AUT i.e., antenna under test in the anechoic chamber, is shown in Figure 6. The detailed discussion and comparative analysis of measured results are provided in the subsequent section.

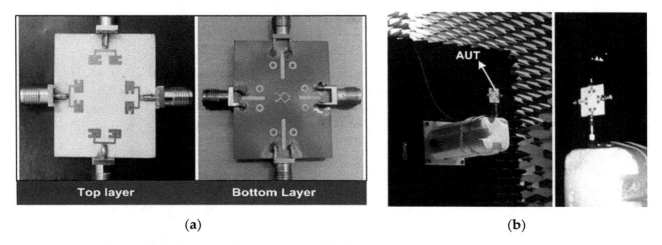

(a)   (b)

Figure 6. (**a**) Fabricated antenna module (**b**) antenna module in measurement.

4.1. Scattering Parameters

The Scattering parameters of the proposed prototype are measured using the Rohde & Schwarz ZVA 40 VNA. Figure 7a,b illustrate the simulated and measured scattering parameter curves for the reported MIMO antenna. It is observed from the reflection coefficient curve in Figure 7a that MIMO Ant1 is covering the 25.5–29.6 GHz, frequency band. Likewise, the other MIMO antennas exhibit nearly similar reflection coefficient curves with slight shift in bands. The maximum measured bandwidth thus achieved for the proposed antenna is 4.1 GHz. The transmission coefficient analysis is shown in Figure 7b. The minimum measured isolation obtained is −17 dB between the Ant3 and Ant4.

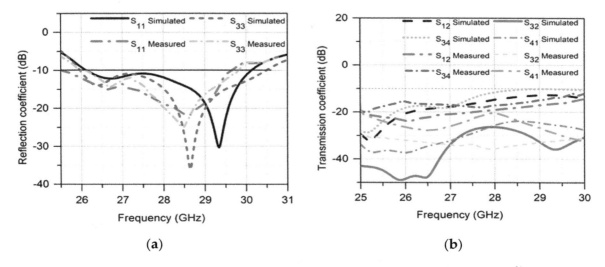

(a)   (b)

Figure 7. Measured and Simulated scattering parameter curves (**a**) Reflection coefficient curves (**b**) Transmission coefficient curves.

Simulated and measured results are exhibiting good coherence. However, insignificant differences are due to fabrication losses or unavoidable use of coaxial cables during the measurement [36,37]. Hence, the obtained measured results possess the suitability of the reported MIMO antenna for future mm-wave 5G applications.

4.2. Radiation Patterns

To understand the radiational behavior of the proposed design, the 2D radiation patterns of antennas were measured using the commercial ORBIT/FR far-field measurement system in an anechoic chamber, as shown in Figure 6b. The far-field measurements were performed in the xz and yz planes with theta range of −90° to 90°. The horn antenna with standard gain of 24 dBi was used for signal transmission. In Figure 8 the simulated and measured 2-D radiation patterns are shown at 27.5 and 28 GHz for Ant1. In the xz plane, the maximum radiation is observed at −35° while in the yz plane, the main beam is directed at −25°. The antenna exhibits overall good performance for both simulated and measured data. However, inconsistencies are observed between simulated and measured results due to fabrication errors and unavoidable cable losses. Also, these sorts of measurement systems are not the most appropriate ones for measuring small antennas, especially in mm wave frequency range, and the effect of the measurement system could affect the results, which in fact creates a discrepancy between simulated and measured results.

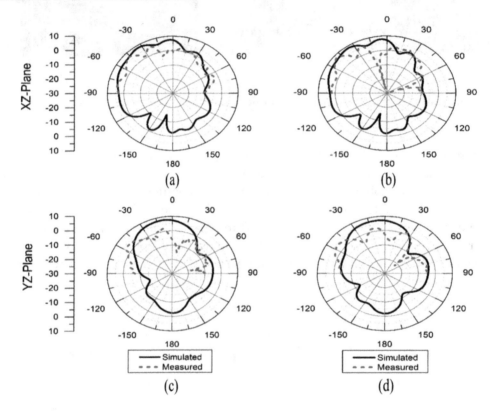

Figure 8. Measured and simulated 2D radiation pattern of antenna 1 (**a**) Radiation pattern at XZ-plane at 27.5 GHz (**b**) Radiation pattern at XZ-plane at 28 GHz (**c**) Radiation pattern at YZ-plane at 27.5 GHz (**d**) Radiation pattern at YZ-plane at 28 GHz.

4.3. Gain and Percentage Efficiency

The proposed design demonstrates a simulated peak gain value of 8.45 dB while peak measured gain is 8.3 dBi for Ant.1. Similarly, the peak antenna efficiency is about 82% as shown in Table 2. Moreover, the antenna exhibited a nearly stable gain with 3dB gain bandwidth from 26–29.97 GHz.

Table 2. Measured and simulated gain and efficiency.

Frequencies (GHz)	Gain at Antenna-1 (dBi)		Percentage Efficiency	
	Simulated	Measured	Simulated	Measured
27.5	8.45	8.3	80	79
28	8.1	8.02	82	80
28.5	8.22	8.1	85	82

5. MIMO Performance Parameters

To ensure the proposed antenna's multi-channel performance was high, the key performance metrics such as ECC, DG, CCL, and MEG were analyzed. Detailed discussion of the parameters is provided below.

5.1. Envelope Correlation Coefficient (ECC)

ECC is one of the key performance parameters of MIMO systems, and it is calculated using Equation (5) [10]. Figure 9a shows the ECC curve of the proposed antenna over frequency, relatively larger values of ECC are shown between antennas 1 and 2 as well as 3 and 4. The overall antenna module ensures that there are correlation values below the practical standard of 0.5.

$$\rho_{eij} = \frac{\left| S_{ii}{}^* S_{ij} + S_{ji}{}^* S_{jj} \right|^2}{\left(1 - |S_{ii}|^2 - S_{ij}{}^2\right)\left(1 - |S_{ji}|^2 - S_{jj}{}^2\right)} \tag{5}$$

5.2. Diversity Gain (DG)

Diversity gain demonstrates "the loss in transmission power when diversity schemes are performed on the module" for the MIMO configuration. The diversity gain is calculated by using Equation (6) given in reference [10]. Figure 9b describes the DG to be approximately 10 dB throughout the band, which ensures good diversity performance of the antenna.

$$DG = 10\sqrt{1 - |\rho_{eij}|^2} \tag{6}$$

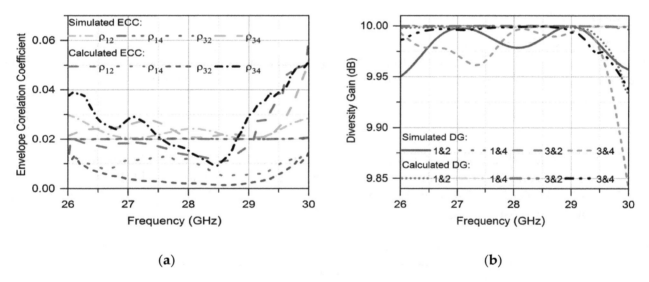

(a)   (b)

Figure 9. (a) ECC of the proposed MIMO antenna array (b) Diversity Gain of the MIMO antenna array.

5.3. Channel Capacity Loss (CLL)

CCL was enlisted among the MIMO performance parameters, thereby providing details of channels capacity losses of the system during the correlation effect. The CCL is calculated numerically by Equations (7)–(10). Figure 10 illustrates that for the proposed MIMO antenna, the obtained CCL is less than the practical standard of 0.4 bit/s/Hz [38] for the entire operating band, which ensures the proposed system's high throughput

$$C(loss) = -log_2 \det(a) \tag{7}$$

where a is the correlation matrix,

$$a = \begin{bmatrix} \sigma_{11} & \sigma_{12} \\ \sigma_{21} & \sigma_{22} \end{bmatrix} \tag{8}$$

$$\sigma_{ii} = 1 - \left(|S_{ii}|^2 - |S_{ij}|^2\right) \tag{9}$$

$$\sigma_{ij} = -\left(S_{ii}{}^{*}S_{ij} + S_{ji}S_{jj}{}^{*}\right) \tag{10}$$

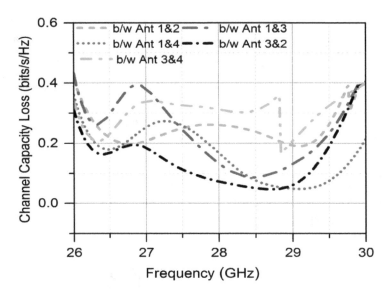

Figure 10. Channel capacity loss analysis of proposed MIMO antenna.

5.4. Mean Effective Gain(MEG)

For diversity, performance analysis mean effective gain is an important parameter and is defined as the mean received power in the fading environment. The mean effective gain is calculated using Equation (11) provided below, and the numerically estimated values are tabulated in Table 3.

$$MEG_i = 0.5\,\mu_{irad} = 0.5\left(1 - \sum_{j=1}^{K}\left|S_{ij}\right|\right) \tag{11}$$

In this equation, K is the number of antennas, i represents antenna under observation, and μ_{irad} is the radiation efficiency. For good diversity performance, the practical standard followed is that MEG should be $-3 \le MEG\ (dB) < -12$, which is therefore validated for the obtained MEG values of all MIMO antennas of the proposed design.

Table 3. The mean effective gain of the reported antenna.

Frequency	Mean Effective Gain (-dB)			
(GHz)	Ant-1	Ant-4	Ant-3	Ant-4
26	7.243892	7.28562	7.623244	7.208327
26.5	6.75436	6.602545	7.359296	6.676395
27	6.798682	6.495418	7.873314	7.096469
27.5	7.002822	6.549647	7.135389	7.211303
28	6.961063	6.652316	6.887685	7.075835
28.5	6.832857	6.634896	7.020392	6.901157
29	6.662024	6.534248	7.630851	7.051805
29.5	6.576111	7.024126	8.36803	7.311751
30	7.101393	8.203815	9.089905	7.458108

6. Comparison with Related Work

A comparison with related works reported in literature is tabulated in Table 4. The comparative analysis with other works shows that the proposed design demonstrates better performance in terms of compactness, bandwidth and gain. Moreover, MIMO performance analysis is provided in detail for the proposed design, and it is observed that the MIMO antenna proposed in this work exhibits better performance as compared to other reported works. Hence, the suitability of the proposed MIMO antenna 5G mm–wave applications was ascertained.

Table 4. Performance comparison with recent state of the art in the literature.

Ref.	Frequency (GHz)	Board Size (mm^3)	No. of Ports	Bandwidth (GHz)	Gain (dBi)	ECC, DG (dB)	CCL Bits/s/Hz
[11]	3.6	150 × 75 × 1.6	8	1.2	2.5	<0.01, Not provided	Not provided
[15]	28	41.3 × 46 × 0.508	4	3.35	13.1	Not provided	Not provided
[25]	24	15 × 19 × 0.254	2	0.8	6	0.24, 9.7	Not provided
[26]	28	Not provided	4	1.5	7.41	Not provided	Not provided
[27]	5.2 & 24	40 × 25 × 0.254	2	0.1 & 0.77	5 &7.37	Not provided	Not provided
[28]	30	48 × 21 × 0.13	2	1	>7	<0.4, Not provided	Not provided
[34]	28	19 × 19 × 7.608	4 (multi-layers)	8	14 dBiC	0.05, Not provided	Not provided
[35]	28	20 × 20 × 0.254	2	0.85	8 dB	0.13, 9.9	Not provided
This work	28	30 × 35 × 0.76	4	4.1	8.3 dB	<0.01, >9.96	<0.4

7. Conclusions

We presented a four-port MIMO antenna for 5G mm-wave applications. Each MIMO antenna of the proposed design is composed of the wideband and a high gain antenna array of two elements. The operating frequency band is covering 25.5–29.6 GHz, having a bandwidth of 4.1 GHz. Moreover, the measured peak gain obtained for the presented design is 8.3 dBi. Furthermore, to ensure the prototype's MIMO performance, DG, ECC, CCL, and MEG were also calculated and observed to be within practically acceptable values. The good radiation characteristics of the reported antenna system certify it for the future devices operating in the 5G mm-wave bands.

Author Contributions: Conceptualization M.K. and S.I.N.; Data curation, M.K, S.I.N. and F.; Formal analysis, M.K., S.I.N., and N.H.; Investigation, M.R., M.J.K. and Y.A.; Methodology M.K. and S.I.N. and M.R.; Project administration, Y.A.; Resources, M.R. and S.S.M. and N.H.; Software, M.J.K; Supervision, S.I.N and Y.A.; Validation, M.K. and S.I.N.; Visualization, N.H. and F. and S.S.M.; Writing–original draft, M.K.; Writing–review & editing, S.I.N., N.H., Y.A. and M.R. All authors have read and agreed to the published version of the manuscript.

References

1. Andrews, J.G.; Buzzi, S.; Choi, W.; Hanly, S.V.; Lozano, A.; Soong, A.C.; Zhang, J.C. What Will 5G Be? *IEEE J. Sel. Areas Commun.* **2014**, *32*, 1065–1082. [CrossRef]

2. Global Mobile Suppliers Association. The Road to 5G: Drivers, Applications, Requirements and Technical Development. Available online: https://www.google.com.hk/url?sa=t&rct=j&q=&esrc=s&source=web& cd=1&cad=rja&uact=8&ved=2ahUKEwimhtnryMPmAhXiGEKHQqMBOcQFjAAegQIAhAC&url= https%3A%2F%2Fwww.huawei.com%2Fminisite%2F5g%2Fimg%2FGSA_the_Road_to_5G.pdf&usg= AOvVaw1RWAb8E8EVfk8xLN0weKBv (accessed on 1 November 2015).

3. Pi, Z.; Khan, F. An introduction to millimeter-wave mobile broadband systems. *IEEE Commun. Mag.* **2011**, *49*, 101–107. [CrossRef]

4. Rappaport, T.S.; Sun, S.; Mayzus, R.; Zhao, H.; Azar, Y.; Wang, K.; Wong, G.N.; Schulz, J.K.; Samimi, M.; Gutierrez, F. Millimeter Wave Mobile Communications for 5G Cellular: It Will Work! *IEEE Access* **2013**, *1*, 335–349. [CrossRef]

5. Thompson, J.; Ge, X.; Wu, H.C.; Irmer, R.; Jiang, H.; Fettweis, G.; Alamouti, S. 5G wireless communication systems: Prospects and challenges part 2 [Guest Editorial]. *IEEE Commun. Mag.* **2014**, *52*, 24–25. [CrossRef]

6. Zhang, J.; Ge, X.; Li, Q.; Guizani, M.; Zhang, Y. 5G Millimeter-Wave Antenna Array: Design and Challenges. *IEEE Wirel. Commun.* **2017**, *24*, 106–112. [CrossRef]

7. Shayea, I.; Abd Rahman, T.; Hadri Azmi, M.; Islam, M.R. Real Measurement Study for Rain Rate and Rain Attenuation Conducted Over 26 GHz Microwave 5G Link System in Malaysia. *IEEE Access* **2018**, *6*, 19044–19064. [CrossRef]

8. Zhao, Q.; Li, J. Rain Attenuation in Millimeter Wave Ranges. In Proceedings of the 2006 7th International Symposium on Antennas, Propagation & EM Theory, Guilin, China, 26–29 October 2006; pp. 1–4. [CrossRef]

9. Rahman, M.; NagshvarianJahromi, M.; Mirjavadi, S.S.; Hamouda, A.M. Compact UWB Band-Notched Antenna with Integrated Bluetooth for Personal Wireless Communication and UWB Applications. *Electronics* **2019**, *8*, 158. [CrossRef]

10. Sharawi, M.S. Printed Multi-Band MIMO Antenna Systems and Their Performance Metrics [Wireless Corner]. *IEEE Antennas Propag. Mag.* **2013**, *55*, 218–232. [CrossRef]

11. Ojaroudi Parchin, N.; Jahanbakhsh Basherlou, H.; Alibakhshikenari, M.; Ojaroudi Parchin, Y.; Al-Yasir, Y.I.A.; Abd-Alhameed, R.A.; Limiti, E. Mobile-Phone Antenna Array with Diamond-Ring Slot Elements for 5G Massive MIMO Systems. *Electronics* **2019**, *8*, 521. [CrossRef]

12. Yashchyshyn, Y.; Derzakowski, K.; Bogdan, G.; Godziszewski, K.; Nyzovets, D.; Kim, C.H.; Park, B. 28 GHz Switched-Beam Antenna Based on S-PIN Diodes for 5G Mobile Communications. *IEEE Antennas Wirel. Propag. Lett.* **2018**, *17*, 225. [CrossRef]

13. Tang, M.; Shi, T.; Ziolkowski, R.W. A Study of 28 GHz, Planar, Multilayered, Electrically Small, Broadside Radiating, Huygens Source Antennas. *IEEE Trans. Antennas Propag.* **2017**, *65*, 6345–6354. [CrossRef]

14. Lin, X.; Seet, B.; Joseph, F.; Li, E. Flexible Fractal Electromagnetic Bandgap for Millimeter-Wave Wearable Antennas. *IEEE Antennas Wirel. Propag. Lett.* **2018**, *17*, 1281–1285. [CrossRef]

15. Yoon, N.; Seo, C. A 28-GHz Wideband 2 × 2 U-Slot Patch Array Antenna. *J. Electromagn. Eng. Sci.* **2017**, *17*, 133–137. [CrossRef]

16. Ta, S.X.; Choo, H.; Park, I. Broadband Printed-Dipole Antenna and Its Arrays for 5G Applications. *IEEE Antennas Wirel. Propag. Lett.* **2017**, *16*, 2183–2186. [CrossRef]

17. Kim, J.; Song, S.C.; Shin, H.; Park, Y.B. Radiation from a Millimeter-Wave Rectangular Waveguide Slot Array Antenna Enclosed by a Von Karman Radome. *J. Electromagn. Eng. Sci.* **2018**, *18*, 154–159. [CrossRef]

18. Dzagbletey, P.A.; Jung, Y. Stacked Microstrip Linear Array for Millimeter-Wave 5G Baseband Communication. *IEEE Antennas Wirel. Propag. Lett.* **2018**, *17*, 780–783. [CrossRef]

19. Khalily, M.; Tafazolli, R.; Xiao, P.; Kishk, A.A. Broadband mm-Wave Microstrip Array Antenna with Improved Radiation Characteristics for Different 5G Applications. *IEEE Trans. Antennas Propag.* **2018**, *66*, 4641–4647. [CrossRef]

20. Zhu, S.; Liu, H.; Chen, Z.; Wen, P. A Compact Gain-Enhanced Vivaldi Antenna Array with Suppressed Mutual Coupling for 5G mmWave Application. *IEEE Antennas Wirel. Propag. Lett.* **2018**, *17*, 776–779. [CrossRef]

21. Briqech, Z.; Sebak, A.; Denidni, T.A. Low-Cost Wideband mm-Wave Phased Array Using the Piezoelectric Transducer for 5G Applications. *IEEE Trans. Antennas Propag.* **2017**, *65*, 6403–6412. [CrossRef]

22. Yu, B.; Yang, K.; Sim, C.; Yang, G. A Novel 28 GHz Beam Steering Array for 5G Mobile Device with Metallic Casing Application. *IEEE Trans. Antennas Propag.* **2018**, *66*, 462–466. [CrossRef]

23. Bang, J.; Choi, J. A SAR Reduced mm-Wave Beam-Steerable Array Antenna with Dual-Mode Operation for Fully Metal-Covered 5G Cellular Handsets. *IEEE Antennas Wirel. Propag. Lett.* **2018**, *17*, 1118–1122. [CrossRef]

24. Ikram, M.; Wang, Y.; Sharawi, M.S.; Abbosh, A. A novel connected PIFA array with MIMO configuration for 5G mobile applications. In Proceedings of the 2018 Australian Microwave Symposium (AMS), Brisbane, Australia, 6–7 February 2018; pp. 19–20. [CrossRef]

25. Iqbal, A.; Basir, A.; Smida, A.; Mallat, N.K.; Elfergani, I.; Rodriguez, J.; Kim, S. Electromagnetic Bandgap Backed Millimeter-Wave MIMO Antenna for Wearable Applications. *IEEE Access* **2019**, *7*, 111135–111144. [CrossRef]

26. Park, J.; Ko, J.; Kwon, H.; Kang, B.; Park, B.; Kim, D. A Tilted Combined Beam Antenna for 5G Communications Using a 28-GHz Band. *IEEE Antennas Wirel. Propag. Lett.* **2016**, *15*, 1685–1688. [CrossRef]

27. Sun, Y.; Leung, K.W. Substrate-Integrated Two-Port Dual-Frequency Antenna. *IEEE Trans. Antennas Propag.* **2016**, *64*, 3692–3697. [CrossRef]

28. Sharawi, M.S.; Podilchak, S.K.; Hussain, M.T.; Antar, Y.M.M. Dielectric resonator based MIMO antenna system enabling millimetre-wave mobile devices. *IET Microw. Antennas Propag.* **2017**, *11*, 287–293. [CrossRef]

29. Yang, B.; Yu, Z.; Dong, Y.; Zhou, J.; Hong, W. Compact Tapered Slot Antenna Array for 5G Millimeter-Wave Massive MIMO Systems. *IEEE Trans. Antennas Propag.* **2017**, *65*, 6721–6727. [CrossRef]
30. Jilani, S.F.; Alomainy, A. Millimetre-wave T-shaped MIMO antenna with defected ground structures for 5G cellular networks. *IET Microw. Antennas Propag.* **2018**, *12*, 672–677. [CrossRef]
31. Shoaib, N.; Shoaib, S.; Khattak, R.Y.; Shoaib, I.; Chen, X.; Perwaiz, A. MIMO Antennas for Smart 5G Devices. *IEEE Access* **2018**, *6*, 77014–77021. [CrossRef]
32. Saad, A.A.R.; Mohamed, H.A. Printed millimeter-wave MIMO-based slot antenna arrays for 5G networks. *AEU Int. J. Electron. Commun.* **2019**, *99*, 59–69. [CrossRef]
33. Jiang, H.; Si, L.; Hu, W.; Lv, X. A Symmetrical Dual-Beam Bowtie Antenna with Gain Enhancement Using Metamaterial for 5G MIMO Applications. *IEEE Photonics J.* **2019**, *11*, 1–9. [CrossRef]
34. Hussain, N.; Jeong, M.; Park, J.; Kim, N. A Broadband Circularly Polarized Fabry-Perot Resonant Antenna Using A Single-Layered PRS for 5G MIMO Applications. *IEEE Access* **2019**, *7*, 42897–42907. [CrossRef]
35. Zhang, Y.; Deng, J.; Li, M.; Sun, D.; Guo, L. A MIMO Dielectric Resonator Antenna with Improved Isolation for 5G mm-Wave Applications. *IEEE Antennas Wirel. Propag. Lett.* **2019**, *18*, 747–751. [CrossRef]
36. Liu, L.; Cheung, S.W.; Weng, Y.F.; Yuk, T.I. *Cable Effects on Measuring Small Planar UWB Monopole Antennas, in Ultra Wideband Current Status and Future Trends*; InTech: London, UK, 2012; Chapter 12.
37. Liu, L.; Weng, Y.F.; Cheung, S.W.; Yuk, T.I.; Foged, L.J. Modeling of cable for measurements of small monopole antennas. In Proceedings of the Loughborough Antennas Propagation Conference (LAPC), Loughborough, UK, 14–15 November 2011; pp. 1–4.
38. Chae, S.H.; Oh, S.; Park, S. Analysis of Mutual Coupling, Correlations, and TARC in WiBro MIMO Array Antenna. *IEEE Antennas Wirel. Propag. Lett.* **2007**, *6*, 122–125. [CrossRef]

5

Site-Specific Propagation Loss Prediction in 4.9 GHz Band Outdoor-to-Indoor Scenario

Kentaro Saito *, Qiwei Fan †, Nopphon Keerativoranan and Jun-ichi Takada

School of Environment and Society, Tokyo Institute of Technology, Tokyo 152-8550, Japan;
key@ap.ide.titech.ac.jp (Q.F.); keerativoranan.n.aa@m.titech.ac.jp (N.K.); takada@tse.ens.titech.ac.jp (J.-i.T.)
* Correspondence: saitouken@tse.ens.titech.ac.jp
† Present address: Marvelous Inc., Tokyo 140-0002, Japan.

Abstract: Owing to the widespread use of smartphones and various cloud services, user traffic in cellular networks is rapidly increasing. Especially, the traffic congestion is severe in urban areas, and effective service-cell planning is required in the area for efficient radio resource usage. Because many users are also inside high buildings in the urban area, the knowledge of propagation loss characteristics in the outdoor-to-indoor (O2I) scenario is indispensable for the purpose. The ray-tracing simulation has been widely used for service-cell planning, but it has a problem that the propagation loss tends to be underestimated in a typical O2I scenario in which the incident radio waves penetrate indoors through building windows. In this paper, we proposed the extension method of the ray-tracing simulation to solve the problem. In the proposed method, the additional loss factors such as the Fresnel zone shielding loss and the transmission loss by the equivalent dielectric plate were calculated for respective rays to eliminate the penetration loss prediction error. To evaluate the effectiveness of the proposed method, we conducted radio propagation measurements in a high-building environment by using the developed unmanned aerial vehicle (UAV)-based measurement system. The results showed that the penetration loss of direct and reflection rays was significantly underestimated in the ray-tracing simulation and the proposed method could correct the problem. The mean prediction error was improved from 7.0 dB to −0.5 dB, and the standard deviation was also improved from 8.2 dB to 5.3 dB. The results are expected to be utilized for actual service-cell planning in the urban environment.

Keywords: Building entry loss; outdoor-to-indoor propagation; penetration loss; propagation loss measurement; ray-tracing simulation; super high frequency band propagation; unmanned aerial vehicle

1. Introduction

Owing to the widespread use of various application services such as video streaming and cloud services, which are accompanied by the advancement of mobile terminals, user traffic in cellular networks is rapidly increasing. Thus, for efficient radio resource usage, service-cell planning is an important issue. In particular, cell planning becomes complex in urban areas because many users are also inside high buildings, which are sometimes higher than the base stations (BSs) as shown in Figure 1. Therefore, three-dimensional (3D) service-cell planning is considered in those areas [1].

Knowledge of the propagation loss characteristics in the outdoor-to-indoor (O2I) environment is important for this purpose. To clarify the building entry loss (BEL) characteristics, radio propagation measurements were conducted by up to the 2 GHz band [2,3], 3.5 GHz band [4], 5 GHz band [5,6], 8 GHz band [7], 10 GHz band [8], and 38 GHz band [9]. It was confirmed that the BEL characteristics changed according to the incident angle of the radio wave to the building. Those statistical characteristics were also adopted in the global standard models such as the COST 231 BEL model [10] and ITU-R P.2109 model [11]. For service-cell planning, it is also needed to predict the propagation loss characteristics in the specific service area. The ray-tracing simulation has been widely utilized for predicting those site-specific propagation channel characteristics [12–14]. In a typical O2I scenario in which the incident radio waves penetrated indoors through building windows, the penetration loss was represented by the diffraction loss of rays at the window edges [15,16]. However, there still existed a significant discrepancy between the measurement and the ray-tracing simulation. One reason for the discrepancy is thought to be that other penetration losses occurred also by appurtenances such as window blinds and window fences which are attached to the window. Another reason is that the penetration loss occurred because the Fresnel zones of the radio waves were partially shielded by the window. In [17,18], the physical optics approximation was used to clarify the penetrating wave characteristics by calculating the re-radiation of the electromagnetic wave on the window surface. However, because it was necessary to divide the window surface into the numerous meshes such that the size was smaller than the wavelength of the carrier wave for the calculation, it was not feasible to apply it to the ray-tracing simulation for the site planning from the viewpoint of calculation cost.

In this paper, we proposed the extension method of ray-tracing simulation to take those additional penetration loss effects into the simulation. The method consists of the Fresnel zone shielding loss calculation and the transmission loss calculation by equivalent dielectric plate. We conducted radio propagation measurements in front of a research building in the university campus to evaluate the effectiveness of the proposed method. We developed a radio measurement system using an unmanned aerial vehicle (UAV) to realize the various O2I scenarios where the radio waves penetrated indoors through the window from various arrival angles [19–21]. In the measurement, the carrier frequency was the 4.9 GHz band for assuming the low super high frequency (SHF) band communication of the 5G cellular system [22,23]. The contribution of our work was to propose the novel extension method of the ray-tracing simulation to predict the penetration loss in the O2I scenario. We proposed to combine the ray-tracing simulation and the theory of diffraction by small holes [24,25] to include the effect of appurtenances such as window blinds and window fence into the simulation for the first time. We showed the effectiveness of the proposed method through exhaustive radio propagation measurements by using the UAV-based measurement system. The mean prediction error was improved from 7.0 dB to −0.5 dB, and the standard deviation was also improved from 8.2 dB to 5.3 dB by the proposed method.

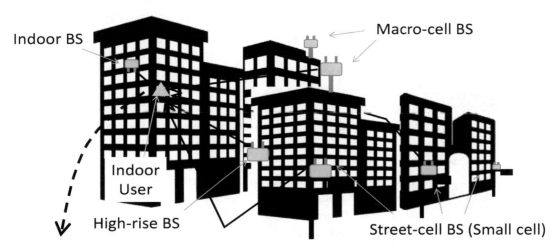

To realize the efficient service-cell planning, the propagation loss characteristics from various outdoor BSs to indoor MS have to be clarified.

Figure 1. The necessity of three-dimensional (3D) service-cell planning.

2. Proposal of Propagation Loss Prediction Method for Outdoor-to-Indoor Scenario

2.1. Proposed Extension Method for Ray-Tracing Simulation

The ray-tracing simulation [12–14] has been widely used for the propagation loss prediction in the site-specific environments for service-cell planning. In the simulation, the trajectories of each radio wave, which are called "rays" are traced by calculating the interactions such as reflection, diffraction, and transmission by interactive objects in the environment. Based on the ray optics approximation, all the objects are modeled by large flat polygon surfaces for the calculation. Therefore, the ray-tracing is especially suitable for the radio propagation simulation in built environments. In the simulation, the electric field strength of the received signal E is calculated as follows

$$
\begin{aligned}
E &= \sum_{i=1}^{I} e_i \\
&= \frac{\lambda}{4\pi} \sum_{i=1}^{I} \left(\frac{\sqrt{P_T G_T(i) G_R(i)} e^{-jkl_{i,m}}}{l_{i,1}} \prod_{n=1}^{N} R_{i,n} \prod_{m=1}^{M} \sqrt{\frac{l_{i,m}}{(l_{i,m} + l_{i,m+1}) l_{i,m+1}}} D_{i,m} e^{-jkl_{i,m+1}} \prod_{p=1}^{P} T_{i,p} \right)
\end{aligned}
\tag{1}
$$

Here, we assume that the propagation channel is represented by the superposition of I rays. e_i represents the received electric field strength of ray $i (1 \le i \le I)$, λ is the carrier wavelength, k is the wave-number $\frac{2\pi}{\lambda}$, P_T is the transmission power, and $G_T(i)$ and $G_R(i)$ are the transmitter (Tx) and receiver (Rx) antenna gains for ray i. The antenna gains are calculated by considering the angle of departure and the angle of arrival of the ray. It is assumed that the ray i suffers the N times reflections, the M times diffractions, and the P times transmissions during the propagation, and it suffered the interaction losses $R_{i,n}$, $D_{i,m}$, and $T_{i,p}$, respectively. The diffraction loss is calculated based on the uniform geometrical theory of diffraction (UTD) [26,27]. $l_{i,m}$ is the total propagation length from the $(m-1)$-th diffraction point to the m-th diffraction point. The pathgain G_P is the gain by the radio propagation and defined as follows.

$$
G_P = \frac{|E|^2}{P_T}
\tag{2}
$$

Although the ray-tracing simulation can be utilized for the propagation loss prediction in the O2I scenarios, there are several problems. We assume in a typical O2I scenario that the rays penetrate indoors through the building windows as shown in Figure 2. In the existing ray-tracing simulation, the penetration loss is modeled by the diffraction loss at window edges, and no penetration loss is taken into account if the trajectory of ray does not cross the window edges by even a little gap. This calculation has a risk to underestimate the penetration loss, because each ray has its Fresnel zone along the propagation path, and the penetration loss occurs if interactive objects shield the zone [28]. This phenomenon is fundamental to understand various penetration loss characteristics through the window, such as the distance dependency between the window and the indoor station and carrier frequency dependency. In addition, many appurtenances such as window blinds and window fences are attached to the window, and they are also thought to cause further penetration loss. However, taking the influence of those appurtenances into the simulation is not straightforward because of their complex structures, which are far from the assumption of the ray-tracing simulation that the objects are modeled by large flat polygon surfaces. Therefore, the influence of appurtenances has been ignored in the current researches, but it can cause further penetration loss underestimation problem.

In this paper, we propose the extension method of the ray-tracing simulation to take those effects into the penetration loss calculation. In our proposal, we don't require the fine 3D environment model for the simulation because there is another difficulty to obtaining accurate models in real environments. Because the very detailed electromagnetic simulation has a high calculation cost and it is not suitable for the service-cell planning, we used the simplified calculation method from the simplified model. The difference between the existing ray-tracing simulation and the proposed method is shown in Figure 2. The proposed method consists of the Fresnel zone shielding loss calculation and the transmission loss calculation by equivalent dielectric plate. In the proposed method, normal ray-tracing simulation was calculated firstly, and the trajectory of each ray was investigated. Next, the Fresnel zone shielding loss $L_{F,i}$ and the transmission loss by appurtenances $L_{T,i}$ of ray i were calculated, and those additional losses were added to the electric field strength E' and pathgain G'_P calculation in (1) and (2).

$$E' = \sum_{i=1}^{I} \frac{e_i}{L_{F,i}L_{T,i}} \tag{3}$$

$$G'_P = \frac{|E'|^2}{P_T} \tag{4}$$

The details for the calculation of $L_{F,i}$ and $L_{F,i}$ are explained in the next subsection.

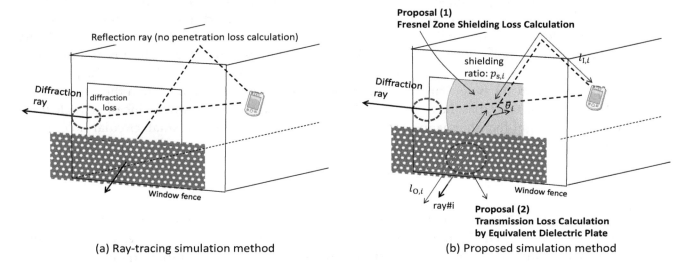

(a) Ray-tracing simulation method   (b) Proposed simulation method

Figure 2. The difference between ray-tracing simulation and the proposed method.

2.2. Fresnel Zone Shielding Loss Calculation

In the Fresnel zone shielding loss calculation, the total propagation distance from the previous indoor diffraction point to the window intersection point $l_{I,i}$ and the distance from the window intersection point to the next outdoor diffraction point $l_{O,i}$ are calculated for ray i as shown in Figure 2 (**b**). Although it is assumed that the ray originated from the indoor Tx and propagated to outdoor Rx in the figure, the same theory can be applied also if the Tx and Rx positions were opposite because the reciprocity was satisfied. Fresnel radius of the ray r_i at the window intersection point is calculated as follows.

$$r_i = \sqrt{\lambda \frac{l_{I,i} l_{O,i}}{l_{I,i} + l_{O,i}}} \tag{5}$$

The cross-section of the Fresnel zone on the window plane is calculated by the incident angle of the ray to the window θ_i. The cross-section becomes an ellipse with the major axis $r_{1,i} = r_i / \cos(\theta_i)$ and minor axis $r_{2,i} = r_i$. Next, the shielding ratio of the first Fresnel zone $p_{s,i}$ is calculated by considering the cross-section shape of the window. In this paper, to simplify the calculation, we approximate the Fresnel zone shielding loss by the inverse of $p_{s,i}$.

$$L_{F,i} = \frac{1}{p_{s,i}} \tag{6}$$

The reason for the simplification is that there is difficulty in obtaining the detailed 3D environment model in real environments, and the rigorous calculation is not always effective regardless of the high calculation cost. Those calculation procedures were performed for all rays except for the rays that diffracted at the window edge. The reason why the diffracted rays were excluded from the calculation is that the Fresnel zone shielding effect of those rays is already included in the diffraction loss calculation. In that case, $L_{F,i}$ is set to 1 in (3).

2.3. Transmission Loss Calculation by Equivalent Dielectric Plate

Many appurtenances such as window blinds and window fences are often attached around the window, and they are thought to affect the penetration loss characteristics. However, the calculation of those effects is not straightforward, because their shapes are thought to be a kind of screen but they have a complicated structure in detail, which is quite different from the plane surface that the ray-tracing simulation assumes. Therefore, the calculation method needed to deal with the effect of the structure in the ray-tracing based simulation. In this paper, we assumed that those screen structures were able to be modeled by the metal plates with periodically perforated small holes, and we calculate the transmission loss by regarding the plate as an equivalent dielectric plate [29,30] based on the theory of diffraction by small holes [24,25]. The calculation model is shown in Figure 3. it is assumed that small round holes are periodically perforated in the metal plate. Δw is the thickness of the plate, and the hole spacing a and the hole diameter d are thought to be enough smaller than the carrier wave wavelength λ. We consider that the transverse electric (TE) wave arrives from the incident angle θ_i'. In the case, it was thought that the magnetic dipole $m_{0,i}$ was induced on each small circular hole by the incident magnetic field $H_{0,i}$.

$$m_{0,i} = \frac{1}{6}\mu_0 d^3 H_{0,i} \tag{7}$$

Here, μ_0 is the permeability of the free space. Another magnetic dipole $m_{1,i}$ was induced also on the opposite side of the plate by the magnetic field, which passed through the hole. The attenuation coefficient by the hall was analyzed numerically by [31], and it is known that $m_{1,i}$ is calculated as follows in case of the round hall.

$$m_{1,i} = -m_{0,i}\exp\left(\frac{-3.682\Delta w}{d}\right) \tag{8}$$

The electric field on the opposite side of the plate was obtained by calculating the re-radiation of the electric field by the induced magnetic dipoles. Here, we approximated the electric field by assuming that the magnetic dipoles on each hole were not coupled, and the phases of radiated waves were the same from the far-field assumption. In that case, the electric field on the opposite side $E_{s,i}$ was thought to be proportional to the hole density $\frac{S}{a^2}$, and $E_{s,i}$ was calculated as follows in the far-field condition.

$$
\begin{aligned}
E_{s,i} &= -\frac{j\omega^2\exp(-jkr)\cos(\theta_i')S}{\pi r a^2 c}m_{1,i} \\
&= \frac{j\omega^2\exp(-jkr)\cos(\theta_i')d^3 S}{6\pi r a^2 c^2}E_{0,i}\exp\left(\frac{-3.682\Delta w}{d}\right)
\end{aligned}
\tag{9}
$$

Here, S is the area of the plate, ϵ_0 is the permittivity of the free space, ω is the angular frequency, r is the distance between the mobile station (MS) and the BS, and $c = \frac{1}{\sqrt{\epsilon_0\mu_0}}$ is the speed of light. We used the relation $H_{0,i} = \sqrt{\frac{\epsilon_0}{\mu_0}}E_{0,i}$. The electric field $E_{d,i}$ when there was no obstacle between the MS and BS can be obtained by calculating the re-radiation of the electric field by the equivalent source on the plane by the Huygens–Fresnel principle.

$$E_{d,i} = \frac{\omega}{2\pi cr}\exp(-jkr)SE_{0,i} \tag{10}$$

The transmission loss can be calculated from the ratio of $E_{d,i}$ and $E_{s,i}$.

$$\frac{|E_{d,i}|}{|E_{s,i}|} = \frac{3a^2\lambda}{2\pi d^3 \cos(\theta_i')} \exp(\frac{3.682\Delta w}{d}) \qquad (11)$$

(11) was slightly modified to improve the estimation accuracy in case of large incident angle θ_i' in [29], and finally, the total transmission loss of ray i $P_{T,i}$ [dB] defined as follows.

$$\begin{aligned} P_{T,i} &= 10\log_{10}[\frac{|E_{d,i}|^2}{|E_{s,i}|^2}] \qquad &(12)\\ &\simeq 10\log_{10}[1 + \frac{1}{4}(\frac{3a^2\lambda}{\pi d^3 \cos(\theta_i')})^2] + \frac{32\Delta w}{d}\\ L_{T,i} &= 10^{P_{T,i}/20} \qquad &(13) \end{aligned}$$

The transmission loss characteristics by a variety of screen structures are modeled by the parameters such as the plate thickness, the hole spacing, and the hole diameter. It might be possible to substitute the different types of screens by the periodically perforated plates also. The parameterizing method of the different types of screens by the equivalent periodically perforated plates will be future work.

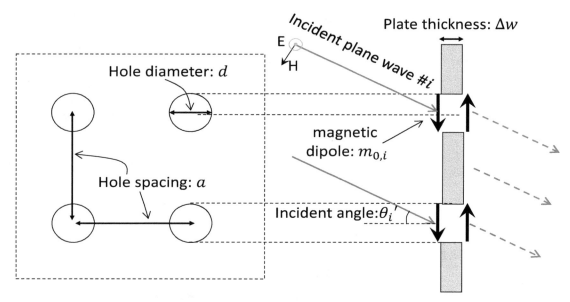

Figure 3. The calculation model of transmission loss by metal plate with small holes.

3. 4.9 GHz Band Radio Measurement for Penetration Loss Characteristics from Window

3.1. Radio Measurement System Using UAV

The diagram of the developed radio propagation measurement system is shown in Figure 4. The radio transceiver was implemented on a universal software radio peripheral (USRP) N210 [32], which is one of the commercial SDR platforms. In the measurement, the Tx was set indoors, and the Rx was mounted on the UAV. These were regarded as the indoor MS and virtual outdoor BS, respectively. The Tx-side USRP was controlled by a laptop personal computer (PC) and sends the continuous wave (CW) signal. The Rx recorded the instantaneous narrowband receiving power continuously during the measurement. The Rx-side USRP was controlled by Raspberry Pi [33], which was a single-board computer to reduce the payload. The photograph of the UAV station is shown in Figure 5. The total payload was less than 1 kg, and it could be operated using the portable battery of the UAV. The total dimensions of the system including the UAV were less than 60 cm × 60 cm × 50 cm, and it was usable for various BS placements in various measurement environments. Since the flight course of the UAV was not correctly controlled as planned because of natural disturbances such as wind conditions, it is crucial to know its actual flight trajectory for the data analysis. The UAV also recorded several sensor outputs as the flight data. We calculated the UAV height from the barometer data, the horizontal position from the GPS data, and the direction from the gyroscope data. Since the measurement and flight data were associated with the time stamp information, it was possible to trace the 3D position of the UAV and obtain the receiving power at each position by the post data processing.

In the data analysis, the average receiving power $\tilde{P}_R(q)$ of the q-th snapshot was defined by 3D moving averaging as follows:

$$\tilde{P}_R(q) = \underset{\substack{|x(q)-x(q')|<\Delta x, |y(q)-y(q')|<\Delta y, \\ |z(q)-z(q')|<\Delta z}}{\text{mean}} [P_R(q')] \tag{14}$$

where, $x_R(q) = [x(q), y(q), z(q)]$ is the UAV position of the q-th snapshot, and $\Delta x = (\Delta x, \Delta y, \Delta z)$ is the 3D window length. (14) gives the average receiving power inside the rectangular prism that size is $\Delta x \times \Delta y \times \Delta z$. The actual window length values that was used for the experiment is shown in Table 1.

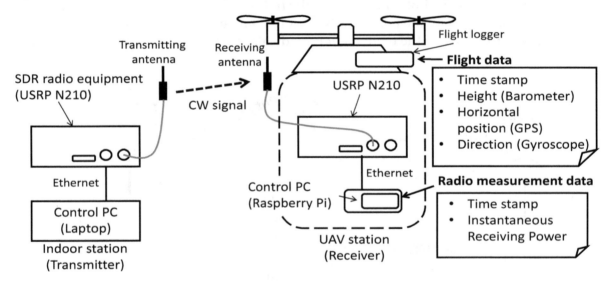

Figure 4. Diagram of developed radio propagation measurement system.

Figure 5. Photo of the UAV station.

Table 1. Measurement parameters.

Radio Equipment	
Center frequency	4.89 GHz
Transmit power	11.5 dBm
Transmit signal	CW
Receiver dynamic range	From -120 dBm to -30 dBm
Tx/Rx antennas	Dipole (2.14 dBi)
Polarization	Vertical polarization
Antenna height (UAV)	From 0 m to 15 m
Indoor floor height	20 m (building), 7 m (cafeteria)
Antenna height (indoor)	1.7 m from floor
Receiver sampling rate	100 Hz
Moving average	2 m (horizontal) 1 m (vertical)
UAV equipment	
Flight controller	DJI NAZA-M v2
Flight recorder	DJI iOSD MARK II
UAV sensor recording rate	100 Hz

3.2. Measurement Method

Radio measurements were conducted in the 4.9 GHz band to investigate penetration loss characteristics from various incident angle conditions from the outdoor BS to the indoor MS through a building window. The measurement map and photos are shown in Figure 6. The Tx was set indoors, and the Rx was mounted on the UAV. These were regarded as the indoor MS and virtual outdoor BS, respectively, because the reciprocity was satisfied in the measurement. The indoor MSs were fixed at three positions marked MS1, MS2, and MS3 in the conference room on the sixth floor. The distances of the positions from the window were 0.4 m, 2.4 m, and 4.4 m, respectively. As shown in 6, a steel window fence was attached outside the window. The window fence consisted of the thick balustrade and the meshed screen with small holes. The floor height was 20 m from the ground, and the MS antenna height was 1.7 m from the floor. The outside measurement areas were grass fields in front of the buildings. The horizontal measurement courses were set parallel to the building external walls, and the course length was 20 m. The measurement plane was 20 m horizontal length by 15 m height in front of the buildings. We obtained the propagation loss profile on the plane to investigate the penetration loss characteristics from various BS positions.

To simplify the measurement procedure, we divided the horizontal course into 2 m intervals. At each measurement point, the BS ascended to approximately 15 m in height and subsequently descended slowly. By repeating the same procedure at all points, the propagation loss characteristics of the 20 m by 15 m measurement plane in front of the building were obtained. Other measurement parameters are summarized in Table 1. Before the measurement, the measured value of the Rx was calibrated by connecting the RF cables between the Tx and the Rx directly. The UAV position was obtained from the flight log, and the moving average of the receiving power was calculated as shown in (14) to eliminate the multi-path fading effect. The antenna radiation pattern of the UAV station was measured in an anechoic chamber in advance, and the antenna elevation directivity effect was canceled from the measured data based on the grazing angle.

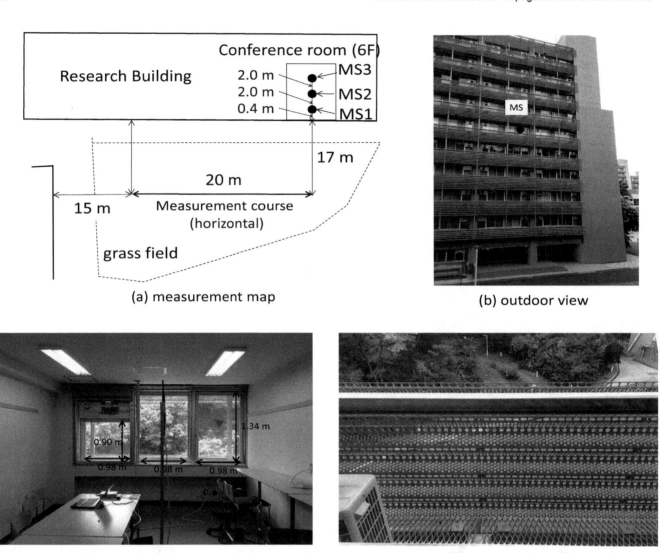

(a) measurement map

(b) outdoor view

(c) indoor view

(d) photo of window fence

Figure 6. Measurement overview ((**a**) measurement map, (**b**) outdoor view, (**c**) indoor view, and (**d**) window fence).

4. Penetration Loss Prediction Results

4.1. Ray-Tracing Simulation Method

The ray-tracing simulation was performed to evaluate the prediction performance of the proposed method. The 3D environment model which was created by Sketch-up is shown in Figure 7. The concrete, the plasterboard, and the steel were selected as the building materials. To simplify the model, all the furniture in the room was excluded, and other buildings around the target building were not taken into account. About the window fence, only the balustrade part was included in the model to calculate the diffraction wave at the balustrade.

Raplab software [34] was used for the ray-tracing engine. In the simulation, the method of imaging algorithm [14] was used, and the maximum number of reflections was three, the maximum number of diffractions was one, and the maximum number of transmissions was one. Dipole antennas were applied on both the BS and MS sides. Although the theoretical radiation pattern was used for the indoor MS, the measured pattern was used for the outdoor BS to consider the influence of the UAV frame on the pattern. The elevation radiation pattern of BS measured in an anechoic chamber is shown in Figure 7c. Because the pattern had an upward trend owing to the UAV frame effect, it is thought that the contribution of the reflection waves from the ground was not significant in the measurement. Therefore, the ground reflection waves were not calculated in this simulation.

Firstly, the normal ray-tracing simulation was performed by the above model. Then, the Fresnel zone shielding loss was calculated from the simulation result and the window geometry as explained in Section 2.2. The metal plate with periodically perforated small holes was assumed for the screen part of the window fence, and the equivalent transmission loss was calculated as explained in Section 2.3 if the rays intersected the fence. Other simulation parameters are summarized in Table 2. The normal ray-tracing simulation result and the proposed method were compared to the measurement result, to evaluate the prediction performance.

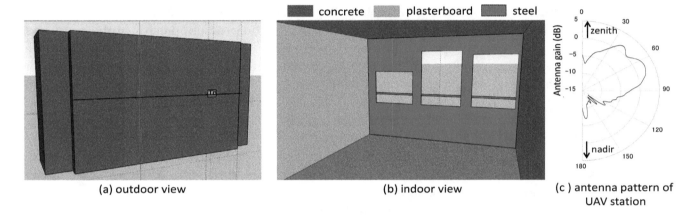

| (a) outdoor view | (b) indoor view | (c) antenna pattern of UAV station |

Figure 7. 3D environment model for the ray-tracing simulation ((**a**) outdoor view, (**b**) indoor view).

Table 2. Simulation parameters.

Ray-Tracing Simulation	
Calculation Method	**Method of Imaging**
Number of reflections	3
Number of diffractions	1
Number of transmissions	1
Material parameters	Concrete: $\epsilon_r = 6.8, \sigma = 2.3 \times 10^{-3} [S/m]$
(ϵ: relative permittivity	Ground: $\epsilon_r = 3.0, \sigma = 1.0 \times 10^{-4} [S/m]$
σ: conductivity)	Metal: $\epsilon_r = 1.0, \sigma = 1.0 \times 10^{7} [S/m]$
	Plaster board: $\epsilon_r = 1.5, \sigma = 3.3 \times 10^{-3} [S/m]$
Equivalent dielectric plate calculation	
Plate width Δw	3 mm
Hole diameter d	20 mm
Hole spacing u	30 mm

4.2. Numerical Results

The measurement and simulation results of the MS1 and MS3 settings are summarized in Figures 8 and 9. Figures 8a and 9a show the vertical–horizontal domain receiving power profiles of the measurements. The MS position is also shown for reference. In either MS settings, although the measured receiving power tended to increase as the BS position approached to the MS position, the receiving power in the MS3 setting was much smaller than the result of MS1 because of more severe penetration loss. Figure 8b showe the receiving power prediction result of the ray-tracing simulation. The receiving power was obviously overestimated at most of the BS positions. To clarify the reason for the discrepancy, the schematic propagation routes of the rays in the simulation are shown in Figure 8c. In the MS1 setting, most of the BS positions were under the line-of-sight (LoS) condition because the indoor MS was located close to the window. Therefore, any penetration loss was not considered in that area. Figure 8d presents the receiving power result of the ray-tracing simulation, but all distraction rays were excluded from the calculation. The figure showed that the area where the receiving power was overestimated in Figure 8 a corresponds to the area of Figure 8d. About the reflection and diffraction rays, although the rays reflected from the ceiling of the room were observed in the area marked (A), the receiving power contribution was not significant because this area overlapped with the LoS area. Because no direct and reflection waves were observed the diffraction waves were dominant in the area marked (B). Therefore, the receiving power significantly decreased in the same area in Figure 8a. The receiving power prediction of the proposed method is shown in Figure 8e. The overestimation of receiving power was corrected well in the proposed method by considering the Fresnel zone shielding loss and the transmission loss by equivalent dielectric plate for the direct and reflection rays. The result indicated that the dominant reasons for the penetration loss were the Fresnel zone shielding of the window and the window fence shielding rather than the diffraction at the window edge in this setting. The result showed the effectiveness of the proposed method. The prediction error of the proposed method compared to the measurement is shown in Figure 8f. The positive value means that the prediction of the proposed method was higher than the measurement. The error was relatively higher in the areas where the direct and reflection waves were dominant, but the error was less than ±5 dB in most of the areas.

The same analysis was performed for the MS3 setting. The measurement result and the ray-tracing simulation result are shown in Figure 9a,b. In the simulation, the receiving power was overestimated in the area marked (A) while it was underestimated in the area marked (B). The reason for these discrepancies was analyzed in Figure 9c,d. Because the distance between the indoor MS and the window increased, most of the BS positions were under the non-Line-of-Sight (NLoS) condition. Therefore, the single and double bounce reflections became a more dominant propagation mechanism. By comparing Figure 9a,d, it can be seen that area (A) corresponded to the area that direct and reflection waves were observed like the MS1 setting. No direct and reflection waves were observed when the BS height was about 5 m, and the diffraction waves were dominant in the area. Especially, the receiving power reduced in area (B) because multiple losses owing to the reflections and the diffractions occurred. On the whole, in the ray-tracing simulation, the diffraction waves were tended to be estimated weaker than the measurements. This problem might be because of the issue of the ray-tracing simulation itself or because of the accuracy of the environment model such as the detailed structures and material parameters of the interactive objects. Further detailed analysis will be future work. In the receiving power prediction of the proposed method shown in 9e, the overestimation issue of the receiving power in the area (A) was fairly corrected. However, the underestimation issue in the area (B) remained.

The prediction errors of the ray-tracing simulation and the proposed method are summarized in Figure 10. The data represent the prediction error CDF of all BS positions. The positive error means that the prediction result was higher than the measurement. The mean and the standard deviation of the error is summarized in Table 3. As described in the previous paragraphs, the ray-tracing simulation had the problem of underestimating the penetration loss. The mean errors were 11.1 dB and 8.0 dB in the MS1 and MS2 settings, respectively. Although the mean error was 0.4 dB in the MS3 setting, it does not mean the correct power was predicted. In actuality, the power of direct and reflecting waves was overestimated and the power of diffraction waves was underestimated, and they happened to balance in this environment. In the proposed method, the mean errors were improved to 0.3 dB and 1.5 dB in the BS1 and BS2 settings, respectively, while the error slightly increased in the BS3 setting because of the inaccuracy of diffraction loss estimation. However, the total prediction error was improved from 7.0 dB to −0.5 dB, and the result proved the effectiveness of the proposed method. About the dispersion of error, the standard deviation was also improved from 8.2 dB to 5.3 dB.

The significance of the result is that our proposal is not a kind of predicted offset that increases or decreases the receiving power in the same way. Because the power correction was executed for each ray by considering the physical propagation mechanism, the method is robust enough to be applied to a variety of radio propagation conditions. The utilization of the proposal for the actual service-cell planning is expected to be future work.

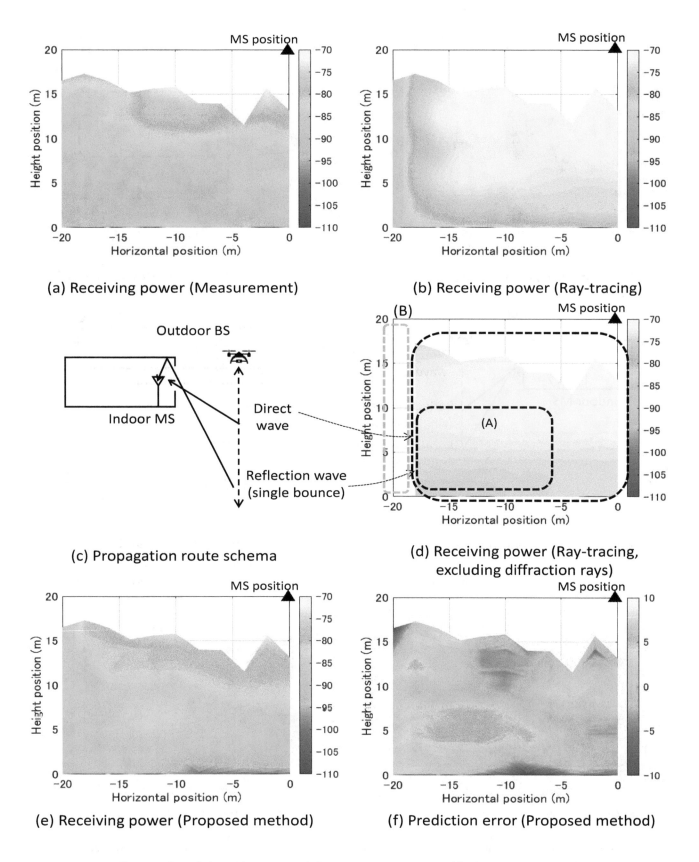

Figure 8. The results of the MS1 setting ((**a**) receiving power profile (measurement), (**b**) receiving power profile (ray-tracing), (**c**) propagation route schema, (**d**) receiving power profile (ray-tracing, excluding diffraction rays), (**e**) receiving power profile (proposed method), and (**f**) prediction error (proposed method)).

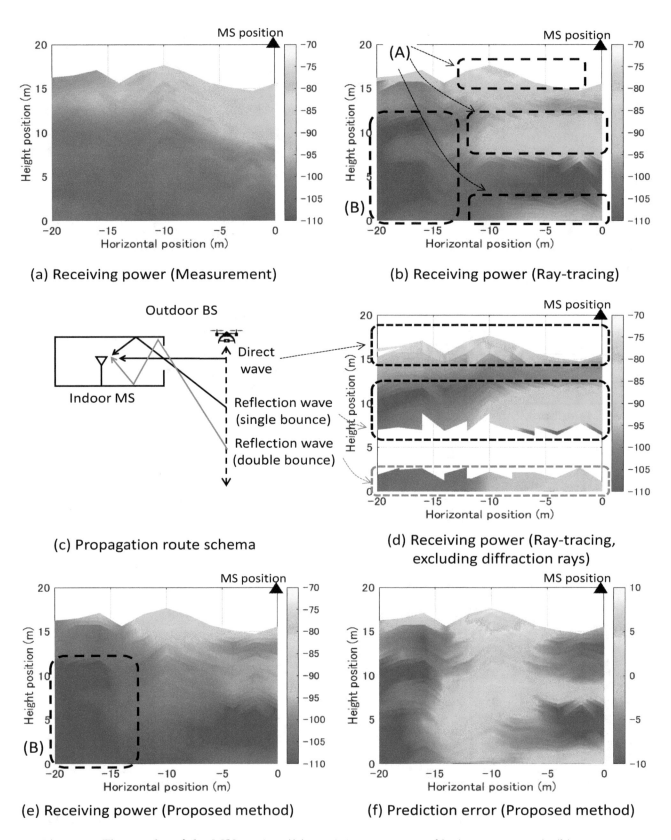

Figure 9. The results of the MS3 setting ((**a**) receiving power profile (measurement), (**b**) receiving power profile (ray-tracing), (**c**) propagation route schema, (**d**) receiving power profile (ray-tracing, excluding diffraction rays), (**e**) receiving power profile (proposed method), and (**f**) prediction error (proposed method)).

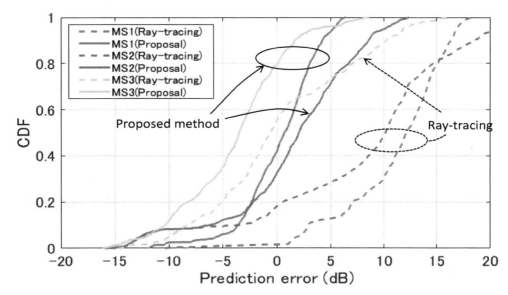

Figure 10. Prediction error CDF (error = simulation − measurement).

Table 3. Prediction error summary (error = simulation − measurement).

	Ray-Tracing				Proposed Method			
	MS1	MS2	MS3	Total	MS1	MS2	MS3	Total
mean (dB)	11.1	8.0	0.4	7.0	0.3	1.5	−3.9	−0.5
standard deviation (dB)	4.8	8.8	6.7	8.2	3.4	6.0	4.8	5.3

5. Conclusions

In this paper, we proposed the extension method of the ray-tracing simulation for a typical O2I scenario in which the incident radio waves penetrate indoors through the building windows. Because only the diffraction loss at window edges is considered in the existing ray-tracing simulation, it has the problem of underestimating the penetration loss. Our proposed method consists of the Fresnel zone shielding loss calculation and the transmission loss calculation by equivalent dielectric plate. In the Fresnel zone shielding loss calculation, the Fresnel zone cross-sections of respective rays on the window plane were evaluated, and the shielding losses were calculated by the shielding ratios of the zones. In the transmission loss calculation by equivalent dielectric plate, we proposed to substitute the screen-type appurtenances such as window blinds and window fences by an equivalent dielectric plate. The transmission loss of the equivalent plate can be calculated based on the theory of diffraction by small holes.

To evaluate the effectiveness of our proposal, we conducted radio propagation measurements in a high-building environment by using the developed UAV based measurement system. The result showed that the penetration losses of direct and reflection rays were underestimated in the ray-tracing simulation. This prediction error became significant as the indoor MS approached the window, and the mean error was 11.1 dB when the distance between the indoor MS and the window was 0.4 m. The total prediction error was 7.0 dB in the ray-tracing simulation. In the proposed method, the mean prediction error was improved to 0.3 dB in that case and the total prediction error was improved to −0.5 dB. The standard deviation of the prediction error was also improved from 8.2 dB to 5.3 dB. The results are expected to be utilized for actual service-cell planning in the urban environment. We found another problem that the diffraction loss was overestimated in the ray-tracing simulation. The solution to the problem will be future work.

Author Contributions: K.S. designed the measurement system and the experiment, and he was in charge of the data analysis and simulation part of the study. Q.F. was in charge of the drone controlling for the experiment. N.K. was in charge of the implementation of radio transceiver on the SDR platform. J.-i.T. supervised the project, and provided advice and support to promote the project.

References

1. Omote, H.; Miyashita, M.; Yamaguchi, R. Measurement of time-spatial characteristics between indoor spaces in different LOS buildings. In Proceedings of the 2015 International Symposium on Antennas and Propagation (ISAP), Hobart, Australia, 9–12 November 2015; pp. 1–4.
2. de Jong, Y.L.C.; Koelen, M.H.J.L.; Herben, M.H.A.J. A building-transmission model for improved propagation prediction in urban microcells. *IEEE Trans. Veh. Technol.* **2004**, *53*, 490–502. [CrossRef]
3. Axiotis, D.I.; Theologou, M.E. An empirical model for predicting building penetration loss at 2 GHz for high elevation angles. *IEEE Antennas Wirel. Propag. Lett.* **2003**, *2*, 234–237. [CrossRef]
4. Chee, K.L.; Anggraini, A.; Kaiser, T.; Kurner, T. Outdoor-to-indoor propagation loss measurements for broadband wireless access in rural areas. In Proceedings of the 5th European Conference on Antennas and Propagation (EUCAP), Rome, Italy, 11–15 April 2011; pp. 1376–1380.
5. Alatossava, M.; Suikkanen, E.; Veli-Matti, J.M.H.; Ylitalo, J. Extension of COST 231 Path Loss Model in Outdoor-to-Indoor Environment to 3.7 GHz and 5.25 GHz. In Proceedings of the 11th International Symposium on Wireless Personal Multimedia Communications, 23–27 September 2008; pp. 1–4.
6. Medbo, J.; Furuskog, J.; Riback, M.; Berg, J.E. Multi-frequency path loss in an outdoor to indoor macrocellular scenario. In Proceedings of the 2009 3rd European Conference on Antennas and Propagation, Berlin, Germany, 23–27 March 2009; pp. 3601–3605.
7. Okamoto, H.; Kitao, K.; Ichitsubo, S. Outdoor-to-Indoor Propagation Loss Prediction in 800-MHz to 8-GHz Band for an Urban Area. *IEEE Trans. Veh. Technol.* **2009**, *58*, 1059–1067. [CrossRef]
8. Roivainen, A.; Hovinen, V.; Tervo, N.; Latva-aho, M. Outdoor-to-indoor path loss modeling at 10.1 GHz. In Proceedings of the 2016 10th European Conference on Antennas and Propagation (EuCAP), Davos, Switzerland, 10–15 April 2016; pp. 1–4. [CrossRef]
9. Imai, T.; Kitao, K.; Tran, N.; Omaki, N.; Okumura, Y.; Nishimori, K. Outdoor-to-Indoor path loss modeling for 0.8 to 37 GHz band. In Proceedings of the 2016 10th European Conference on Antennas and Propagation (EuCAP), Davos, Switzerland, 10–15 April 2016 ; pp. 1–4. [CrossRef]
10. COST 231. *COST Action 231—Digital Mobile Radio Towards Future Generation Systems—Final Report*; Office for Official Publications of the European Communities: Luxembourg, 1999.
11. ITU-R Recommendation P. 2109. *Prediction of Building Entry Loss*; 2017. Available online: https://www.itu.int/rec/R-REC-P.2109/en (accessed on 21 November 2019).
12. McKown, J.W.; Hamilton, R.L. Ray tracing as a design tool for radio networks. *IEEE Netw.* **1991**, *5*, 27–30. [CrossRef]
13. Seidel, S.Y.; Rappaport, T.S. Site-specific propagation prediction for wireless in-building personal communication system design. *IEEE Trans. Veh. Technol.* **1994**, *43*, 879–891. [CrossRef]
14. Costa, E. Ray tracing based on the method of images for propagation simulation in cellular environments. In Proceedings of the Tenth International Conference on Antennas and Propagation (Conf. Publ. No. 436), Edinburgh, UK, 14–17 April 1997; Volume 2, pp. 204–209. [CrossRef]
15. Rodriguez, I.; Nguyen, H.C.; Sorensen, T.B.; Zhao, Z.; Guan, H.; Mogensen, P. A novel geometrical height gain model for line-of-sight urban micro cells below 6 GHz. In Proceedings of the 2016 International Symposium on Wireless Communication Systems (ISWCS), Poznan, Poland, 20–23 September 2016; pp. 393–398. [CrossRef]
16. Inomata, M.; Sasaki, M.; Onizawa, T.; Kitao, K.; Imai, T. Effect of reflected waves from outdoor buildings on outdoor-to-indoor path loss in 0.8 to 37 GHz band. In Proceedings of the 2016 International Symposium on Antennas and Propagation (ISAP), Okinawa, Japan, 24–28 October 2016; pp. 62–63.
17. Hasegawa, K.; Taga, T. A proposal of double aperture field method, and its experimental confirmation. In Proceedings of the 2015 International Workshop on Electromagnetics: Applications and Student Innovation Competition (iWEM), Hsinchu, Taiwan, 16–18 November 2015; pp. 1–2. [CrossRef]

18. Imai, T.; Okumura, Y. Study on hybrid method of ray-tracing and physical optics for outdoor-to-indoor propagation channel prediction. In Proceedings of the 2014 IEEE International Workshop on Electromagnetics (iWEM), Sapporo, Japan, 4–6 August 2014; pp. 249–250. [CrossRef]

19. Saito, K.; Fan, Q.; Keerativoranan, N.; Takada, J. Vertical and Horizontal Building Entry Loss Measurement in 4.9 GHz Band by Unmanned Aerial Vehicle. *IEEE Wirel. Commun. Lett.* **2019**, *8*, 444–447. [CrossRef]

20. Saito, K.; Fan, Q.; Keerativoranan, N.; Takada, J. 4.9 GHz Band Outdoor to Indoor Propagation Loss Analysis in High Building Environment Using Unmanned Aerial Vehicle. In Proceedings of the 2019 13th European Conference on Antennas and Propagation (EuCAP), Krakow, Poland, 31 March–5 April 2019; pp. 1–4.

21. Saito, K.; Fan, Q.; Keerativoranan, N.; Takada, J. 4.9 GHz Band Outdoor-to-Indoor Radio Propagation Measurement by an Unmanned Aerial Vehicle. In Proceedings of the 2018 IEEE International Workshop on Electromagnetics:Applications and Student Innovation Competition (iWEM), Nagoya, Japan, 29–31 August 2018. [CrossRef]

22. 3GPP TS38.913 V14.3.0. Access, Evolved Universal Terrestrial Radio. Study on Scenarios and Requirements for Next Generation Access Technologies. 2017. Available online: https://www.3gpp.org/DynaReport/38-series.htm (accessed on 21 November 2019).

23. 3GPP TS38.101-1 16.1.0 Access, Evolved Universal Terrestrial Radio. NR: User Equipment (UE) Radio Transmission and Reception; Part 1: Range 1 Standalone. 2019. Available online: https://www.3gpp.org/DynaReport/38-series.htm (accessed on 21 November 2019).

24. Bethe, H.A. Theory of Diffraction by Small Holes. *Phys. Rev.* **1944**, *66*, 163–182. [CrossRef]

25. Culshaw, W. Reflectors for a Microwave Fabry-Perot Interferometer. *IRE Trans. Microw. Theory Tech.* **1959**, *7*, 221–228. [CrossRef]

26. McNamara, D.A.; Pistorius, C.W.I.; Malherbe, J.A.G.*Introduction to the Uniform Geometrical Theory of Diffraction*; Artech House on Demand: Boston, MA, USA, 1990.

27. Kouyoumjian, R.G.; Pathak, P.H. A uniform geometrical theory of diffraction for an edge in a perfectly conducting surface. *Proc. IEEE* **1974**, *62*, 1448–1461. [CrossRef]

28. Molisch, A.F. *Wireless Communications*, 2nd ed.; Wiley Publishing: Hoboken, NJ, USA, 2011.

29. Otoshi, T.Y. A Study of Microwave Leakage through Perforated Flat Plates (Short Papers). *IEEE Trans. Microw. Theory Tech.* **1972**, *20*, 235–236. [CrossRef]

30. Yamamoto, S.; Hamano, A.; Hatakeyama, K.; Iwai, T. EM-wave transmission characteristic of periodically perforated metal plates. In Proceedings of the 2016 IEEE 5th Asia-Pacific Conference on Antennas and Propagation (APCAP), Kaohsiung, Taiwan, 26–29 July 2016; pp. 7–8. [CrossRef]

31. McDonald, N.A. Electric and Magnetic Coupling through Small Apertures in Shield Walls of Any Thickness. *IEEE Trans. Microw. Theory Tech.* **1972**, *20*, 689–695. [CrossRef]

32. Ettus Research. Universal Software Radio Peripheral N210. Available online: https://www.ettus.com/product/details/UN210-KIT (accessed on 21 November 2019).

33. The Raspberry Pi Foundation. Raspberry Pi 2 Model B. Available online: https://www.raspberrypi.org/products/raspberry-pi-2-model-b/ (accessed on 21 November 2019).

34. Kozo Keikaku Enginerring Inc. RapLab, Radio Wave Propagation Analysis Tool. Available online: https://www.kke.co.jp/en/solution/theme/raplab.html/ (accessed on 21 November 2019).

6

Coplanar Stripline-Fed Wideband Yagi Dipole Antenna with Filtering-Radiating Performance

Yong Chen [1], Gege Lu [2], Shiyan Wang [2] and Jianpeng Wang [2,*]

[1] School of Physics and Electronic Electrical Engineering, Huaiyin Normal University, Huaian 223300, China; yongchen@hytc.edu.cn

[2] Ministerial Key Laboratory of JGMT, Nanjing University of Science and Technology, Nanjing 210094, China; lugege_njust@163.com (G.L.); nustwang@163.com (S.W.)

*   Correspondence: eejpwang@njust.edu.cn

Abstract: In this article, a wideband filtering-radiating Yagi dipole antenna with the coplanar stripline (CPS) excitation form is investigated, designed, and fabricated. By introducing an open-circuited half-wavelength resonator between the CPS structure and dipole, the gain selectivity has been improved and the operating bandwidth is simultaneously enhanced. Then, the intrinsic filtering-radiating performance of Yagi antenna is studied. By implementing a reflector on initial structure, it is observed that two radiation nulls appear at both lower and upper gain passband edges, respectively. Moreover, in order to improve the selectivity in the upper stopband, a pair of U-shaped resonators are employed and coupled to CPS directly. As such, the antenna design is finally completed with expected characteristics. To verify the feasibility of the proposed scheme, a filtering Yagi antenna prototype with a wide bandwidth covering from 3.64 GHz to 4.38 GHz is designed, fabricated, and measured. Both simulated and measured results are found to be in good agreement, thus demonstrating that the presented antenna has the performances of high frequency selectivity and stable in-band gain.

Keywords: CPS (coplanar stripline); Yagi antenna; filtering-radiating performance; frequency selectivity

1. Introduction

There is an increasing demand for RF front end to possess much more potential characteristics for application in modern wireless communication systems, such as compact structure, low cost, high efficiency, multiple functions, and so on. It is well known that both antennas and filters are two key components in the RF front end as they play important roles in whole communication systems [1–7]. If the antenna and filter can be integrated into one module, which possesses not only the radiation characteristics but also the filtering function, the extra matching network between these two components can be removed and the footprint of whole system will be reduced efficiently. In this context, antennas with filtering performance have been attracting more and more attention [8–10].

Antennas with unidirectional radiation are much more practical in some modern wireless communication systems [8–14], such as missiles, aircrafts, and vehicles. To accommodate to this tendency, Yagi antennas have been widely used as a kind of classical structure since its original design and operating principles were first described by Uda and Yagi [15,16]. Quite recently, filtering Yagi antennas have been proposed and investigated [17–19]. In [17], a filtering quasi-Yagi antenna was designed by using cascade strategy. Multimode balun bandpass filter was directly integrated into the antenna so as to achieve filtering performance. In [19], the principle from filter to antenna was adopted. Yagi structure here acted as the last-stage resonator of a filter. However, antennas designed by these two kinds of methods are bulky. Actually, the Yagi structure can exhibit the filtering performance itself. As demonstrated in [20], the out-of-band gain suppression of Yagi antenna can be improved by optimizing the length and spacing of directors and reflectors, while narrowing the operating bandwidth compared with the conventional counterpart. Meanwhile, it is well known that the CPS structure is much appreciated by engineering according to its advantages in greatly simplifying the differential-fed network for unidirectional radiation antenna and convenient integration with the active circuits and monolithic microwave integrated circuits [21–25].

The main motivation of this article is to propose a CPS-fed wideband Yagi dipole antenna with filtering-radiating performance. The intrinsic filtering performance of the Yagi structure has been utilized here to produce two radiation nulls emerging at both lower and upper passband edges, respectively. To overcome the narrow operation bandwidth caused by this filtering scheme, an open-circuited half-wavelength resonator is introduced between CPS and driven dipole. As such, both the gain selectivity and the operating bandwidth have been enhanced simultaneously. It is demonstrated that the introduced resonator herein serves as a first-order resonator. Moreover, in order to improve the selectivity of the upper passband edge, a pair of U-shaped resonators are employed and coupled to CPS directly. Finally, an antenna prototype with operation frequency band covering from 3.64 GHz to 4.38 GHz is fabricated and measured. All results are observed as being in good agreement, thereby verifying the validity of this design.

2. Design of the Proposed Antenna

Figure 1 illustrates the configuration of the proposed filtering Yagi dipole antenna, which is fabricated on a polytetrafluoroethylene (PTFE) substrate with a relative permittivity of 2.2, thickness of 1 mm, and dimensions of 30×42 mm^2. The filtering antenna is composed of four parts: CPS for differential signal excitation, a pair of U-shaped resonators symmetrically coupled to the CPS, two-element radiators consisting of a driven element and a reflector, as wel as a folded open-circuited half-wavelength resonator inserted between the feedline and driver. All the four parts are printed on the top side of the substrate while no metal parts exist on its bottom to ensure the operating environment of dipole structure. The antenna structure is symmetrical with respect to the reference line along the middle axis of CPS along the x-axis.

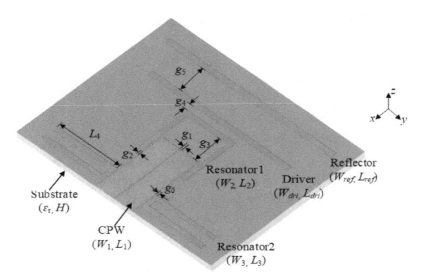

Figure 1. Geometry of the proposed filtering Yagi dipole antenna in 3D view.

2.1. Modified Dipole with Bandwidth Improved (Type A)

Firstly, the frequency selectivity of dipole antenna is investigated. For clear illustrating, both the conventional dipole antenna and the new dipole structure named type A are presented and depicted in Figure 2a,b, respectively. It should be mentioned that these two dipole antennas are designed on the same substrate and operate at the same frequency. As indicated in Figure 2b, for the new dipole structure, an open-circuited half-wavelength resonator has been introduced as one filtering element between the feedline and dipole. By virtue of this scheme, both the selectivity and bandwidth of dipole antenna are expected to be improved. Comparative results including reflection coefficients and normalized realized gains of antennas are shown in Figure 3. It can be observed that by inserting the half-wavelength resonator, an additional resonance point appears at about 3.65 GHz, naturally resulting in enhanced bandwidth. Besides, for the type A antenna, its in-band gain becomes flatter and the out-of-band suppression is better than the conventional counterpart, which means an improved selectivity.

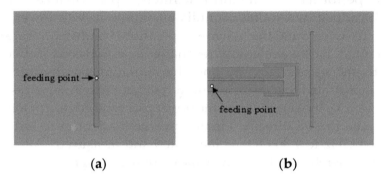

(a)   **(b)**

Figure 2. Dipole antennas. (**a**) Conventional. (**b**) Proposed type A.

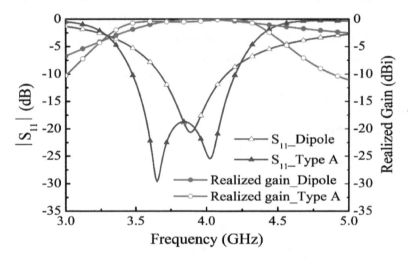

Figure 3. Comparison about reflection coefficients and realized gains between the conventional dipole antenna and proposed Type A.

2.2. Study of the Filtering Performance of Yagi Antenna

Herein, the filtering performance of Yagi structure itself is discussed. A typical three-element Yagi structure, including both reflector and director, is utilized and shown in Figure 4. By optimizing the parameters, a filtering-radiating Yagi antenna can be obtained. Table 1 tabulates the detailed dimensions of traditional and optimized Yagi antenna. A comparison about normalized realized gain and reflection coefficient curves versus frequency between them is provided in Figure 5. It can be easily observed that after the optimization process, the roll-off performance for both the realized gain and |S11| curves have been improved remarkably, revealing the desired filtering performance. In fact, the filtering property, especially out-of-band gain suppression is here caused by the enhanced loaded Q-factor, which is influenced by the distance between the driver and parasitic elements and will be higher under small element spaces [10]. The described optimization exactly reduces the distances among elements, thus increasing the loaded Q-factor. Besides, it is seen that not only the gain of the optimized one becomes flatter, but also the upper and lower radiation nulls are generated. One thing should be mentioned that radiation nulls are produced by the intrinsic characteristic of the Yagi antenna. It is known that the parasitic element working as director or reflector is determined by the phase condition between the currents on the driver and the element. If the phase of current on parasitic element is prior to that of the driver, the parasitic one will act as a director. Similarly, if the phase of the current on the parasitic element is delayed to that of driver, a reflector can be achieved. Besides, there is no doubt that phase condition is related to the operating frequency. As such, the radiation pattern and beam direction will turn 180 degrees at some certain frequencies where the reflector or director change their role. In this context, when the front-to-back ratio reaches its minimum, radiation nulls appear. The radiation nulls at the upper and lower band edges are caused by the director and reflector, respectively. It should also be noted that the resonant frequency will also be changed slightly with different distance values as the distances will affect the coupling condition for the Yagi dipole antenna [26–32]. To verify the aforementioned statement, the radiation patterns at these two radiation nulls are displayed in Figure 6. It can be clearly found that beam direction has been reversed. The optimization process changes the distances among elements, which determines coupling and phase condition. Therefore, the location of radiation nulls can be controlled.

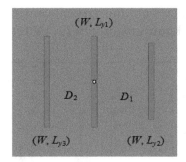

Figure 4. Configuration of a three-element Yagi antenna.

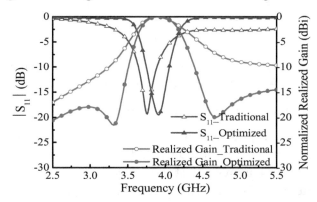

Figure 5. Comparison about normalized realized gain and reflection coefficient curves versus frequency between the traditional and optimized Yagi antenna.

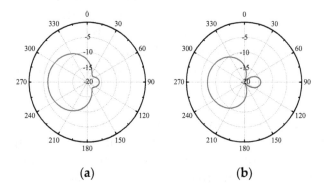

(a)   (b)

Figure 6. E planes of the optimized Yagi operating at two radiation nulls. (a) 3.32 GHz. (b) 4.66 GHz.

Table 1. Dimensions of Yagi antenna in Figure 4

Type	Parameters	W	L_{y1}	L_{y2}	L_{y3}	D_1	D_2
Traditional	Values(mm)	2	30	31.22	26.33	18.98	15.3
Optimized	Values(mm)	2	30	27.55	33.67	2.02	3.06

In addition, as indicated in Table 1, the most difference between traditional and optimized antennas lies in the length of parasitic element and distance from it to the driven one. In this context, the relationship between the occurring frequency of lower radiation null and the spacing from reflector to driven element D_2, as wel as the length of reflector L_{y3} are selected for investigation and shown in Figure 7. It can be found that when D_2 remains unchanged, increasing L_{y3} will move down the frequency location of radiation null. Moreover, with a fixed L_{y3}, increasing D_2 can also result in a variation of the frequency of radiation null. Herein, according to the above discussions, both the length of the parasitic element and the distance from the driven element to parasitic one in Yagi structure has been properly selected to obtain good gain selectivity for better filtering performance.

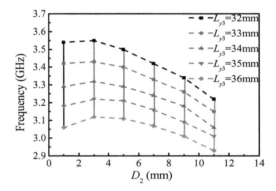

Figure 7. Relationship between the occurring frequency of lower radiation null and the spacing from reflector to driven element D_2, as wel as the length of reflector L_{y3}.

2.3. Realization of the Proposed Filtering Yagi Dipole Antenna

Based on the aforementioned dipole antenna with improved selectivity and enhanced bandwidth, a Yagi dipole antenna named type B is presented by adding a reflector in the dipole structure (Type A), as shown in Figure 8a. As we have proved that the reflector can reflect electromagnetic waves and enhance the directivity of antenna. Also, it actually can improve the out of band suppression. Figure 9 sketches the realized gain of proposed Type B. From the gain results, it is found that two radiation nulls appear at the lower and upper passband edges; nevertheless, the suppression of the upper stopband is not very good. This phenomenon is caused by the fact that the introduced folded resonator hardly influences the radiation of driver and cannot work as a director due to its limited length. This is exactly the reason why the filtering performance of Type B at the upper band edge is worse than the optimized Yagi antenna. Limited by the CPS feedline, the director has not been utilized in this proposed structure. Moreover, inside the operation band, there also exist two resonance points, so as to maintain a wide bandwidth. Compared with type A antenna, the filtering characteristic of type B is enhanced. Undoubtedly, the increment on filtering performance is achieved by intrinsic property of Yagi antenna.

To further improve the suppression in the upper stopband, a pair of U-shaped resonators are deployed and directly coupled to the CPS feedline. Until now, the implementation of wideband Yagi dipole filtering antenna—i.e., antenna type C—is finally accomplished, as shown in Figure 8b. Figure 9 depicts its realized gain. It is found that one additional radiation null has been achieved at the near edge of operating frequency band, thus manifesting an improved gain selectivity than the one caused by the reflector. Herein, the introduced U-shaped folded line functions as an open-circuited half-wavelength resonator corresponding to a specific frequency; thus, signals at this frequency cannot pass through the feedline but couple to the U-shaped resonator. One thing should be mentioned that the U-shaped folded line is not same as those reported structures with notch since ground is nonexistent. The introduced U-shaped line radiates here and works just like a folded dipole antenna, which possesses an omnidirectional radiation pattern. As such, the omnidirectional radiation caused by U-shaped line worsens the front-to-back ratio, leading to an additional radiation null.

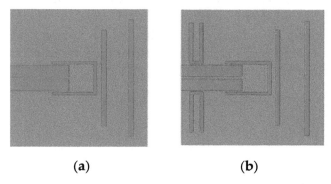

(a)   (b)

Figure 8. Configuration of proposed Yagi dipole antennas. (**a**) Type B. (**b**) Type C.

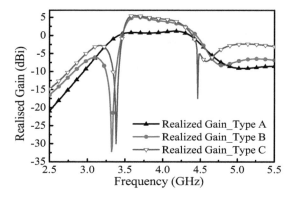

Figure 9. Comparison about reflection coefficients and normalized realized gains between the conventional dipole antenna and proposed Type A.

3. Fabrication and Experimental Results

Figure 10 shows the photograph of fabricated antenna prototype. The dimensions of the antenna prototype are: W1 = 3.8 mm, W2 = 1 mm, W3 = 0.8 mm, Wdri = 1 mm, Wref = 1 mm, L1 = 13 mm, L2 = 28.4 mm, L3 = 27.4 mm, L4 = 13 mm, Ldri = 30 mm, Lref = 36 mm, g1 = 0.2 mm, g2 = 0.3 mm, g3 = 5 mm, g4 = 1 mm, g5 = 5 mm, and g6 = 0.3 mm. The CPS feedline is achieved by a pair of parallel coupled lines which are respectively connected with the inner and outer conductor of SMA, thus forming a balun structure. Besides, the flange of SMA is selected with a compact size for reducing its influence on radiation. The proposed antenna is easily fabricated and assembled.

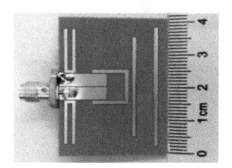

Figure 10. Photograph of the fabricated antenna prototype.

Measured and simulated reflection coefficients and gains of the fabricated antenna prototype are provided in Figure 11. It can be found that the measured operating frequency band covering from 3.64 GHz to 4.38 GHz (18.5%) is little higher than the simulated one. Moreover, these two gain curves have the same tendency with the variation of frequency. The emerged radiation nulls at specific frequency locations agree well with our expectations. Compared with simulated result, the measured one also slightly moves toward a higher frequency, which corresponds to the results of reflection coefficients. Figure 12 depicts the measured and simulated radiation patterns of fabricated antenna prototypes operating at 3.66 GHz and 4.13 GHz, which means the resonant frequencies inside the working band. Measured patterns match well with the simulated ones, except the measured cross-polarization is a little higher than the simulated one.

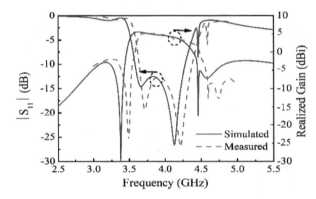

Figure 11. Measured and simulated reflection coefficients and realized gains of the fabricated antenna prototype.

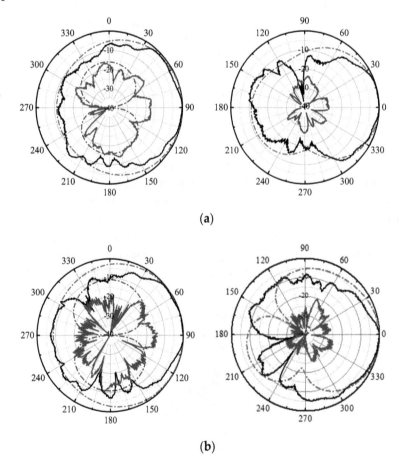

(b)

Figure 12. Measured and simulated radiation patterns of fabricated antenna prototype operating at (**a**) 3.66 GHz. (**b**) 4.13 GHz.

To highlight the advantages of this work, the performances in comparison with other reported counterparts are summarized in Table 2. It can be concluded from this table that our presented work exhibits good properties of compact size, wider bandwidth, simple structure of one-layer circuit, as wel as an improved gain selectivity.

Table 2. Comparison among filtering Yagi antennas

Ref.	Bandwidth (%)	Central Frequency	Design Principle	Realized Gain (dBi)	Size (λ_0)	Number of Radiation Null	Number of Layer
[17]	16.7	1.72 GHz	Cascade	5.9	$0.6 \times 0.6 \times 0.007$	0	3
[18]	5.5	1.82 GHz	Cascade	5.1	$0.75 \times 0.6 \times 0.007$	0	3
[19]	25	5.2 GHz	From filter to antenna	7.0	$0.65 \times 0.62 \times 0.009$	1	1
[20]	8.7	0.95 GHz	From antenna to filter	10.6	$0.55 \times 0.15 \times$ N.A	2	1
Our work	18.5	4.01 GHz	From antenna to filter	5.8	$0.56 \times 0.4 \times 0.013$	3	1

4. Conclusions

In this article, a wideband filtering-radiating Yagi dipole antenna is presented and investigated. With resorting to the CPS differential feed network, the design complexity has been effectively reduced. Then, by introducing an open-circuited half-wavelength resonator between CPS and dipole, a new dipole antenna with improved selectivity and extended bandwidth is proposed. Afterwards, the intrinsic filtering performance of Yagi structure has been studied and implemented on the dipole antenna. Meanwhile, a pair of U-shaped resonators are employed and coupled to CPS directly so as to accomplish the filtering radiation purpose. An antenna prototype has been fabricated and tested. Measured results demonstrate a wide bandwidth covering from 3.64 GHz to 4.38 GHz (18.5%) and three desired radiation nulls. Since no filtering and matching networks are involved, the structure of the proposed antenna is very simple. In addition, this antenna has the features of low profile, compact size, and easy fabrication, which will make it a good candidate for modern 5G wireless communication systems.

Author Contributions: Methodology, investigation, writing—original draft, and data curation, Y.C.; Conceptualization and project administration, G.L.; Resources, S.W.; Supervision, writing—review and editing, J.W. All authors have read and agreed to the published version of the manuscript.

Acknowledgments: The authors wish to express their thanks for the support provided by the National Natural Science Foundation of China (grant no. 61771247).

References

1. Pozar, D.M. *Microwave Engineering*, 2nd ed.; Wiley: New York, NY, USA, 1998.

2. Yang, N.; Caloz, C.; Wu, K. Co designed CPS UWB filter-antenna system. In Proceedings of the 2007 IEEE Antennas and Propagation Society International Symposium, Honolulu, HI, USA, 9–15 June 2007; Volume 6, pp. 1433–1436.

3. Shen, P.; Qi, Y.; Wang, X.; Zhang, W.; Yu, W. A 2 × 2 MIMO Throughput Analytical Model for RF Front End Optimization. *J. Commun. Inf. Netw.* **2020**, *5*, 194–203.

4. Wu, W.; Ker, M.; Chen, S.; Chen, J.; Linten, D.; Groeseneken, G. RF/High-Speed I/O ESD Protection: Co-optimizing Strategy Between BEOL Capacitance and HBM Immunity in Advanced CMOS Process. *IEEE Trans. Electron. Devices* **2020**, *67*, 2752–2759. [CrossRef]

5. Harati, P.; Kallfass, I. Analogue feed-forward carrier recovery for millimetre-wave broadband wireless links. *IET Microw. Antennas Propag.* **2020**, *14*, 366–373. [CrossRef]

6. Tabarani, F.; Boccia, L.; Calzona, D.; Amendola, G.; Schumacher, H. Power-efficient full-duplex K/Ka-band phased array front-end. *IET Microw. Antennas Propag.* **2020**, *14*, 268–280. [CrossRef]

7. Garcia, J.C.B.; Kamoun, M.; Sibille, A. Complexity Adaptive Spatial Processing of ESPAR Antenna Systems. *IEEE Trans. Wirel. Commun.* **2020**, *19*, 3700–3711. [CrossRef]

8. Jian, F.L.; Zhi, N.C.; Duo, L.W.; Gary, Z.; Yan-Jie, W. Dual-beam filtering patch antennas for wireless communication application. *IEEE Trans. Antennas Propag.* **2018**, *66*, 3730–3734.

9. Zhang, G.; Wang, J.; Wu, W. Wideband balun bandpass filter explored for a balanced dipole antenna with high selectivity. *Electron. Lett.* **2016**, *52*, 1153–1155. [CrossRef]

10. Deng, J.; Hou, S.; Zhao, L.; Guo, L. A reconfigurable filtering antenna with integrated bandpass filters for UWB/WLAN applications. *IEEE Trans. Antennas Propag.* **2018**, *66*, 401–404. [CrossRef]

11. Sohail, M.F.; Leow, C.Y.; Won, S. Energy-Efficient Non-Orthogonal Multiple Access for UAV Communication System. *IEEE Trans. Veh. Technol.* **2019**, *68*, 10834–10845. [CrossRef]

12. Kim, J.; Lee, H.; Chong, S. Super-MAC Design for Tightly Coupled Multi-RAT Networks. *IEEE Trans. Commun.* **2019**, *67*, 6939–6951. [CrossRef]

13. Capovilla, C.E.; Casella, I.R.; Costa, F.F.; Luz-de-Almeida, L.A.; Sguarezi Filho, A.J. DFIG-Based Wind Turbine Predictive Control Employing a Median Filter to Mitigate Impulsive Interferences on Transmitted Wireless References. *IEEE Trans. Ind. Appl.* **2019**, *55*, 4091–4099. [CrossRef]

14. Wu, C.; Wang, Y.; Yin, Z. Realizing Railway Cognitive Radio: A Reinforcement Base-Station Multi-Agent Model. *IEEE Trans. Intell. Transp. Syst.* **2019**, *20*, 1452–1467. [CrossRef]

15. Yagi, H. Beam transmission of the ultra short waves. *Proc. IRE* **1928**, *16*, 715–741. [CrossRef]

16. Liu, S.; Raad, R.; Theoharis, P.; Tubbal, F.A. Printed Yagi Antenna for CubeSat with Multi-Frequency Tilt Operation. *Electronics* **2020**, *9*, 986. [CrossRef]

17. Tang, H.; Chen, J.X.; Chu, H.; Zhang, G.Q.; Yang, Y.J.; Bao, Z.H. Integration design of filtering antenna with load-insensitive multilayer balun filter. *IEEE Trans. Compon. Packag. Manuf. Technol.* **2016**, *6*, 1408–1416. [CrossRef]

18. Shi, J.; Wu, X.; Chen, Z.N.; Qing, X.; Lin, L.; Chen, J.; Bao, Z. A compact differential filtering Quasi-Yagi antenna with high frequency selectivity and low cross-polarization levels. *IEEE Antennas Wirel. Propag. Lett.* **2015**, *14*, 1573–1576. [CrossRef]

19. Deng, H.W.; Xu, T.; Liu, F. Broadband pattern reconfigurable filtering microstrip antenna with Quasi-Yagi structure. *IEEE Antennas Wirel. Propag. Lett.* **2018**, *17*, 1127–1131. [CrossRef]

20. Wang, Z.P.; Hall, P.S.; Gardner, P. Yagi antenna with frequency domain filtering performance. In Proceedings of the 2012 IEEE International Symposium on Antennas and Propagation, Chicago, IL, USA, 8–14 July 2012; pp. 1–2.

21. Liang, Z.; Yuan, J. Compact dual-wideband multi-mode printed quasi-Yagi antenna with dual-driven elements. *IET Microw. Antennas Propag.* **2020**, *14*, 662–670. [CrossRef]

22. Byeon, C.W.; Eun, K.C.; Park, C.S. A 2.65-pJ/Bit 12.5-Gb/s 60-GHz OOK CMOS Transmitter and Receiver for Proximity Communications. *IEEE Trans. Microw. Theory Tech.* **2020**, *68*, 2902–2910. [CrossRef]

23. Yang, L.; Zhuang, J. Compact quasi-Yagi antenna with enhanced bandwidth and stable high gain. *Electron. Lett.* **2020**, *56*, 219–220. [CrossRef]

24. Sarkar, D.; Mikki, S.M.; Antar, Y.M.M. Poynting Localized Energy: Method and Applications to Gain Enhancement in Coupled Antenna Systems. *IEEE Trans. Antennas Propag.* **2020**, *68*, 3978–3988. [CrossRef]

25. Alekseytsev, S.A.; Gorbachev, A.P. The Novel Printed Dual-Band Quasi-Yagi Antenna with End-Fed Dipole-Like Driver. *IEEE Trans. Antennas Propag.* **2020**, *68*, 4088–4090. [CrossRef]

26. Zhou, Z.; Wei, Z.; Tang, Z.; Yin, Y. Design and Analysis of a Wideband Multiple-Microstrip Dipole Antenna with High Isolation. *IEEE Antennas Wirel. Propag. Lett.* **2019**, *18*, 722–726. [CrossRef]

27. He, K.; Gong, S.; Gao, F. A Wideband Dual-Band Magneto-Electric Dipole Antenna with Improved Feeding Structure. *IEEE Antennas Wirel. Propag. Lett.* **2014**, *13*, 1729–1732. [CrossRef]

28. Liu, Y.; Yi, H.; Wang, F.; Gong, S. A Novel Miniaturized Broadband Dual-Polarized Dipole Antenna for Base Station. *IEEE Antennas Wirel. Propag. Lett.* **2013**, *12*, 1335–1338. [CrossRef]

29. Wang, Z.; Wu, J.; Yin, Y.; Liu, X. A Broadband Dual-Element Folded Dipole Antenna with a Reflector. *IEEE Antennas Wirel. Propag. Lett.* **2014**, *13*, 750–753. [CrossRef]

30. Lehmensiek, R.; de Villiers, D.I.L. Constant Radiation Characteristics for Log-Periodic Dipole Array Antennas. *IEEE Trans. Antennas Propag.* **2014**, *62*, 2866–2869.

31. Tao, J.; Feng, Q.; Liu, T. Dual-Wideband Magnetoelectric Dipole Antenna with Director Loaded. *IEEE Antennas Wirel. Propag. Lett.* **2018**, *17*, 1885–1889. [CrossRef]

32. Yin, H.; Wang, M.; Wang, J.; Wu, W. A compact wideband filtering quasi-yagi antenna. In Proceedings of the 2017 Sixth Asia-Pacific Conference on Antennas and Propagation (APCAP), Xi'an, China, 16–19 October 2017; pp. 1–3.

7

A Low-Cost CPW-Fed Multiband Frequency Reconfigurable Antenna for Wireless Applications

Tayyaba Khan [1], MuhibUr Rahman [2,*], Adeel Akram [1], Yasar Amin [1,3] and Hannu Tenhunen [3,4]

[1] Department of Telecommunication Engineering, University of Engineering and Technology, Taxila 47050, Pakistan

[2] Department of Electrical Engineering, Polytechnique Montreal, Montreal, QC H3T 1J4, Canada

[3] Department of Electronic Systems, Royal Institute of Technology (KTH), Isafjordsgatan 26, SE 16440 Stockholm, Sweden

[4] Department of Information Technology, TUCS, University of Turku, Turku 20520, Finland

* Correspondence: muhibur.rahman@polymtl.ca

Abstract: A novel, cedar-shaped, coplanar waveguide-fed frequency reconfigurable antenna is proposed. The presented antenna uses low-cost FR4 substrate with a thickness of 1.6 mm. Four PIN diodes are inserted on the antenna surface to variate the current distribution and alter the resonant frequencies with different combinations of switches. The proposed antenna is fabricated and measured for all states, and a good agreement is seen between measured and simulated results. This antenna resonates within the range of 2 GHz to 10 GHz, covering the major wireless applications of aviation service, wireless local area network (WLAN), worldwide interoperability for microwave access (WiMAX), long distance radio telecommunications, and X-band satellite communication. The proposed antenna works resourcefully with reasonable gain, significant bandwidth, directivity, and reflection coefficient. The proposed multiband reconfigurable antenna will pave the way for future wireless communications including WLAN, WiMAX, and possibly fifth-generation (5G) communication.

Keywords: reconfigurable antennas; PIN diodes; wireless applications; fifth-generation (5G) communication

1. Introduction

The antenna is considered a vital component for every wireless communication system nowadays. To meet the requirements of different applications in one device, multiband antennas, wideband antennas, and reconfigurable antennas have served this purpose. These individual devices can have many uses, e.g., for global positioning systems (GPS), global system for mobile communication (GSM), Bluetooth, wireless local area network (WLAN), etc. Antenna size and cost are two significant factors when designing any sort of antenna.

Reconfigurable antennas are found to be the best solution, as they can be reconfigured to resonate on different frequencies to provide various functions. Reconfigurable antennas also offer unique advantages, such as flexibility and compactness. Reconfigurable antennas can modify their properties such as frequency, polarization, and radiation pattern with changing environmental conditions or varying system requirements in a controlled manner [1]. A comprehensive and detailed review regarding microstrip reconfigurable antennas has been provided by Shakhirul et al. [2]. Different modern techniques for reconfiguration have been analyzed and the pros and cons are discussed. Methods of improving the switching capability and achieving operation at different frequency bands are discussed and investigated.

Frequency reconfigurable antennas are advantageous for diverse applications, as the need for the bandwidth of spread-spectrum signals is decreased. This is because all operating frequencies are not supposed to be covered by a wireless application simultaneously. As a result, the functionality of the antenna is improved by keeping the size small and less complicated [3]. In frequency reconfigurable antennas, low cost, miniaturized size, and use for various applications are other necessary characteristics that result in their incorporation in maximum modern wireless communication systems [4].

Many types of techniques for feeding have been reported, but the coplanar waveguide (CPW) feeding technique has been the most preferable. It reduces the complexity in design by employing the antenna patch and the ground plane on the same side of the substrate.

Several researchers have attained frequency reconfiguration by changing the operative length of the antenna by removing or adding part of its length by using various approaches, such as varactors or variable capacitors [5]. Varactor diodes are used to redirect surface current, therefore letting a smooth change in frequency with altering capacitance [6]. The authors presented a varactor-based reconfigurable antenna with the capability to operate in the tuning range from 890 to 1500 MHz with good impedance matching [7]. This antenna is very promising because the multiple smaller patches are connected by means of varactors which permit polarization agility, frequency tuning, and phase shifting. Likewise, Korosec et al. [8] presented a different multisubpatch microstrip antenna loaded with varactors for solving the reconfiguration problem.

A cedar-shaped reconfigurable antenna was designed for operation in WiMAX, Bluetooth, GPS, and WLAN bands [9]. Both PIN and varactor diodes are used in combination to achieve the reconfigurability. Three pairs of varactor diodes were utilized for achieving such an objective. Due to the utilization of so many varactor diodes with different variations, it becomes very lossy as the varactors used were from two different manufacturers, including SMV1211 from SKYWORKS and 1SV325 from TOSHIBA. Also, the overall size of the antenna is very bulky, and the total board size is 6.5×6 cm^2. However, varactor diodes are considered undesirable as they are nonlinear and exhibit a narrow tuning range [10]. Likewise, the authors proposed a varactor-based antenna with a tuning range from 2.34 to 2.68 GHz [11]. Such tuning is achieved by changing the varactor diode capacitance between 12.33 pF (0 V) to 1.30 pF (15 V). Babakhani et al. [12] reported a tunable wideband frequency with polarization reconfiguration. The antenna performed in a continuously switching frequency range from 1.17 to 1.58 GHz by loading the varactor between the patch and the ring.

Reconfiguration using RF-MEMS gave lower loss and increased Q factor compared to PIN diodes and varactors [13]. Amongst these switching techniques, PIN (Positive-Intrinsic-Negative) diodes are considered to be low cost with sound isolation, easy fabrication, and low insertion loss for optimal performance. They are very consistent and compact as they tend to increase switching speeds and decrease resistive capacitance in both states, i.e., ON and OFF. CPW feed slot and folded slot antennas have been reconfigured to alter the resonant frequency by switching the slot length [14].

Previous reported works have utilized bias lines or vias for activation of PIN diodes, which make the circuit complicated and degrade the performance of the overall antenna system. Therefore, in this work, the biasing circuit is integrated into the same antenna to make it simple, and no vias are used. The effective length of the antenna is controlled by PIN diodes.

Han et al. [15] reported L-shaped and U-shaped slots using three pin diodes for long term evolution (LTE) and WLAN applications, whereas its substrate has an antenna element on both sides. A pattern and frequency reconfigurable microstrip-based antenna is presented that has three operational states using five pin diodes; unidirectional at 5.4 GHz, omnidirectional at 2.4 GHz, and both unidirectional and omnidirectional functionalities operate simultaneously [16]. Similarly, three pin diodes are implanted on the ground plane of a frequency reconfigurable antenna to control switching states [17]. Its radiating patch is designed as a square for WLAN, WiMAX, and Bluetooth applications. The FR4-based frequency reconfigurable antenna presented by Liu et al. [18] resonates between different modes, i.e., narrow band, ultra-wideband, and dual-band. The ground is slotted with four pin diodes that result in switching. Five PIN diodes are embedded in the microstrip frequency reconfigurable antenna, with six states switchable between 2.2 GHz to 4.75 GHz [19]. However, there are three drawbacks in the design; complexity due to the large number of switches; large size; and low impedance bandwidth. It must be noted that the concept of achieving wideband performance from the tapered structure is based on [20,21], where the authors utilized a printed tapered monopole antenna (PTMA) and investigated the ultra-wideband (UWB) behavior. They showed their step by step approach to bandwidth dispensation from corresponding rectangular to tapered and then from circular/cylindrical to tapered structures. They also compare the bandwidth performance in each case.

Therefore, in this paper, a CPW-fed, compact, cedar-shaped, multiband frequency reconfigurable antenna is proposed. The main advantage of this antenna is its ability to tune to different resonant frequencies. The length of the radiating patch is changed by the lumped elements, mainly PIN diodes in this design, to achieve different resonant frequencies ranging from 2 GHz to 10 GHz.

The paper is arranged in the following manner. Section 2 gives step by step guidelines and the approach used in designing the proposed antenna. Section 3 investigates and provides the refined antenna's simulated vs. measured results in different scenarios. Section 4 compares the proposed work with the recently published state of the art designs in the literature, which is followed by a conclusion section.

2. Step by Step Design Guidelines and Approach

2.1. Antenna Geometry

The corresponding geometrical dimensions of the proposed multiband reconfigurable antenna is displayed in Figure 1. This antenna utilizes a low-cost FR4-Epoxy substrate with loss tangent $tan\ \delta = 0.02$, dielectric constant $\varepsilon_r = 4.4$, and thickness of 1.6 mm. The overall dimensions of the antenna are 40×60 mm². The antenna design modeling is carried out in a commercially available FEM (Finite element method)-based simulator, ANSYS HFSS. Moreover, we have also validated our results on time domain-based simulator CST (FDTD) before fabrication and measurement.

To achieve improved impedance matching and acceptable gain, the shape of the main radiating patch resembles the cedar tree and the rectangular slots are placed over an equilateral triangle. The antenna is designed in such a way to resonate on different frequency bands, using four PIN diodes, named 2, 3, 4, and 5, with a lumped element boundary in a reserved slot of 1 mm. The width of the CPW feedline is 1 mm to excite the antenna with the characteristic impedance of 50 Ω. This value of CPW feedline is evaluated by using Equations (1)–(6) [22].

$$k' = \sqrt{1 - k^2} \tag{1}$$

$$k_1 = \frac{S_c}{S_c + 2W} \tag{2}$$

$$k_2 = \frac{\sinh \frac{\pi a}{2h}}{\sinh \frac{\pi b}{2h}} \tag{3}$$

$$\frac{K(k)}{K'(k)} = \begin{cases} \dfrac{\pi}{\ln\left[\dfrac{2\left(1+\sqrt{k'}\right)}{1-\sqrt{k'}}\right]} & (0 \le k \le 0.707 \\[3ex] \dfrac{\ln\left[\dfrac{2\left(1+\sqrt{k'}\right)}{1-\sqrt{k'}}\right]}{\pi} & (0.707 \le k \le 1 \end{cases} \tag{4}$$

$$\varepsilon_{eff} = 1 + \frac{\varepsilon_r - 1}{2} \frac{K(k_2)}{k'(k_2)} \frac{k'(k_1)}{K(k_1)} \tag{5}$$

$$Zo_{cp} = \frac{30\pi}{\sqrt{\varepsilon_{eff}}} \frac{k'(k_1)}{K(k_1)} \tag{6}$$

The parameters used in these equations are described as follows: S_c is the width of the central conductor, Zo_{cp} is the characteristic impedance of the CPW line, W is the separation between the central conductor and ground plane, ε_r is the relative dielectric constant of the substrate, ε_{eff} is an effective dielectric constant, h is the thickness of the substrate, K_1, and K_2 can be calculated by utilizing S_c, W, h, a, and b as defined before. $K(k)$ is defined as the integral of first order with argument k or complementary argument k'.

Table 1 displays the optimized overall dimensions of the antenna. Transmission line theory is used to calculate the effective length for a specific resonating frequency using Equation (7) taken from [23].

$$L_r = \frac{c}{4 f_r \sqrt{\frac{\varepsilon_r + 1}{2} + \frac{\varepsilon_r - 1}{2}\left(1 + \frac{12h}{W}\right)^{-0.5}}} \tag{7}$$

where h is the thickness of the substrate, c is the speed of light, L_r is the length of the resonant frequency, and ε_r is the relative permittivity of the substrate.

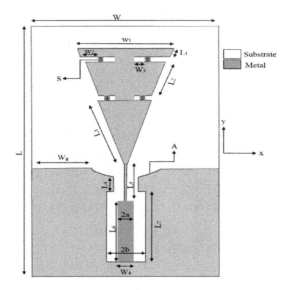

Figure 1. Geometrical dimensions of the proposed antenna.

Table 1. Dimensions of the proposed antenna (all values are in mm).

Parameter	Value	Parameter	Value	Parameter	Value	Parameter	Value
L	60	W_2	3.2	L_1	2.62	L_5	8.35
W	40	W_3	2.93	L_2	8.41	L_6	15.8
W_g	12.25	W_4	3.05	L_3	18.39	L_7	18.3
W_1	20	S	1	L_4	4.51	A	5.8
S_c	1	h	1.6	a	1.525	b	3.535

2.2. Design Methodology

Figure 2 shows the steps to achieve the cedar-shaped antenna at different operating modes with its corresponding S_{11}, which basically represents how much power is reflected from the antenna, and hence is known as the reflection coefficient [22]. The minimum value of $|S11|$ is 0 (best case and corresponding dB value is $-\infty$), the maximum value of $|S11|$ is 1 (worst case and corresponding dB value is 0), and the compromised/acceptable value of $|S11|$ is 0.33 (corresponding dB value is almost -10). As can be seen in our simulations, the corresponding S_{11}-parameter value is less than -10 dB in all passbands. Parametric analysis of the antenna is performed based on the corresponding S_{11}. To perform it, first, an occupied equilateral triangle is designed as a basic structure with a CPW feedline. In this case, the electrical length is large, and the current is flowing through the whole antenna, thus making it resonate at multiple frequencies, i.e., 2.7 GHz, 4.3 GHz, and 6.2 GHz. Secondly, a strip is defined at the upper side of the equilateral triangle which results in a frequency shift at 3 GHz. In the last case, the lower part of the equilateral triangle is disjoined from the previous design to obtain a higher frequency, i.e., 8.2 GHz, as the current path becomes shorter compared to other cases. All these variations with corresponding reflection coefficient plots are shown in Figure 1.

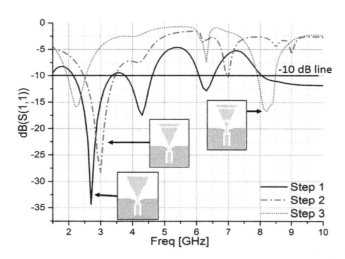

Figure 2. Various steps of the proposed antenna with the corresponding S_{11}.

The ultimate proposed design is obtained by the incorporation of four PIN diodes (2, 3, 4, 5) at specified locations. The diodes are termed as S with 1 mm reserved slots between the upper and middle portion of the equilateral triangle. This results in frequencies ranging from 2 GHz to 10 GHz. A flow chart demonstrating the steps taken to achieve the desired design is shown in Figure 3. The corresponding radiation pattern, including both principal planes for the fundamental structure, is shown in Figure 4.

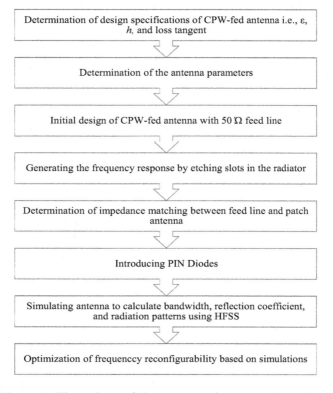

Figure 3. Flow chart of the antenna design methodology.

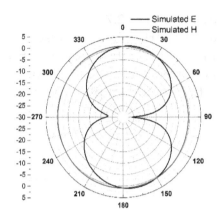

Figure 4. Simulated radiation pattern of the proposed antenna at 2.7 GHz.

2.3. Switch Design

To confer electronic reconfigurability, model shunt switches were substituted with PIN diodes. These PIN diodes are suitable for reconfiguration owing to their reliability, low cost, high switching speed, small size, and decreased capacitance and resistance in the ON and OFF states. Figure 5a,b illustrates both the ON and OFF states in that order. The reactive components C_t and L_s produce the packaging impact, while others develop the electronic traits in the ON and OFF states of the diode junction. Referring to the datasheet of PIN diode SMP 1320-079, the lumped characteristics for the ON state are $L_s = 0.7$ nH, $R_s = 0.9$ Ω. In the OFF state (reverse biased), R_p is given a higher value than the reactance C_t because of packaging; therefore, it is ignored from the equivalent diode model.

It is noteworthy that during the simulation process in HFSS, a strip element of size around 1 mm × 1 mm was used to characterize the PIN diode, and its equivalent circuit in the "ON" and "OFF" states is as shown in Figure 5a,b, respectively. The reason for using a PIN diode in this work is because the PIN diode has better RF switching capability [23]. Notably, removing this strip element during the simulation to indicate the "OFF" state has also been investigated, and the results are very similar to applying the lumped element as in Figure 5. Thus, in the following simulation, the latter technique was used. To maintain good DC isolation in practical scenarios, four RF chokes (inductors with 1 µH) must be used in the DC biasing circuit, and they need to be linked to a two separate two-position dual inline package (DIP) switch and a low-power lithium battery (approximately 3 volts DC and a continuous standard load current of 0.2 mA). However, it must be noted that the DC supply is to be protected from RF. DC blocking is done using an appropriate capacitor and RF blocking is done using an RF choke. Values of capacitance and inductance are based on the operating frequency.

For clarity and simplicity, we employed a simple strip that can turn on and off the switch for demonstration. For further use in practical applications, the biasing circuit of Case 3 is provided, which is very simple to predict. The switch is connected to the 3 V battery and the two terminals are considered, one with a via hole and the other through an RF choke of 1 µH. A strip has been placed on the back side for via connection through the battery and can be soldered in the same way as in [24,25]. A similar procedure has been adopted for Case 4. All these cases were first studied in a simulation environment in order to determine the possible states. In fabrication, different antennas are designed for each case in order to clearly distinguish them and so that the reader may understand the step by step approach. All cases are measured, and the results are provided in each case. Only states that have an independent feed are considered and placed in the manuscript. A canonical model of Case 3 is also provided in Figure 5c for further illustration.

In the proposed antenna design, modeling of the PIN diode in HFSS has been done by employing two sets of lumped RLC boundary conditions, as shown in Figure 6. L is the first portion of this RLC model and the second is a parallel combination of R_p and C_t for the OFF state or R_s for ON state, respectively. Ansoft HFSS (High Frequency Structure Simulator) is used for modeling of the PIN diodes, which provides robustness regarding reconfiguration among several resonating frequency bands.

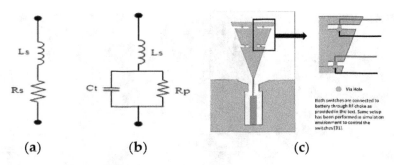

(a)   (b)   (c)

Figure 5. Positive-Intrinsic-Negative (PIN) diode equivalent circuit: (**a**) ON state. (**b**) OFF state, (**c**) Canonical biasing model of Case 3.

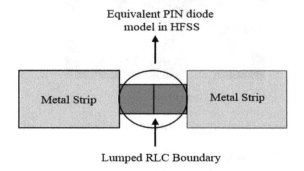

Figure 6. Modeling of the PIN diode in Ansoft HFSS (High Frequency Structure Simulator).

3. Results and Discussions

For validation of the performance parameters of the designed antenna, a prototype was fabricated for all four cases, as shown in Figure 7, and measured. A vector network analyzer was used for measurement of the reflection coefficient and VSWR. The antenna radiation pattern is measured using the anechoic chamber installed in the NUST research center for microwave and millimeter waves (RIMMS). The chamber can characterize antennas in the frequency range from 0.8 GHz to 40 GHz. The anechoic chamber is equipped with a near-field planner scanner and a far-field tower to measure the radiation pattern of a given antenna under test (AUT). The measurement software has the capability to transform the near-field data to far-field data for plotting antenna radiation patterns in 3D. A snapshot of the measurement setup in the anechoic chamber of the proposed antenna is shown in Figure 8a while the zoom in view is shown in Figure 8b.

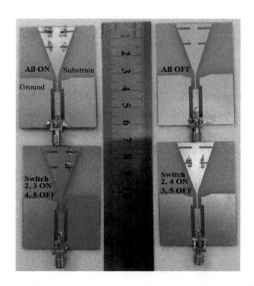

Figure 7. Fabricated prototype of the antenna showing different states.

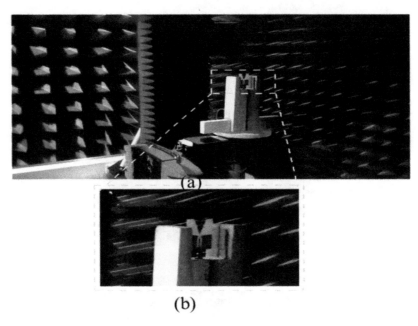

Figure 8. Measurement setup (**a**) Proposed antenna in the anechoic chamber; (**b**) Zoom in view.

The general states of the achieved bands and positions relating to the various configurations of the PIN diodes are depicted in Table 2. The electrical length of the presented antenna is characterized by the switch's ON/OFF states that radiate at a particular frequency band. In the Table '0' represents the OFF state, whereas '1' represents the ON state. The best possible results out of the 16 tested states were chosen for the manuscript. The four states mentioned in the paper have outstanding characteristic parameters, including VSWR less than 2, wide bandwidth, reasonable gain, and excellent radiation efficiency at all operating frequencies. The reflection coefficient at all states is below −10 dB, which confirms that the antenna operates properly. The antenna parameters were also theoretically calculated using the below-stated Equations (8)–(12) [26–29]. The calculated values are approximately the same as of the experimentally obtained values.

$$VSWR = \frac{1 + |\Gamma|}{1 - |\Gamma|} \tag{8}$$

$$L_{RT} = -20 \log_{10} |\Gamma| \tag{9}$$

$$\Gamma = \frac{Z_a - Z_o}{Z_a + Z_o} \tag{10}$$

$$Z_a = R_a + jX_a \tag{11}$$

$$R_a = R_r + R_L \tag{12}$$

The variables from the above equations are described as follows: Γ is reflection coefficient, L_{RT} is the return loss, Z_a is input impedance of the antenna, R_L is the loss resistance, and R_r is the radiation resistance of the antenna.

Along with above-mentioned antenna features, effective aperture is one of the critical parameters to calculate because the surface current and field intensity at the aperture of the antenna is not uniform. The effective aperture of the antenna is theoretically calculated from Equation (13) [30]. To obtain gain and directivity, D values for the proposed antenna were determined. η_{ap} is the aperture efficiency, A_p is the physical aperture, and A_e is the effective aperture of the antenna.

$$D = \frac{4\pi}{\lambda^2} A_e = \frac{4\pi}{\lambda^2} \eta_{ap} A_p \tag{13}$$

It should be noted that physical aperture A_p is always less than the effective aperture A_e.

Table 2. Different configurations of pin diodes and their corresponding operating bands.

Case	Diodes				Frequency Bands	Operating Frequency (GHz)	No. of Bands
	S2	S3	S4	S5	(GHz)		
1	1	1	1	1	2.2–3.4, 3.8–4.7, 7.8–8.4, 9.2–9.7	2.7, 4.4, 8.1, 9.4	4
2	0	0	0	0	2.1–3.3, 3.6–4.3, 5.2–5.46, 6.4–6.8, 8.4–8.8	2.7, 4.1, 5.4, 6.6, 8.6	5
3	1	0	1	0	2.2–3.3, 3.7–4.4, 5.2–5.6, 6.7–7.2, 8.4–9.2	2.7, 4.1, 5.4, 6.9, 8.8	5
4	1	1	0	0	2.2–3.3, 3.7–4.5, 8.1–8.8	2.7, 4.2, 8.4	3

3.1. Case 1: (All Switches ON)

The measured and simulated S_{11} in Case 1 are shown in Figure 9. All the switches are ON in this case; thus, current pass through the longest path and all parts of the antenna radiates. In this mode, the antenna resonates at four different operating bands from 2.2 GHz to 9.7 GHz. This band has almost same radiation pattern, i.e., the omnidirectional pattern in H-plane and bidirectional in E-plane, as all its frequency bands, so a radiation pattern that is appropriate for all frequencies is selected for inclusion in the paper. This antenna is an excellent choice for integration into portable devices for wireless communication. Figure 10 shows the radiation pattern for all switches in the ON state.

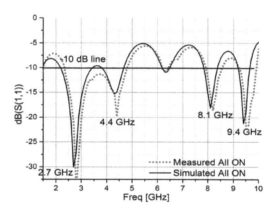

Figure 9. Measured and simulated S_{11} for all switches ON.

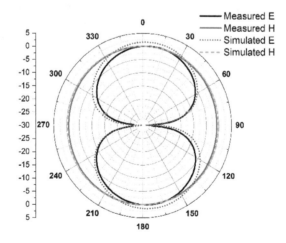

Figure 10. Measured and simulated radiation pattern for all switches ON.

The first band is from 2.2 GHz to 3.4 GHz with a mid-operating frequency of 2.7 GHz and an impedance bandwidth of 1200 MHz. It covers the WLAN standards of 2.4 GHz to 2.484 GHz and WiMAX standards of 2.5 GHz to 2.69 GHz. The second band is from 3.8 GHz to 4.7 GHz with an operating frequency of 4.4 GHz and an impedance bandwidth of 900 MHz. This band covers a part of the S-band. The S-band is used for weather radar, surface ship radar, and some communications satellites (WLAN, Bluetooth, ZigBee, radio astronomy, microwave devices/communications, mobile phones, GPS, amateur radio).

The C-band is used for long-distance radio telecommunications. Mostly 4.4 GHz is used for defense communication. The third band achieved is from 7.8 GHz to 8.4 GHz with an operating frequency of 8.1 GHz and an impedance bandwidth of 600 MHz. This band covers an essential part of the X-band for wireless communication. It includes the military demand for satellite uplinks and the mobile satellite sub-band from 7.9 GHz to 8 GHz for naval and land mobile satellite earth stations. The military requirement for earth exploration satellite (downlink) purposes is in the band from 8 GHz to 8.4 GHz. The last band, in this case, starts from 9.2 GHz and ends at 9.7 GHz with an operating frequency of 9.4 GHz and an impedance bandwidth of 500 MHz. It is used for satellite communications, radar, terrestrial broadband, space communications, and amateur radio. The summarized results of this case are presented in Table 3.

Table 3. Summarized results of Case 1.

Freq. (GHz)	2.7	4.4	8.1	9.4
RL (dB)	−30	−15	−18	−21
BW (MHz)	1200	900	600	500
Gain (dBi)	1.8	1.9	2.01	1.41
Dir. (dB)	1.7	2.1	2.2	1.54
Eff. (%)	90	91	88	91
VSWR	1.09	1.2	1.03	0.9
App.	WLAN/ WiMAX	S/C-Band	Fixed/ Mobile Satellite	Satellite/ Radar

3.2. Case 2: (All Switches Off)

The simulated and measured S_{11} in Case 2 are depicted in Figure 11. All the switches are OFF in this case; thus, current passes through the shortest path, and part of the antenna resonates. In this mode, the antenna resonates at five different operating bands with an impedance bandwidth of 2.1 GHz to 8.8 GHz. The radiation pattern of this case shows the same behavior, with an omnidirectional pattern in the H-plane and bidirectional in the E-plane as shown in Figure 12.

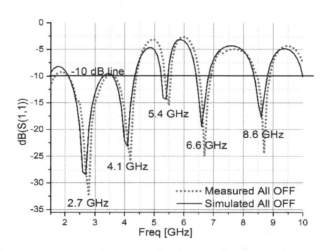

Figure 11. Measured and simulated S_{11} for all switches OFF.

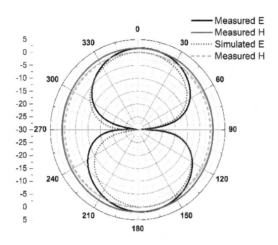

Figure 12. Measured and simulated radiation pattern for all switches OFF.

The first band achieved, in this case, is from 2.1 GHz to 3.3 GHz with an operating frequency of 2.7 GHz and an impedance bandwidth of 700 MHz. This band is used in earth exploration satellites, radio astronomy, and space research services whereas the 2.7 GHz to 2.9 GHz band is used by aviation services. The band from 3.1 GHz to 3.3 GHz band is used in by radiolocation services. This band is allocated for both federal and nonfederal use.

The second band achieved is from 3.6 GHz to 4.3 GHz with an operating frequency of 4.1 GHz and an impedance bandwidth of 700 MHz. This band is useful for WiMAX applications. The third band achieved is from 5.2 GHz to 5.46 GHz with an operating frequency of 5.4 GHz and an impedance bandwidth of 260 MHz. This band can be used mainly for 5.15 GHz to 5.825 GHz standard WLAN applications. The fourth band achieved is from 6.4 GHz to 6.8 GHz with an operating frequency of 6.6 GHz and an impedance bandwidth of 400 MHz. This band is used for long-distance radio telecommunications. The last band of this case covers 8.4 GHz to 8.8 GHz with an operating frequency at 8.6 GHz and an impedance bandwidth of 400 MHz. This band is used for X-band Satellite communication. Table 4 presents the summarized results of this case.

Table 4. Summarized results of Case 2.

Freq.(GHz)	2.7	4.1	5.4	6.6	8.6
RL (dB)	−29	−23	−14	−19	−18
BW (MHz)	1200	700	260	400	400
Gain (dBi)	1.55	1.7	0.64	0.94	2.34
Dir. (dB)	1.62	1.9	0.8	1.10	2.6
Eff. (%)	95	88	77	85	87
VSWR	1.2	1.02	1.03	1.011	1.04
App.	Aviation Service	WiMAX	WLAN	Long Dist. Radio Comm.	X-Band Sat. Comm.

3.3. Case 3: (Switches 2 and 4 ON, Switches 3 and 5 OFF)

The measured and simulated S_{11} in Case 3 are shown in Figure 13. Switches 2 & 4 are ON in this case, whereas switches 3 & 5 are OFF; thus, current passes through the upper part of the antenna only. In this mode, the antenna resonates at five different operating bands with an impedance bandwidth of 2.2 GHz to 9.2 GHz. The radiation pattern of this case shows the same behavior, having an omnidirectional pattern in the H-plane and a bidirectional pattern in the E-plane, as shown in Figure 14.

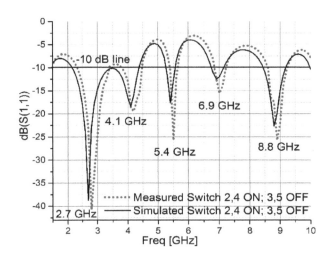

Figure 13. Measured and simulated S$_{11}$ for switches 2 and 4 ON, switches 3 and 5 OFF.

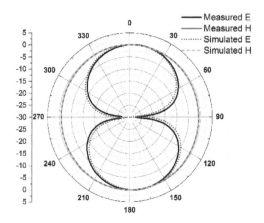

Figure 14. Measured and simulated radiation pattern for switches 2 and 4 ON, switches 3 and 5 OFF.

In this case, the first three bands are similar to the previous cases but with slight differences. The fourth band ranges from 6.7 GHz to 7.2 GHz with an operating frequency of 6.9 GHz and an impedance bandwidth of 500 MHz. This band supports fixed satellite communication and WiMAX applications. The last band covers 8.4 GHz to 9.2 GHz with an operating frequency of 8.8 GHz and an impedance bandwidth of 800 MHz. This band is used for satellite and space communications. Table 5 shows the summarized results of this case.

Table 5. Summarized results of Case 3.

Freq. (GHz)	2.7	4.1	5.4	6.9	8.8
RL (dB)	−39	−18	−17	−13	−23
BW (MHz)	1100	700	400	500	800
Gain (dBi)	1.35	1.7	0.65	1.53	2.5
Dir. (dB)	1.55	1.98	0.80	1.7	2.9
Eff. (%)	87	85	82	89	88
VSWR	1.4	1.09	1.02	1.04	0.9
App.	S-band	Aeronautical Radio Navi.	WiMAX	FixedSat./ WiMAX	Sat./Space Comm.

Regarding Case 2 and Case 3, the applications of both cases are different since the variation in frequency is very minor, i.e., Case 2 antennas can be used for aviation services, WiMAX, WLAN, long distance radio communication, and X-band satellite communication, while Case 3 antennas can be used for S-band applications, WiMAX, aeronautical radio navigation services, and X-band satellite communication.

3.4. Case 4: (Switches 2 and 3 ON, Switches 4 and 5 OFF)

The measured and simulated S_{11} in Case 4 are shown in Figure 15. Switches 2 & 3 are ON in this case, whereas switches 4 & 5 are OFF; thus, current passes through the right part of the antenna only. In this mode, the antenna resonates at three different operating bands with an impedance bandwidth of 2.2 GHz to 8.8 GHz. The radiation pattern of this case shows the same behavior, having an omnidirectional pattern in the H-plane and bidirectional pattern in the E-plane, as shown in Figure 16.

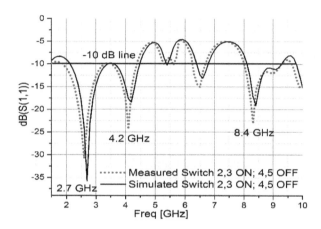

Figure 15. Measured and simulated S_{11} for switches 2 and 3 ON, switches 4 and 5 OFF.

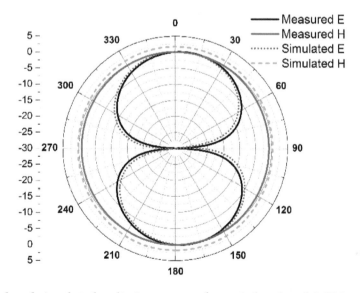

Figure 16. Measured and simulated radiation pattern for switches 2 and 3 ON, switches 4 and 5 OFF.

The first band in this state ranges from 2.2 GHz to 3.3 GHz with an operating frequency of 2.7 GHz and an impedance bandwidth of 1100 MHz. This band is used in the aviation service, maritime service, and the radiolocation service. The second band ranges from 3.7 GHz to 4.5 GHz with an operating frequency of 4.2 GHz and an impedance bandwidth of 800 MHz. This band is used for long-distance radio telecommunications. The last band ranges from 8.1 GHz to 8.8 GHz with an operating frequency of 8.4 GHz and an impedance bandwidth of 700 MHz. This band covers a productive part of the X-band wireless application. The summarized results of this case are given in Table 6.

Table 6. Summarized results of Case 4.

Freq. (GHz)	2.7	4.2	8.4
RL (dB)	−35	−17	−19
BW (MHz)	1100	800	700
Gain (dBi)	1.43	1.80	1.70
Dir. (dB)	1.52	2.02	2.02
Eff. (%)	94	89	84
VSWR	1.4	0.9	1.02
App.	Maritime/ Radiolocation Service	Long Distance Comm.	X-band/Satcom App.

3.5. Proposed Antenna Gain and Efficiency (%) at Different Stages

The gain (dBi) and percentage efficiency of the proposed antenna is also taken into account and considered at all four cases. The brief description of the gain is tabulated in Table 7 which shows the gain of the proposed antenna in all four cases at their resonance frequencies. It is seen that the proposed antenna possesses an acceptable gain. Also, a brief description of the antenna efficiency (%) is tabulated in Table 8 and plotted in Figure 17, which shows the percentage radiation efficiency of the proposed antenna in all four cases at their resonance frequencies. It shows that antenna efficiency (%) is very stable in each case at the resonance frequency.

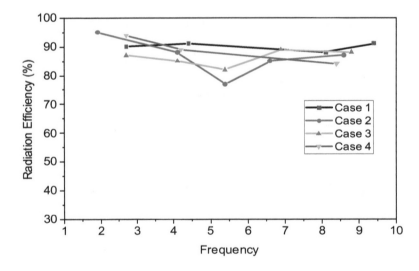

Figure 17. Radiation efficiency (%) of the proposed antenna in all cases.

Table 7. Gain (dBi) in all four cases at resonance frequencies.

	Case 1		Case 2		Case 3		Case 4
Freq.	Gain (dBi)	Freq.	Gain (dBi)	Freq.	Gain (dBi)	Freq.	Gain (dBi)
2.7	1.8	1.9	1.55	2.7	1.35	2.7	1.43
4.4	1.9	4.1	1.7	4.1	1.7	4.2	1.80
8.1	2.01	5.4	0.64	5.4	0.65	8.4	1.70
9.4	1.41	6.6	0.94	6.9	1.53	-	-
-	-	8.6	2.34	8.8	2.5	-	-

Table 8. Radiation Efficiency (%) in all four cases at resonance frequencies.

	Case 1		Case 2		Case 3		Case 4	
Freq.	Efficiency (%)	Freq.	Efficiency (%)	Freq.	Efficiency (%)	Freq.	Efficiency (%)
2.7	90	1.9	95	2.7	87	2.7	94
4.4	91	4.1	88	4.1	85	4.2	89
8.1	88	5.4	77	5.4	82	8.4	84
9.4	91	6.6	85	6.9	89	-	-
-	-	8.6	87	8.8	88	-	-

4. Comparison with Other State of the Art Designs

The proposed antenna was compared with the recent state of the art to highlight the novelty of this work. The comparison is summarized in Table 9. It can be seen that the proposed antenna is advantageous in terms of maximum passbands, its wide reconfigurable behavior, low-cost substrate, and overall size.

Table 9. Performance comparison with other designs in the literature.

Ref.	[9]	[15]	[16]	[17]	[18]	[19]	[31]	[32]	Proposed Work
Area (mm^2)	3900	675	1852.3	400	1720	1892	337.5	2852	2400
Substrate	FR4	RO4350B	RO4350	FR4	FR4	PET	FR4	FR4	FR4
Thickness (mm)	1.55	0.8	1.5	0.8	1.6	0.1	0.8	N/A	1.6
No. of resonances	6	6	4	3	3	3	2	2	12
No. of switches	6	3	5	3	4	1	N/A	N/A	4
BW at diff. resonance bands (MHz)	1400 to 4600	100; 120; 280; 220; 100; 320	690; 300; 740; 620	210; 400; 580	500; 380; 800	160; 180; 270	1575; 244	245; 525; 575	1200; 900; 600; 500; 700; 260; 400; 400; 500; 800; 800; 700

5. Conclusion

A novel cedar-shaped multiband frequency reconfigurable antenna is proposed for WLAN, WiMAX, X-band satellite, and many other wireless communication systems. The broad tunability of operating bands is achieved by employing switches to alter the effective electrical length of the radiating element of the antenna. The proposed antenna is studied and optimized in different reconfigurable scenarios and validated through measurements. The proposed antenna is studied in terms of important antenna parameters, including reflection coefficient, antenna gain, and radiation pattern, through simulation as well as measurements. This antenna radiates efficiently at all the desired bands. Design simplicity, compactness, and reconfigurability are the features that make this antenna a good choice for future wireless communication applications.

Author Contributions: Conceptualization, T.K.; Data curation, T.K.; Formal analysis, T.K. and M.R.; Investigation, M.R., A.A. and Y.A.; Methodology, T.K. and M.R.; Project administration, Y.A.; Resources, A.A.; Software, A.A.; Supervision, A.A. and Y.A.; Validation, Y.A. and H.T.; Visualization, H.T.; Writing—original draft, T.K.; Writing—review & editing, M.R.

Acknowledgments: The work was financially supported by Vinnova (The Swedish Governmental Agency for Innovating Systems) and University of Engineering and Technology Taxila, Pakistan through the Vinn Excellence centers program and ACTSENA research group funding, respectively. The fabrication of the antenna has been carried out at National Institute of Electronics (NIE) Islamabad, Pakistan, whereas antenna testing and measurement was done at Research Institute for Microwave and Millimeter-Wave Studies (RIMMS), National University of Science and Technology (NUST), Islamabad, Pakistan.

References

1. Yuan, Z.; Chang-Ying, W. An approach for optimizing the reconfigurable antenna and improving its reconfigurability. In Proceedings of the IEEE International Conference on Signal Processing, Communications, and Computing (ICSPCC), Hong Kong, China, 5–8 August 2016; pp. 1–5.

2. Shakhirul, M.S.; Jusoh, M.; Lee, Y.S.; Husna, C.R.N. A Review of Reconfigurable Frequency Switching Technique on Micostrip Antenna. *J. Phys. Conf. Ser.* **2018**, *1019*, 012042. [CrossRef]

3. Hannula, J.-M.; Holopainen, J.; Viikari, V. Concept for frequency reconfigurable antenna based on distributed transceivers. *IEEE Antennas Wirel. Propag. Lett.* **2017**, *16*, 764–767. [CrossRef]

4. Abdulraheem, Y.I.; Oguntala, G.A.; Abdullah, A.S.; Mohammed, H.J.; Ali, R.A.; Abd-Alhameed, R.A.; Noras, J.M. Design of frequency reconfigurable multiband compact antenna using two PIN diodes for WLAN/WiMAX applications. *IET Microw. Antennas Propag.* **2017**, *11*, 1098–1105. [CrossRef]

5. Rahman, M.; NagshvarianJahromi, M.; Mirjavadi, S.S.; Hamouda, A.M. Compact UWB Band-Notched Antenna with Integrated Bluetooth for Personal Wireless Communication and UWB Applications. *Electronics* **2019**, *8*, 158. [CrossRef]

6. Li, T.; Zhai, H.; Li, L.; Liang, C. Frequency-reconfigurable bow-tie antenna with a wide tuning range. *IEEE Antennas Wirel. Propag. Lett.* **2014**, *13*, 1549–1552.

7. Korošec, T.; Ritoša, P.; Vidmar, M. Varactor-tuned microstrip-patch antenna with frequency and polarisation agility. *Electron. Lett.* **2006**, *42*, 1015–1017. [CrossRef]

8. Korosec, T.; Naglic, L.; Tratnik, J.; Pavlovic, L.; Batagelj, B.; Vidmar, M. Evolution of varactor-loaded frequency and polarization reconfigurable microstrip patches. In Proceedings of the IEEE Asia-Pacific Microwave Conference, Melbourne, VIC, Australia, 5–8 December 2011; pp. 705–708.

9. Madi, M.A.; Al-Husseini, M.; Kabalan, K.Y. Frequency Tunable Cedar-Shaped Antenna for WIFI and Wimax. *Prog. Electromagn. Res. Lett.* **2018**, *72*, 135–143. [CrossRef]

10. Xie, P.; Wang, G.; Li, H.; Liang, J. A dual-polarized two-dimensional beam-steering fabry–pérot cavity Antenna with a reconfigurable partially reflecting surface. *IEEE Antennas Wirel. Propag. Lett.* **2017**, *16*, 2370–2374. [CrossRef]

11. Muthuvel, S.K.; Choukiker, Y.K. Frequency tunable circularly polarized antenna with branch line coupler feed network for wireless applications. *Int. J. RF Microw. Comput. Eng.* **2019**, *29*, e21784. [CrossRef]

12. Babakhani, B.; Sharma, S. Wideband frequency tunable concentric circular microstrip patch antenna with simultaneous polarization reconfiguration. *IEEE Antennas Propag. Mag.* **2015**, *57*, 203–216. [CrossRef]

13. Singh, R.; Slovin, G.; Xu, M.; Schlesinger, T.E.; Bain, J.A.; Paramesh, J. A Reconfigurable Dual-Frequency Narrowband CMOS LNA Using Phase-Change RF Switches. *IEEE Trans. Microw. Theory Tech.* **2017**, *65*, 4689–4702. [CrossRef]

14. Nguyen-Trong, N.; Piotrowski, A.; Hall, L.; Fumeaux, C. A frequency- and polarization-reconfigurable circular cavity antenna. *IEEE Antennas Wirel. Propag. Lett.* **2017**, *16*, 999–1002. [CrossRef]

15. Han, L.; Wang, C.; Chen, X.; Zhang, W. Compact frequency reconfigurable slot antenna for wireless applications. *IEEE Antennas Wirel. Propag. Lett.* **2018**, *15*, 1795–1798. [CrossRef]

16. Li, P.K.; Shao, Z.H.; Wang, Q.; Cheng, Y.J. Frequency and pattern reconfigurable antenna for multi standard wireless applications. *IEEE Antennas Wirel. Propag. Lett.* **2015**, *14*, 333–336. [CrossRef]

17. Borhani, M.; Rezaei, P.; Valizade, A. Design of a reconfigurable miniaturized microstrip antenna for switchable multiband systems. *IEEE Antennas Wirel. Propag. Lett.* **2016**, *15*, 822–825s. [CrossRef]

18. Liu, X.; Yang, X.; Kong, F. A frequency-reconfigurable monopole antenna with switchable stubbed ground structure. *Radio Eng. J.* **2015**, *24*, 449–454. [CrossRef]

19. Majid, H.A.; Rahim, M.K.A.; Hamid, M.R.; Ismail, M.F. A compact frequency-reconfigurable narrowband microstrip slot antenna. *IEEE Antennas Wirel. Propag. Lett.* **2012**, *11*, 616–619. [CrossRef]

20. Verbiest, J.R.; VandenBosch, G.A.E. A novel small-size printed tapered monopole antenna for UWB WBAN. *IEEE Antennas Wirel. Propag. Lett.* **2006**, *5*, 377–379. [CrossRef]

21. Azim, R.; Islam, M.T.; Misran, N. Compact tapered-shape slot antenna for UWB applications. *IEEE Antennas Wirel. Propag. Lett.* **2011**, *10*, 1190–1193. [CrossRef]

22. Pozar, D.M. *Microwave Engineering*; John Wiley & Sons: New York, NY, USA, 2009.

23. Fallahpour, M.; Ghasr, M.T.; Zoughi, R. Miniaturized reconfigurable multiband antenna for multiradio wireless communication. *IEEE Trans. Antennas Propag.* **2014**, *62*, 6049–6059. [CrossRef]

24. Bernhard, J.T. Reconfigurable antennas. *Synth. Lect Antennas* **2007**, *2*, 1–66. [CrossRef]

25. Sung, Y.; Jang, T.; Kim, Y.-S. A reconfigurable microstrip antenna for switchable polarization. *IEEE Microw. Wirel. Compon. Lett.* **2004**, *14*, 534–536. [CrossRef]

26. Milligan, T.A. *Modern Antenna Design*; Wiley-IEEE Press: Hoboken, NJ, USA, 2006; ISBN 978-0-471-72060-7.

27. Behera, D.; Dwivedy, B.; Mishra, D.; Behera, S.K. Design of a CPW fed compact bow-tie microstrip antenna with versatile frequency tunability. *IET Microw. Antennas Propag.* **2018**, *12*, 841–849. [CrossRef]

28. Balanis, C.A. *Antenna Theory, Analysis and Design*, 2nd ed.; John Wiley & Sons: New York, NY, USA, 2016.

29. Huang, Y.; Boyle, K. *Antennas from Theory to Practice*; Wiley Press: Hoboken, NJ, USA, 2008; ISBN 978-0-470-51028-5.

30. Amin, Y.; Chen, Q.; Tenhunen, H.; Zheng, L.-R. Performance-optimized quadrate bowtie RFID antennas for cost-effective and eco-friendly industrial applications. *Prog. Electromagn. Res.* **2012**, *126*, 49–64. [CrossRef]

31. Xu, Z.; Ding, C.; Zhou, Q.; Sun, Y.; Huang, S. A Dual-Band Dual-Antenna System with Common-Metal Rim for Smartphone Applications. *Electronics* **2019**, *8*, 348. [CrossRef]

32. Azeez, H.I.; Yang, H.-C.; Chen, W.-S. Wearable Triband E-Shaped Dipole Antenna with Low SAR for IoT Applications. *Electronics* **2019**, *8*, 665. [CrossRef]

8

Radio Network Planning towards 5G mmWave Standalone Small-Cell Architectures

Georgia E. Athanasiadou [1,*], **Panagiotis Fytampanis** [1], **Dimitra A. Zarbouti** [1], **George V. Tsoulos** [1], **Panagiotis K. Gkonis** [2] **and Dimitra I. Kaklamani** [3]

[1] Wireless & Mobile Communications Lab, Department of Informatics and Telecommunications, University of Peloponnese, 22131 Tripolis, Greece; pfytampanis@uop.gr (P.F.); dzarb@uop.gr (D.A.Z.); gtsoulos@uop.gr (G.V.T.)

[2] General Department, National and Kapodistrian University of Athens, Sterea Ellada, 34400 Dirfies Messapies, Greece; pgkonis@uoa.gr

[3] Intelligent Communications and Broadband Networks Laboratory, School of Electrical and Computer Engineering, National Technical University of Athens, 9 Heroon Polytechneiou str, Zografou, 15780 Athens, Greece; dkaklam@mail.ntua.gr

* Correspondence: gathanas@uop.gr

Abstract: The 5G radio networks have introduced major changes in terms of service requirements and bandwidth allocation compared to cellular networks to date and hence, they have made the fundamental radio planning problem even more complex. In this work, the focus is on providing a generic analysis for this problem with the help of a proper multi-objective optimization algorithm that considers the main constraints of coverage, capacity and cost for high-capacity scenarios that range from dense to ultra-dense mmWave 5G standalone small-cell network deployments. The results produced based on the above analysis demonstrate that the denser the small-cell deployment, the higher the area throughput, and that a sectored microcell configuration can double the throughput for ultra-dense networks compared to dense networks. Furthermore, dense 5G networks can actually have cell radii below 400 m and down to 120 m for the ultra-dense sectored network that also reached spectral efficiency 9.5 bps/Hz/Km^2 with no MIMO or beamforming.

Keywords: radio network planning; 5G wireless communications systems; mmWave small cells

1. Introduction

Radio planning is an essential task for wireless networks that mainly refers to calculating the number, location and configuration of the radio network nodes. In the early days, since no prior network infrastructure existed, this task considered only the estimation of the number of base stations (BSs) and their locations, i.e., the BS location problem [1–4]. In addition, for early cellular networks, radio network planning was split into two separate tasks: coverage and capacity. Nevertheless, as it was shown in several publications (e.g., [5–7]), capacity and coverage planning are not un-correlated. On the contrary, they are inter-related in 3G, 4G and now 5G wireless networks and hence, they must be treated together.

The 5G networks introduce really different elements from the previous generations, mainly due to virtualization and service-based architecture. Among other things, they are designed for considerably higher data rates, very large numbers of connected Internet of Things (IoT) devices and low latency while providing adaptive means for network scalability and flexibility. The number of the 5G radio frequency bands is targeted to be higher than in previous generations of cellular networks, more specifically multiple mmWave bands [8]. Also, massive Multiple Input Multiple Output (MIMO) and hybrid beamforming are core techniques to achieve the targeted high data rates and the large number of devices [9].

Typically the cell deployment architectures can be classified into standalone and overlay architectures. In the context of 5G, the first refers to a network deployment that consists of mere mmWave small cells, while the latter refers to the deployment of mmWave small cells on top of the existing macro-relay networks in the form of hierarchical or mixed cell structures. In the overlay architecture, the existing (pre5G) macrocell layer is mainly for coverage as well as mobility and signaling problems originating from the mmWave small-cell layer, which exists for capacity boosts. The much wider bandwidth as well as the beamforming/MIMO capabilities, together with the reduced access-link distances, give the mmWave small cells the capability to substantially increase the system capacity. It has to be mentioned here that another advantage of the overlay network architecture is the separation of control and message transmissions. All the control signaling is supported via the existing macrocells and high data rate transmissions go through the mmWave small-cell network.

5G radio planning is now of utmost importance since not only does it require cost-optimized deployments capable of handling a variety of demand constraints, but also since it then affects the optimal placement of the core network elements, e.g., for achieving low latency values.

In order to tackle these issues, this article presents a generic analysis that allows for evolutionary radio network planning towards 5G mmWave standalone small-cell architectures. Due to the complex nature of the problem, a multi-objective optimization algorithm is appropriately modified in order to provide solutions in the context of the multiple constraints and specifics of a 5G mmWave network with particular emphasis in the 28 GHz band. The three main constraints of coverage, capacity and cost are studied for a range of cases that reflect scenarios from dense to ultra-dense network deployments in order to achieve high capacity.

The optimization algorithm used in this paper is the evolution of the algorithm previously used for the analysis of Long-Term Evolution (LTE) systems [7]. In this study, the analysis is focused on network dimensioning for the mmWave band at 28 GHz. Hence, the propagation module is modified accordingly and a path loss model based on results of field trials at the 28 GHz band is applied. For this analysis only microcells are considered with tri-sector directional antennas, as well as omnidirectional antenna patterns. Moreover, the system module is upgraded to encompass 5G rate calculations, while the core analysis is focused on the impact of high-density microcell deployments on the spectral efficiency and throughput efficiency of these systems.

The remainder of the paper is organized as follows: Section 2 starts with the formulation of the radio network planning optimization problem and then explains the developed simulation methodology in terms of the employed propagation model, the necessary 5G system characteristics and the planning optimization method. Section 3 presents the results for different 5G standalone small-cell network deployments, namely dense, very dense and ultra-dense. Finally, concluding remarks are provided in Section 4.

2. Problem Formulation and Simulation Methodology

For the study of dense networks at high frequencies, two types of serving radio network nodes were considered in the planning process of our analysis: omnidirectional and tri-sectored microcells. When sectorized scenarios were considered, at each site three gNBs were co-located (reuse pattern 1 $\times$ 3 $\times$ 1). During the planning process, all three sectors or only a subset of them could be activated, according to the required signal strength in an area and/or the need to limit the overall interference of the system. Although the algorithm was developed to also examine tri-sectored macrocells with relay nodes, they were not considered here since the focus was to analyze the capacity capabilities of ultra-dense standalone small-cell architectures. Note that each cell can also have MIMO capabilities.

Control Nodes (CNs) were used in our study in order to represent throughput requirements as well as coverage constraints. Specifically, K CNs were distributed in the area under study and each CN was associated with a rate requirement, i.e., $R_{CN,min}$. Introduced by system link budget calculations, the received power at each CN should always exceed a minimum power threshold.

Furthermore, the optimization process was employed on discrete locations rather than on continuous x-y space, an approach already applied and tested for previous generation network deployments [2,5,6]. This way the complexity of the NP-hard problem of continuous space calculations was avoided [10], but also, this was closer to the actual planning process employed by network operators where existing sites are most certainly reused in any future network layout.

The outline of the radio network planning process is shown in Figure 1. A more detailed description is provided in the next sections.

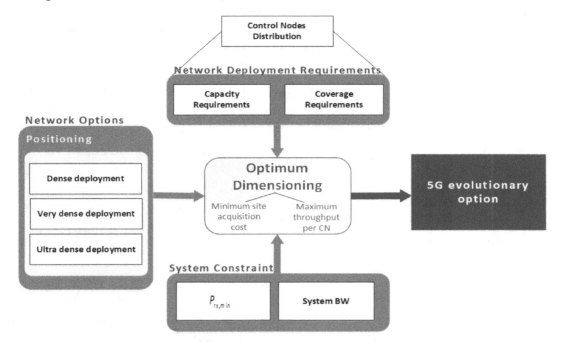

Figure 1. Radio network planning outline.

2.1. Scenario: Network Deployment Requirements

Herein, our planning process was employed in a rectangular area with dimensions 1 Km × 1 Km. Due to the area characteristics, the requirements/constraints of the planning process were defined against a set of 25 equi-spaced CNs (separation distance of 200 m). This grouping of CNs represented the high-capacity requirements in a small area, i.e., a demanding network planning operational scenario that requires a dense network deployment.

Since the trade-off between performance and network density was our main concern, a clear preexisting network layout was considered and new transmitting sites were explored, i.e., microcells with various density deployments were examined. Scenarios with 30 (5 × 6 grid), 100 (10 × 10 grid) and 400 (20 × 20 grid) candidate microcells uniformly distributed in the area were studied in the analysis below in order to investigate the capabilities of a dense network. However, it should be pointed out that the locations of the candidate network sites were an external parameter to our platform, thus could be defined by any interested stakeholder.

2.2. The Planning Method

The planning analysis consisted of three pillars:

1. Propagation analysis
2. System analysis
3. Planning optimization

2.2.1. Propagation Analysis

Various studies have been conducted in projects like METIS, MiWEBA, ITU-R M, mmMAGIC etc. in order to deploy channel models in higher frequency bands, such as millimeter waves, to support 5G use cases [8,11–13]. The path loss model that was adopted here was the Alpha-Beta-Gama (ABG) empirical path loss model:

$$PL(dB) = 10\,a\,log_{10}(d) + \beta + 10\gamma log_{10}(f_{GHz}) \tag{1}$$

The model has three parameters to describe mean path loss over frequency and distance, which derive from the statistical process of measurements. In our analysis, at urban environments at 28 GHz, these parameters took the following values [13]:

$$a = 3.5,\ \beta = 24.4,\ \gamma = 1.9 \tag{2}$$

The propagation losses of this model vs. distance for f = 28 GHz are depicted in Figure 2 along with the losses of the WINNER model for microcells at the 2 GHz band of LTE [14].

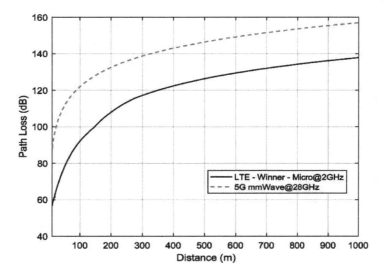

Figure 2. Path loss vs. distance at 28 GHz and 2 GHz bands.

Table 1 summarizes typical values used in the performed simulations for generic mmWave operational scenarios.

We considered the downlink of a 5G network since it is usually the bottleneck for such a system. The Signal to Interference plus Noise Ratio (SINR) at the kth CN is:

$$SINR_k = \frac{P_{tx}^b A_0^b A_{\phi \to k}^b G_{b,k}^{ch}}{\sum_{i \in \{B \setminus b\}} P_{tx}^i A_0^i A_{\phi \to k}^i G_{i,k}^{ch} + P_N} \tag{3}$$

where b is the serving gNB for the kth CN, P_{tx}^b and A_0^b are the transmitted power and maximum antenna gain for the bth gNB, respectively, while $A_{\phi \to k}^b$ is the normalized antenna gain of the bth gNB towards the kth CN and $G_{b,k}^{ch}$ is the channel gain between them. Similarly, for the interference calculation, $A_{\phi \to k}^i$ is the normalized antenna gain of the ith interfering gNB towards the served CN. Lastly, P_N is the noise power at the receiver. The micro gNBs employ either omnidirectional antennas ($A^b = 1$) or directional 120° sector antennas [15] with A_0 = 18 dBi (see Table 1) and $A^b = -min(12(\phi/\phi_{3dB})^2, 25)$, where $\phi_{3dB} = 70^o$. It has to be noted here that the Effective Isotropic Radiated Power (EIRP) of the transmitting BS antennas remains the same with sector and omnidirectional antennas, i.e., 60 dBm.

Table 1. Simulation parameters.

Simulation Parameter	Value
Central Frequency	28 GHz
Transmitter EIRP and User Antenna Gain	60 dBm
Sector Antenna Gain (3GPP pattern for 120° sector)	18 dBi
Omnidirectional Antenna Gain	4 dBi
Total cable losses	2 dB
Channel Bandwidth	100 MHz
Aggregated Bandwidth (16 Carriers)	1600 MHz
Resource Blocks (RBs)	132
Minimum Received Power $P_{rx, min}$	−5 dBm

2.3. 5G System Module

The rate requirements set at each CN are to be met by 5G system options [16–18], hence, a generic 5G rate calculation module was used in order to associate SINR calculations at each CN with the maximum throughput offered by the system. As previously mentioned, the system module that was originally designed upon the LTE-A system's specifications [7] is now extended to consider 5G system options in FR2 bands, i.e., n257 in the mmWave range around the 28 GHz. The system module considers a wideband channel of 100 MHz with subcarrier spacing of 60 KHz, a combination that leads to 132 resource blocks [19]. The carrier aggregation technique was also considered with 16 New Radio (NR) carriers [20], the maximum allowable by the 5G NR Radio Access Technology (RAT). The total bandwidth increased to 1.6 GHz, while the total number of available RBs was 2112 (16 NR carriers × 132 RBs per carrier). The module used the measured SINR at the CN side to estimate the required number of RBs to offer the capacity requirement.

Figure 3 provides an insight on the rate offered from a 5G system as configured herein, when compared to a typical LTE-A system with 20 MHz of bandwidth and five component carriers leading to 100 MHz of aggregated bandwidth. Both systems were considered without MIMO spatial-multiplexing schemes since an eight-layer transmission can be supported by both.

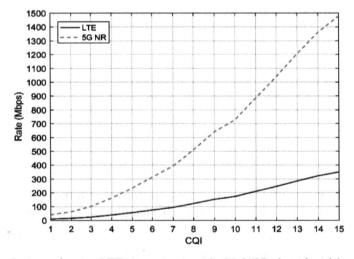

Figure 3. Rate calculations for an LTE-A system with 20 MHz bandwidth and five aggregated component carriers compared to the 5G NR system employed in this work.

2.3.1. The Planning Optimization Algorithm

Our goal herein was to propose a network layout of optimally located microcells that keep the deployment cost to a *minimum* while at the same time specific coverage and capacity criteria are met.

Since in this analysis we only considered microcells, the deployment cost of each node was the same, C, and hence, the total cost of a network deployment was directly proportional to the number of nodes deployed.

Equations (4)–(7) are formulating this optimization problem.

$$B_m^{OPT} \leftarrow \begin{array}{c} \text{argmin} \\ B\prime_m \in \mathbf{B} \end{array} \left(\sum_{b \in B'_m} C_b \right) \tag{4}$$

Subject to:

$$P_{rx}^k \geq P_{rx,min}, \ \forall k \in K \tag{5}$$

$$\sum_{k \in K_g} N_{RB,k} \leq RB_{max}, \ \forall g \in G\prime \tag{6}$$

$$R_k \geq R_{CN,min}, \ \forall k \in K \tag{7}$$

where B'_m is the set of candidate microsites that are under examination, C_b is the cost of the b microsite (herein, $C_b = C$), K is the fixed set of CNs defined in Section 2.1, P_{rx}^k is the received power of the k CN, $G\prime$ is the set gNBs associated with the B'_m sites (as previously mentioned, when sectors are considered, each site comprises three gNBs), K_g is the set of CNs currently served by g gNB, RB_{max} is the maximum allowed resource blocks of a gNB, R_k is the rate offered by the system to the k CN and finally $R_{CN,min}$ is the capacity requirement. Note that the different candidate solutions that were examined (B) were derived in the context of the Genetic Algorithm (GA) methodology as it is described in the following.

The solution to the optimization problem defined earlier was the B_m^{OPT} set of microsites that entails the *minimum total cost* [7] of network deployment while at the same time meets the constraints set by Equations (5)–(7). Herein, the optimization problem was tackled with a multi-objective evolutionary algorithm [10] and specifically with the NSGA-II (Non-sorting Dominated Genetic Algorithm-II). For this, the problem was analyzed and translated into a three-dimensional fitness function, $F = \{F(1), F(2), F(3)\}$. Each dimension corresponded to an objective that needed to be *minimized*. Equations (8)–(10) define the fitness function to be minimized.

$$F(1) = \left| B'_m \right| \tag{8}$$

$$F(2) = \left| K' \right|, \ K' = \left\{ k \in K \middle| N_{RB,k} \to 0 \right\} \tag{9}$$

$$F(3) = \sum_{g \in G\prime} \left\{ \sum_{k \in K_g} N_{RB,k} - RB_{max} \right\} \tag{10}$$

The first objective, $F(1)$, corresponds to the weighted network cost, the second, $F(2)$, is the total number of CNs that do not meet their requirements, either capacity or coverage, while $F(3)$ is the excess RBs needed at a network level.

The NSGA-II algorithm imitates evolutionary mechanisms and employs biological operations found in nature for generating optimal solutions. It starts with a random selection of an initial population (sets of microsites) and by iteratively using operators such as selection, evaluation and crossover or mutation evolves into optimal solutions. When NSGA-II reaches an optimal pareto front, i.e., it cannot be further improved (average pareto spread and distance criteria were used [6]), then a series of solutions corresponding to that pareto front is derived.

3. Radio Network Planning Results

Uniformly distributed CNs in the service area represented the area coverage and capacity criterion. An initial throughput requirement was allocated at each CN, which gradually increased. At relatively low capacity requirements, microcells with omnidirectional antenna patterns were uniformly placed by the optimization algorithm, as seen in Figure 4a, in order for each microcell to serve all neighbor CNs (shown in Figure 4a with black stars). The SINR distribution (Figure 4b) also followed this regular pattern. Note that although the received power fell sharply with distance due to the relatively high frequency, the SINR did not fall as rapidly due to the equivalent reduction of the interference from other cells.

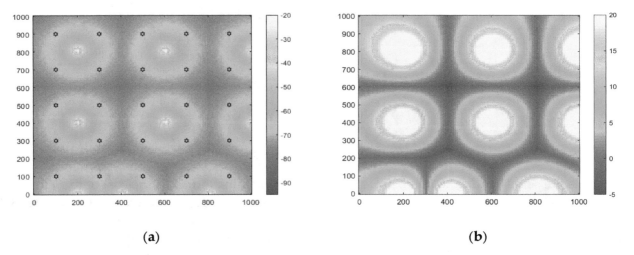

(a)   (b)

Figure 4. (a) Received power (dBm) and (b) SINR distribution in the service area for a dense omnidirectional micro gNB deployment.

Microcells with omnidirectional patterns were then replaced by tri-sectored microcells, with each sector being a separate gNB in order to offer increased spatially selective throughput. Note that the EIRP of the transmitting BS antennas remained the same for sector and omnidirectional antennas (60 dBm), while a sector could be switched off, if it was not needed, in order to keep interference levels low. Hence, in the results depicted in Figure 5, there was a microcell with only one operating sector, and another microcell with two transmitting sectors. The rest of the microcells were placed as far apart as possible in order for all three sectors to be transmitting and covering a wider area.

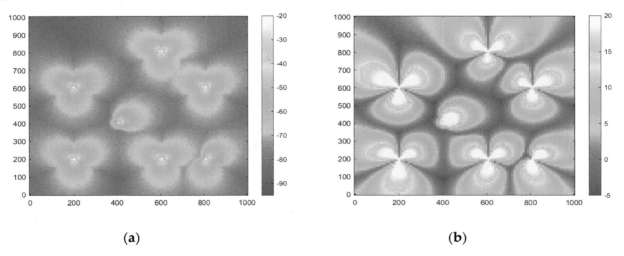

(a)   (b)

Figure 5. (a) Received power (dBm) and (b) SINR distribution in the service area for a dense tri-sector micro gNB deployment.

As the throughput requirements increased, denser grids of microcells were deployed. As can be seen in Figure 6a, for high-capacity requirements with the ultra-dense networks, only microcells close to the CNs were transmitting, with only one sector pointing directly towards the corresponding CN. As a consequence, in Figure 6b, the positions not close to a sector antenna pointing directly towards them had very low SINR due to increased interference from the dense transmitting antenna grid. However, due to the optimization algorithm, the antenna position was such that there was no CN in a position of low SINR. With such optimal placement, the maximum throughput for the 25 CNs in the area reached as high as almost 16 Gbps/Km^2.

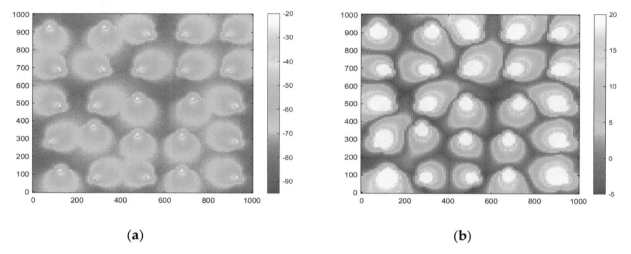

(a)   (b)

Figure 6. (a) Received power (dBm) and (b) SINR distribution in the service area for an ultra-dense tri-sector micro gNB deployment.

Overall, comparing the performance of the various network configurations that resulted from the optimization algorithm, it was evident that the denser the microcell deployment, the higher the area throughput (see Figure 7). Moreover, with tri-sector microcells, the throughput almost doubled at the ultra-dense networks with respect to that of dense networks, either with omnidirectional or sector antennas. At these scenarios, as seen in Figure 6, most of the times each microcell had only one sector antenna operating and hence, the advantage with respect to the omnidirectional deployment was due to interference reduction to adjacent cells.

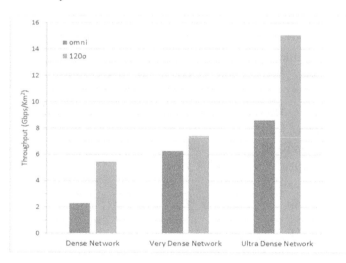

Figure 7. Throughput per Km^2 for different network densities with omnidirectional and tri-sector micro gNBs.

Figure 8 provides an insight into how dense 5G networks with standalone small-cell architectures can actually get. Even for the omnidirectional reference scenario, the cell radius did not get higher than 400 m, while for the ultra-dense network the cell radius further decreased down to 120 m. The spectral efficiency achieved now was 9.5 bps/Hz/Km^2, approximately three times lower than the 5G target of 30 bps/Hz, although no MIMO/beamforming was considered in our generic analysis. If this is also considered, then it becomes obvious that the spectral efficiency goal will be easily exceeded.

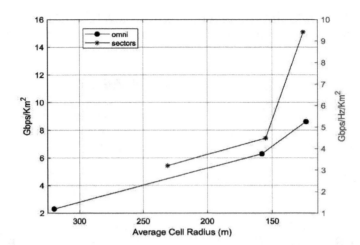

Figure 8. Throughput (and spectral efficiency per Km^2) associated with the cell radius for the omnidirectional and sectorized scenarios and the dense/very-dense/ultra-dense scenarios (from left to right).

4. Conclusions

The ever-increasing demand for new wireless services, together with the evolution of cellular networks, has led to the 5G era. The new network generation introduces major changes and, as a consequence, significant challenges in terms of radio planning. This work focused on providing a generic analysis for this problem, emphasizing the need for dense standalone small-cell networks in the mmWave band of 28 GHz, in order to achieve the capacity goals of 5G. The complex nature of the problem led us to employ a properly modified multi-objective optimization algorithm in order to provide solutions in the context of the multiple constraints and specifics of a 5G mmWave network. The three main constraints of coverage, capacity and cost were studied for a range of cases that reflected scenarios from dense to ultra-dense network deployments in order to achieve high capacity.

The produced results showed that the denser the microcell deployment, the higher the area throughput and also that sectored microcells double the throughput for ultra-dense networks compared to dense networks since better interference control is achieved through the operation of one, instead of two or three, sector antennas. In terms of how dense 5G networks with standalone small-cell architectures can actually get, the analysis showed cell radii below 400 m that reached 120 m for the ultra-dense sectored network. Finally, the spectral efficiency reached 9.5 bps/Hz/Km^2 with no MIMO/beamforming, which leads to the conclusion that the 5G spectral efficiency goal can be easily exceeded if these techniques are also exploited.

Author Contributions: Conceptualization, G.E.A., D.A.Z., G.V.T., P.K.G. and D.I.K.; methodology, G.E.A., D.A.Z. and P.F.; validation, G.E.A., P.F. and D.A.Z.; formal analysis, P.F.; investigation, G.E.A., D.A.Z. and G.V.T.; writing—original draft preparation, G.E.A, D.A.Z. and G.V.T.; writing—review and editing, G.E.A, D.A.Z., G.V.T., P.F., P.K.G. and D.I.K. All authors have read and agreed to the published version of the manuscript.

References

1. Hurley, S. Planning effective cellular mobile radio networks. *IEEE Trans. Veh. Technol.* **2002**, *51*, 243–253. [CrossRef]
2. Molina, A.; Athanasiadou, G.; Nix, A. The automatic location of base-stations for optimised cellular coverage: A new combinatorial approach. Proceedings of the Vehicular Technology Conference. In *1999 IEEE 49th*; IEEE: Houston, TX, USA, 1999; pp. 606–610.
3. Amaldi, E.; Capone, A.; Malucelli, F. Radio planning and coverage optimization of 3G cellular networks. *Wirel. Netw.* **2007**, *14*, 435–447. [CrossRef]

4. Lakshminarasimman, N.; Baskar, S.; Alphones, A.; Willjuice Iruthayarajan, M. Evolutionary multiobjective optimization of cellular base station locations using modified NSGA-II. *Wirel. Netw.* **2011**, *17*, 597–609. [CrossRef]

5. Athanasiadou, G.E.; Zarbouti, D.; Tsoulos, G.V. Automatic location of base-stations for optimum coverage and capacity planning of LTE systems. In *The 8th European Conference on Antennas and Propagation (EuCAP 2014)*; IEEE: Houston, TX, USA, 2014; pp. 2077–2081.

6. Valavanis, I.K.; Athanasiadou, G.E.; Zarbouti, D.A.; Tsoulos, G.V. Multi-Objective Optimization for Base-Station Location in Mixed-Cell LTE Networks. In Proceedings of the 10th European Conference on Antennas and Propagation (EuCAP), Davos, Switzerland, 10–15 April 2016.

7. Athanasiadou, G.; Tsoulos, G.; Zarbouti, D.; Valavanis, I. Optimizing Radio Network Planning Evolution Towards Microcellular Systems. *Wirel. Pers. Commun.* **2019**, *106*, 521–534. [CrossRef]

8. Rappaport, T.S.; Xing, Y.; MacCartney, G.R.; Molisch, A.F.; Mellios, E.; Zhang, J. Overview of Millimeter Wave Communications for Fifth-Generation (5G) Wireless Networks—With a Focus on Propagation Models. *IEEE Trans. Antennas Propag.* **2017**, *65*, 6213–6230. [CrossRef]

9. Tsoulos, G.V. Smart antennas for mobile communication systems: Benefits and challenges. *Electron. Commun. Eng. J.* **1999**, *11*, 84–94. [CrossRef]

10. Deb, K.; Pratap, A.; Agarwal, S.; Meyarivan, T. A fast and elitist multiobjective genetic algorithm: NSGA-II. *IEEE Trans. Evol. Comput.* **2002**, *6*, 182–197. [CrossRef]

11. *3GPP TR 38.900 v14.2.0, LTE; 5G; Study on Channel Model for Frequency Spectrum Above 6 GHz*; ETSI: 650 Route des Lucioles, F-06921 Sophia Antipolis CEDEX, France, 2017.

12. Piersanti, S.; Annoni, L.; Cassioli, D. Millimeter waves channel measurements and path loss models. In Proceedings of the 2012 IEEE International Conference on Communications (ICC), Ottawa, ON, Canada, 10–15 June 2012; pp. 4552–4556.

13. Sun, S.; Rappaport, T.S.; Rangan, S.; Thomas, T.A.; Ghosh, A.; Kovacs, I.Z.; Rodriguez, I.; Koymen, O.; Partyka, A.; Jarvelainen, J. Propagation Path Loss Models for 5G Urban Micro- and Macro-Cellular Scenarios. In Proceedings of the IEEE 83rd Vehicular Technology Conference–Spring, Nanjing, China, 7 May 2016.

14. Pekka, K.; Juha, M.; Hentilä, L. *WINNER II - D1.1.2 - Channel Models*. IST-4-027756 WINNER II. 2007. Available online: http://www.ero.dk/6799B797-53CC-417E-B6FA-059A8E6AF350?frames=no& (accessed on 1 February 2020).

15. *3GPP TR 36.814; Technical Specification Group Radio Access Network; Evolved Universal Terrestrial Radio Access (E-UTRA);Further Advancements for E-UTRA Physical Layer Aspects (Release 9)*; 3GPP: 650 Route des Lucioles, Sophia Antipolis Valbonne, France, 2010.

16. Salem, A.A.; El-Rabaie, S.; Shokair, M. A Proposed Efficient Hybrid Precoding Algorithm for Millimeter Wave Massive MIMO 5G Networks. *Wirel. Pers. Commun.* **2020**, 1–16. [CrossRef]

17. Sharma, V.; You, I.; Leu, F.; Atiquzzaman, M. Secure and efficient protocol for fast handover in 5G mobile Xhaul networks. *J. Netw. Comput. Appl.* **2018**, *102*, 38–57. [CrossRef]

18. Liu, L.; Zhou, Y.; Vasilakos, A.V.; Tian, L.; Shi, J. Time-domain ICIC and optimized designs for 5G and beyond: A survey. *Sci. China Inf. Sci.* **2019**, *62*, 1–28. [CrossRef] [PubMed]

19. *3GPP TS 38.104, "5G; NR; Base Station (BS) Radio Transmission and Reception"*; ETSI: 650 Route des Lucioles, F-06921 Sophia Antipolis CEDEX, France, 2019.

20. *3GPP TR 21.915, "Digital cellular telecommunications system (Phase 2+) (GSM); Universal Mobile Telecommunications System (UMTS); LTE; 5G; Release description; Release 15*; ETSI: 650 Route des Lucioles, F-06921 Sophia Antipolis CEDEX, France, 2019.

9

High-Gain Planar Array of Reactively Loaded Antennas for Limited Scan Range Applications

Ronis Maximidis [1,*], Diego Caratelli [1,2], Giovanni Toso [3] and A. Bart Smolders [1]

[1]   Department of Electrical Engineering, Electromagnetics Group, Eindhoven University of Technology, 5600 MB Eindhoven, The Netherlands; diego.caratelli@antennacompany.com (D.C.); A.B.Smolders@tue.nl (A.B.S.)

[2]   Department of Research and Development, The Antenna Company, 5656 AE Eindhoven, The Netherlands

[3]   Antenna and Sub-Millimeter Waves Section, European Space Agency, ESA/ESTEC, 2200 AG Noordwijk, The Netherlands; giovanni.toso@esa.int

*   Correspondence: R.T.Maximidis@tue.nl

Abstract: This paper proposes a novel high-gain antenna element that can be used in antenna arrays that only require a limited scan range. Each high-gain antenna element uses a linear sub-array of highly-coupled open-ended waveguides. The active central element of this sub-array is directly fed, while the remaining passive waveguides are reactively loaded. The loads are implemented by short-circuits positioned at various distances from the radiating aperture. The short-circuit positions control the radiation pattern properties and the scattering parameters of the array. The proposed sub-array antenna element is optimized in the presence of the adjacent elements and provides a high gain and a flat-top main lobe. The horizontal distance between the sub-array centers is large in terms of wavelengths, which leads to limited scanning capabilities in the E-plane. However, along the vertical axis, the element spacing is around 0.6 wavelength at the central frequency that is beneficial to achieve a wider scan range in the H-plane. We show that the sub-array radiation pattern sufficiently filters the grating lobes which appear in the array factor along the E-plane. To demonstrate the performance of the proposed array configuration, an array operating at 28.0 GHz is designed. The designed array supports scan angles up to ±7.5° along the E-plane and ±24.2° along the H-plane

Keywords: phased arrays; high-gain array element; sub-array; mm-wave; flat-top pattern

1. Introduction

To support the growing demand for high data rates, the telecommunication industry is constantly moving to higher frequencies where larger operational bandwidths are available [1]. However, the propagation loss of radio-wave signals also increases significantly at millimeter-wave frequencies, thus severely limiting the communication range. One way to deal with the increased path losses is to compensate them by increasing the directivity using antenna arrays with electronic beamforming [2,3]. An alternative solution consists in the use of multibeam antennas which reduce the overall system footprint and can enhance, significantly, the system capacity [4]. However, the cost of electronic components required for the development of such a complex antenna system increases dramatically as the frequency of operation becomes higher. Therefore, only small-size antenna arrays with moderate directivities have been developed until now.

On the other hand, there are plenty of applications that require a very high antenna gain but only over a limited field of view (FoV). Examples in this respect are given by point-to-point or point-to-multipoint backhaul links and antennas for sub-urban coverage. For such applications, one can use array elements characterized by a large aperture and, therefore, high directivity [5]. This allows reducing the number of radiating elements needed to achieve the required gain levels.

In turn, this translates into reduced design complexity and manufacturing costs of the feeding network. However, the use of radiating elements with large apertures leads to the appearance of grating lobes in the visible region. The distance between the main lobe and the grating lobes is inversely proportional to the distance between sub-array centers. Therefore, the inter-element spacing should be selected in such a way to keep the grating lobes outside the field of view for all scan angles. To this end, the aperture size should be restricted to certain dimensions. A useful design approach to overcome this drawback, and which is implemented in this research study, is based on the use of overlapping radiating apertures, i.e., the use of the same aperture for multiple feed points. In this way, we can achieve the required levels of gain while maintaining the overall array compactness. Even though overlapping techniques allow keeping grating lobes outside the FoV, the presence of grating lobes still has, per se, certain adverse effects on the side-lobe characteristics of the overall array. More specifically, in transmit mode, they can result in energy leakage along undesired directions that, in turn, can cause a waste of power and potential electromagnetic interferences. In the receive mode, grating lobes can cause the reception of spurious radio signals from undesired directions and, thereby, a degradation of the signal-to-interference ratio. Therefore, appropriate measures have to be taken so as to minimize the detrimental impact of grating lobes. The solution, adopted in this paper, is to filter the grating lobes by the proper shaping of the embedded radiation pattern featured by the individual array element [5,6].

A number of techniques have been developed to reduce the distance between radiating apertures. One can use a complex feeding network to distribute the energy of a port to multiple radiating apertures, enlarging, in this way, the size of each radiating element. Nonetheless, the feeding network becomes very complex even for a small number of array elements; therefore, this technique has found limited applications, mostly in space [7,8]. Another approach consists in using partially reflective surfaces placed above the array structure [9–11]. However, the reflective surfaces increase the height of the structure and reduce its mechanical robustness.

In our previous work [6], we proposed a technique to reduce the separation between the phase centers of a one-dimensional array so as to avoid the use of complex overlapping networks or reflective surfaces. In the proposed design approach, each radiating aperture consists of a linear sub-array composed of one directly fed open-ended waveguide radiator and a large number of short-circuited highly-coupled waveguide elements. The energy is delivered to the passive waveguides by free-space coupling and the sub-array structure is designed in such a way to deliver energy to the passive elements of the adjacent sub-arrays as well. As a result, an effective overlapping between sub-arrays can be realized without the use of any additional power distribution network. The positions of the short circuits define the impedance experienced by the electromagnetic field coupled to the passive waveguides. Due to the low loss of the waveguides, said impedance is mainly reactive. Therefore, by optimizing the positions of short circuits, we can control the wave radiation process by the individual passive waveguide and, in this way, shape the embedded sub-array radiation pattern as well as control the active scattering parameters of the overall array structure.

In the present research study, we apply the technique proposed in [6] to the design of a planar antenna array composed of linear sub-arrays. The embedded radiation pattern of the realized antenna sub-array provides a limited scan range along the sub-array main axis (E-plane) but a substantially larger coverage along the orthogonal axis (H-plane). The array presented in this paper operates at Ka-band at 28.0 GHz, where several emerging wireless applications are being developed [12] but can be easily scaled to any desired frequency.

This paper presents the following new scientific contributions:

- Extension of the overlapping technique based on free-space coupling to design a planar array. The optimized array features high gain and requires a much smaller number of active elements as compared to traditional phased-array systems.
- A theoretical framework to optimize and analyze the proposed class of arrays for any specific application.

The paper is organized as follows. In Section 2, we describe the design procedure, and introduce the mathematical model used for the array optimization; the interested reader can find a more detailed derivation of the model in [6]. In Section 3.1, we present an overlapping sub-array structure featuring an embedded flat-top radiation pattern. Next, in Section 3.2 we use the designed subarray as a building block for the development of a high-gain antenna array. The paper closes with the main conclusions and future directions.

2. Method

Figure 1 shows a schematic of a reactively loaded antenna array consisting of N_a directly-fed waveguides and N_p reactively loaded waveguides. The employed design procedure is general; therefore, the active and passive elements can be randomly distributed across the array.

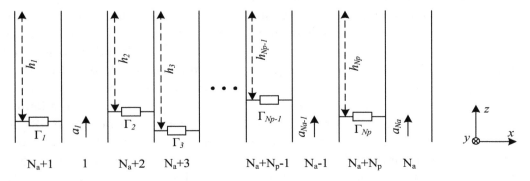

Figure 1. Schematic view of a reactively loaded waveguide antenna array.

The total electric field $\vec{E}(\theta, \varphi)$ radiated by such a structure is given from [6]:

$$\vec{E}(\theta, \varphi) = \mathbf{a}^T \cdot \left[\vec{\mathbf{E}}_a(\theta, \varphi) + \underline{\mathbf{S}}_c \cdot \left(\left[(\underline{\mathbf{\Gamma}}_p \cdot e^{-2\underline{\mathbf{\gamma}}_p \cdot \mathbf{h}_p})^{-1} - \underline{\mathbf{S}}_p \right]^{-1} \right)^T \cdot \vec{\mathbf{E}}_p(\theta, \varphi) \right] \tag{1}$$

with the superscript ^{T} denoting the matrix transposition; the subscripts a and p refer to active and passive elements, respectively. Besides, θ and φ are the conventional spherical angles.

In Equation (1), the vector $\mathbf{a}$ contains the excitation coefficients of the directly-fed array elements, and $\vec{\mathbf{E}}_a(\theta, \varphi)$, $\vec{\mathbf{E}}_p(\theta, \varphi)$ are the vectors containing the electromagnetic field contributions radiated by the active and passive elements, respectively. The matrices $\underline{\mathbf{S}}_p$ and $\underline{\mathbf{S}}_c$ are blocks of the scattering matrix $\underline{\mathbf{S}}$, they are relevant to the entire structure and describe the interaction between the passive waveguides and the coupling between the passive and directly fed waveguides, respectively. Finally, $\underline{\mathbf{\Gamma}}_p$, $\underline{\mathbf{\gamma}}_p$, and $\mathbf{h}_p$ are diagonal matrices containing the reflection coefficient of the load, the complex propagation constant, and the length of the passive waveguides, respectively.

In [6], it has been demonstrated that the scattering matrix of the active elements can be calculated as follows:

$$\underline{\mathbf{S}}'_a = \underline{\mathbf{S}}_a + \underline{\mathbf{S}}_c \cdot \left[(\underline{\mathbf{\Gamma}}_p \cdot e^{-2\underline{\mathbf{\gamma}}_p \cdot \mathbf{h}_p})^{-1} - \underline{\mathbf{S}}_p \right]^{-1} \cdot \underline{\mathbf{S}}_c^T \tag{2}$$

where $\underline{\mathbf{S}}_a$ is the block matrix of $\underline{\mathbf{S}}$ describing the interaction between active elements.

On the basis Equations (1) and (2), the design of the proposed type of arrays can be performed using the following procedure:

1. Decide on the array topology and the minimum number of sub-arrays N_S useful to approximate a large array environment for the central sub-array.
2. Decide on the number of elements per sub-array, N_E.
3. Obtain the scattering matrix and the embedded element patterns for all the elements of the $(N_S \times N_E) \times (N_S \times N_E)$ array.

4. Choose an optimization algorithm and select the optimization parameters.
5. Define the objective function on the basis of Equaitons (1) and (2).
6. After optimization, increase the array size by adding sub-arrays until there is no change in the embedded radiation pattern of the central sub-array.
7. Use the obtained pattern as the embedded element pattern for the design of arrays of any size.

As it is apparent from Equations (1) and (2), the main optimization parameters are $(\underline{\Gamma}_p, \underline{\gamma}_p, \underline{h}_p)$, which define the electrical length of the passive waveguides and their loading profile. The proposed design procedure has been implemented in Matlab [13].

3. Design

3.1. Sub-Array Design

In this paper, we design a planar array built up from linear sub-arrays. The sub-arrays have to support a wide scan range in the H-plane with a reduced scanning capability in the E-plane. The design starts from the structure of Figure 2a, where a linear antenna sub-array of $N_E = 13$ open-ended waveguides is placed in an $N_A = 9 \times 3$ array configuration.

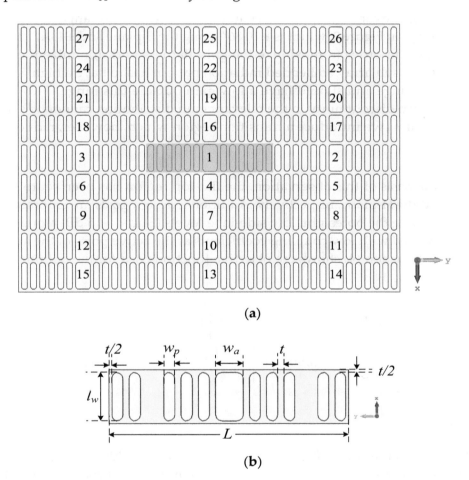

(a)

(b)

Figure 2. (a) Planar $N_A = 9 \times 3$ array configuration using high-gain sub-arrays of open-ended waveguides. The central sub-array is highlighted in green, whereas the relevant geometrical parameters are reported in (b).

The length of the sub-array, $L = 2.653\lambda_0$, with λ_0 denoting the free-space wavelength, was chosen in such a way as to achieve a FoV of about $\pm 8°$. The choice of the length, $l_w = 0.6\lambda_0$, of the waveguides is governed by the cut-off condition of the fundamental TE_{01} mode, that is $l_w > 0.5\lambda_0$. The width of the passive waveguides, $w_p = 0.12\lambda_0$, the thickness of the walls, $t = 0.0679\lambda_0$, as well as the relevant rounding radius of 0.8 mm were chosen so as to simplify the manufacturing process. Finally, the width of the central waveguide, $w_a = 0.3\lambda_0$, is optimized in a way to achieve good impedance matching characteristics. At the initial design step, the length of all the waveguides is set to $0.1\lambda_g$, where λ_g denotes the guided wavelength of the fundamental TE_{01} mode of the passive waveguides. At the working frequency, $f_0 = 28.0$ GHz, and for the specified waveguide dimensions, we get $\lambda_g = 21.104$ mm.

In a linear array, the angular position of the grating lobes is a function of the separation d between antenna elements and the scan angle θ_0, and is given by:

$$\sin\theta_p = \sin\theta_0 + \frac{p\lambda_0}{d} \tag{3}$$

where $p = \pm1, \pm2, \ldots$ is chosen in such a way as to define an angle with real value, as per the condition $|\sin\theta_p| \leq 1$. The maximum scan angle of the array factor before the appearance of the first grating lobe in the field of view is $\theta_p = -\theta_0$.

Using Equation (3), we can evaluate the maximum theoretical scan range in the two main planes of the array of sub-arrays in Figure 2, which is $\pm10.86°$ along the E-plane and $\pm48.47°$ along the H-plane. Therefore, theoretically, in order to maximize the gain flatness and minimize the radiation to the undesired directions, the sub-array embedded radiation pattern has to have a high gain flat-top shape in the calculated scan range and should have a low gain outside this range. However, a radiation pattern with such a jump discontinuity in the relevant gain distribution cannot be synthesized using a limited-size array, unless we introduce a transition region which, however, reduces the maximum scan range.

Therefore, on the basis of this observation, the flat-top region has been set to $\pm8.0°$ and $\pm40.0°$ along the E-plane and the H-plane, respectively. Additional design goals are enforced so as to achieve maximal directivity along the boresight, low side-lobe levels (SLL < -15 dB) along the E-plane, and high return loss combined with low coupling coefficients at the active ports of the array at the central operating frequency. Equation (4) shows the objective function used for the optimization:

$$\Psi_O(f_0, \mathbf{h}_p) = \sum_{j=1}^{N_a} \left| S'_{1j,a}(f_0, \mathbf{h}_p) \right| - D(f_0, \mathbf{h}_p, \theta_0, 0°) + \left| \int_0^{\theta_h} \left[D(f_0, \mathbf{h}_p, \theta, 0°) - D(f_0, \mathbf{h}_p, \theta_0, 0°) \right] d\theta \right|$$

$$+ \left| \int_0^{\theta_e} \left[D(f_0, \mathbf{h}_p, \theta, 90°) - D(f_0, \mathbf{h}_p, \theta_0, 90°) \right] d\theta \right| - 10 \left\{ \left| D(f_0, \mathbf{h}_p, \theta_c, 90°) - D(f_0, \mathbf{h}_p, \theta_0, 90°) \right| \tag{4}$$

$$- \min_{\theta \in [\theta_{SL}, 90°]} \frac{D(f_0, \mathbf{h}_p, \theta_e, 90°)}{D(f_0, \mathbf{h}_p, \theta, 90°)} \right\},$$

where the first term is to minimize $S'_{1j,a}$, $j = 1, 2, \ldots, N_a$, and, in this way, optimize the scattering parameters of the central sub-array. The second term maximizes the broadside directivity. The next two terms are introduced to maximize the flat-top regions, with $\theta_h = 40°$ and $\theta_e = 8°$, along the H-plane and E-plane, respectively. Finally, the last two terms are useful to minimize the directivity outside the field of view along the E-plane, the relative directivity at the critical angle $\theta_c = 14°$, and the SLL for $\theta \in [\theta_{SL}, 90°]$ with $\theta_{SL} = 20°$. The designed sub-array is symmetrical with boresight radiation pattern characteristics along the z-axis. Therefore, we can confine the optimization process to the angular range $\varphi \in [\theta_{SL}, 90°]$. Figure 3 shows graphically the defined angular regions.

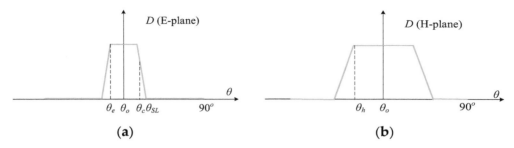

Figure 3. Graphical representation of the array radiation pattern mask along the (**a**) E-plane and (**b**) H-plane.

Utilizing the objective function defined in Equation (4), the design procedure is turned into the solution of the following minimization problem:

$$\underset{0 \le \underline{\mathbf{h}}_p \le 0.5\lambda_g}{\text{argmin}} \; \Psi_O\left(f_0, \underline{\mathbf{h}}_p\right) \tag{5}$$

Due to the symmetry of the considered sub-array structure, the number of design parameters is equal to half the number of passive waveguides embedded in each individual sub-array.

The optimization procedure starts with the calculation of the embedded element patterns and scattering matrix of an array of $9 \times (3 \times 13)$ elements. To this end, a suitable full-wave electromagnetic solver is used [14]. The obtained results are then imported into an in-house developed dedicated software that implements the proposed design procedure. Next, using Equations (1) and (2), the objective function of Equation (4) is implemented. Then, the Global Search Optimization routine embedded in Matlab [13] is adopted to address the minimization problem described by Equation (5). The Global Search Optimization technique combines a global pattern-search-based optimizer with a local gradient-based optimization algorithm. Details of the Global Search Optimization algorithm can be found in [15].

The optimal positions of the short-circuits along the passive waveguides are listed in Table 1, where the waveguides have been numbered starting from the closest one to the central active element. A cross-section of the resulting array is shown in Figure 4.

Table 1. Passive waveguides lengths of the optimized structure.

Section Length	h1	h2	h3	h4	h5	h6
Value in λg	0.196	0.205	0.193	0.0000	0.098	0.206

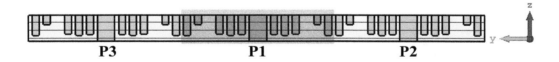

Figure 4. Cross-section of three sub-arrays along the sub-array main axis. The waveguides are shown in blue.

The electromagnetic characteristics of the designed sub-array structure have been successfully validated by using a full-wave solver [14]. After the optimization procedure, the number of sub-arrays has been gradually increased until the radiation pattern of the central sub-array does not change significantly any longer. Figure 5 shows the three main cuts of the embedded radiation pattern of the central sub-array when integrated in 9×3, 11×5, and 15×9 array configurations.

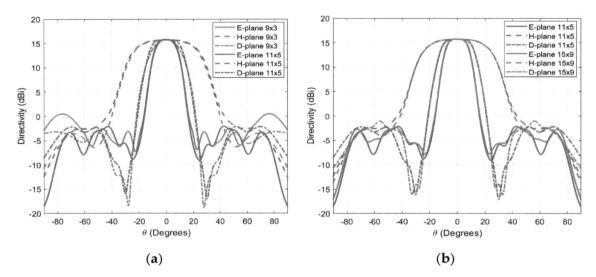

(a)   **(b)**

Figure 5. Embedded element pattern comparison of the central subarrays: (**a**) 9×3 and 11×5; (**b**) 11×5 and 15×9 at the design frequency $f_0 = 28.0$ GHz.

In Figure 5, one can notice a non-negligible change in the embedded radiation patterns along the E- and D-planes when the array size is increased from 9×3 to 11×5 sub-array elements. More specifically, even though the half-power beam-width does not vary along the H-plane, it differs significantly along the other planes, by $0.3°$ along the E-plane and by $1.3°$ along the D-plane. Furthermore, a large deviation of about 7.8 dB is observed in the sidelobe level along the E-plane. No meaningful performance variations occur for further increases of the array size, i.e., to 15×9 elements. As a matter of fact, the maximal difference inside the half-power beam-width along the various planes is about 0.1 dB, whereas the deviation in terms of the sidelobe level is negligible. An additional investigation has been performed on the active reflection coefficient of the central sub-array element in the 9×3, 11×5, and 15×9 array topologies. As it appears in Figure 6, the active reflection coefficients featured by the central elements of the 11×5 and 15×9 arrays show rather similar behavior but differ quite significantly from the one obtained for the 9×3 array. On the basis of the reported results, the radiation pattern of the central element of the 11×5 array has been selected as the embedded sub-array pattern, whose main parameters are detailed in Table 2.

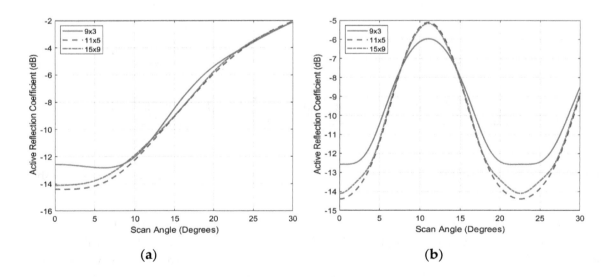

(a)   **(b)**

Figure 6. Active reflection coefficient versus scan angle θ_0 of the central element for (**a**) $\varphi = 0°$ and (**b**) $\varphi = 90°$ at the design frequency $f_0 = 28.0$ GHz.

Table 2. Embedded radiation pattern characteristics of the 11 × 5 array.

Cuts	D(0°) (dBi)	BW(−3 dB) (°)	BW(−0.5 dB) (°)	SLL (dB)
H-plane	15.7	±25.7	±14.9	−18.2
E-plane	15.7	±9.3	±5.9	−17.8
D-plane	15.7	±12.9	±8.5	−17.8

From Figure 5b and Table 2, it can be concluded that the scan range along the E-plane is close to the optimization objective, whereas a gap is observed in relation to the scan range along the H-plane. The sub-array directivity for $\theta_0 = 0°$ is 15.7 dBi, which is 2.3 dB higher than the nominal directivity calculated on the basis of the physical aperture area A_{ap} of the individual sub-array, that is:

$$D_{ap} = \frac{4\pi}{\lambda_0^2} A_{ap} = 13.4 \text{ dBi} \tag{6}$$

We can conclude that the sub-array directivity has been enhanced thanks to the overlapping with adjacent sub-arrays. This increase in directivity can be translated in a 41% reduction of the required transmit power or, equivalently, a 23% reduction in the number of required sub-arrays. In order to provide an insight into the process that is responsible for the enhancement of the directivity, the electric field distribution over the radiating aperture of the 11 × 5 array is reported in Figure 7 when only the central sub-array is active and the other sub-arrays are terminated on matched loads.

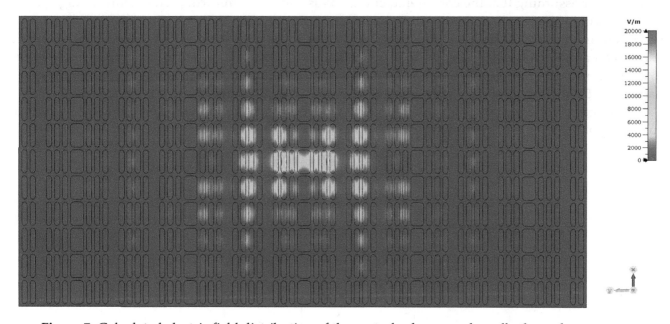

Figure 7. Calculated electric field distribution of the central sub-array when all other sub-arrays are matched in the 11 × 5 array at the design frequency $f_0 = 28.0$ GHz.

From Figure 7, it can be seen that the electric field propagates in a very effective way towards adjacent sub-arrays virtually in all directions. Effectively, all the sub-arrays contribute, to some extent, to the electromagnetic field radiation process. This mechanism enables the sub-array overlapping and, from there, the mentioned directivity enhancement and capability of synthesizing complex-shaped radiation pattern. It should be noted however that, while propagating, the electromagnetic field gradually decays and eventually displays a limited intensity along the edge elements of the array. This is the reason why a further increase in the number of sub-arrays (beyond the size of 11 × 5) does not significantly affect the central sub-array characteristics.In Figure 8, the embedded radiation pattern of the central sub-array of the 11 × 5 element array is presented in the uv-plane, where u = sinθcosφ and v = sinθsinφ.

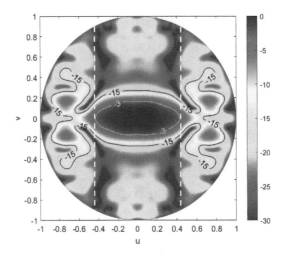

Figure 8. Embedded element pattern of the optimized sub-array at the center of the 11 × 5 array, displayed in the uv-plane, where u = sinθcosφ and v = sinθsinφ. The black and white solid lines encircle areas with a level higher than −15 dB and −3 dB compared to the maximum value. All presented results are obtained at the design frequency f_0 = 28.0 GHz.

Figure 8 shows that the areas where the SLL is larger than −15 dB are outside the region defined by the vertical dotted lines. This region identifies the angular sector where the grating lobes would occur assuming that the considered sub-array is used in a uniform planar array configuration. The performance of such an array can be estimated by analyzing the embedded sub-array pattern presented in Figure 8. Along the H-plane, the grating lobes are outside the visible region over the entire scan range. However, along the E-plane, the grating lobes are inside the visible region. In order to achieve a 10 dB grating-lobe rejection level, we have to limit the scan range to ±8.7°. On the other hand, if a 15 dB grating-lobe rejection level is enforced, the scan range shrinks further to ±6.5°.

The active voltage standing wave ratio (VSWR) versus frequency and scan angle of the central sub-array of the 11 × 5 array configuration is shown in Figure 9.

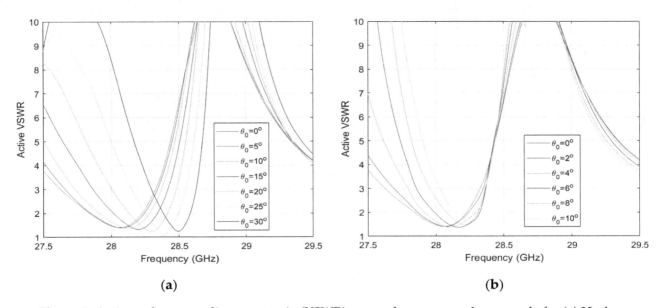

(a)   (b)

Figure 9. Active voltage standing wave ratio (VSWR) versus frequency and scan angle for (a) H-plane and (b) E-plane of the central sub-array of the 11 × 5 array.

Figure 9 shows that the designed sub-array is intrinsically matched, though the relevant impedance matching band shifts with the scan angle. The maximum scanning range can be obtained on the basis of application-specific requirements for active VSWR and bandwidth. In case, where broadband behavior is needed, the integration of a suitable impedance matching network can be explored. The design of such a matching network is outside the scope of this work.

3.2. Array Design

In this section, we use the designed sub-array to realize an antenna array that features a rotationally symmetric mainbeam with a high directivity level of at least 46 dBi. To achieve these goals, a planar array of 65×19 elements is required. The simplest approach to the design of such an array would be to adopt the same embedded radiation pattern of the central sub-array analyzed in the previous section for all the array elements. However, this approach would not account for the fact that the edge elements experience different surroundings and, therefore, are characterized by different embedded radiation patterns. A more accurate design approach consists in the use of the embedded radiation pattern of the edge elements of the small array as an approximation of the one relevant to the edge elements of the large array, as illustrated schematically in Figure 10.

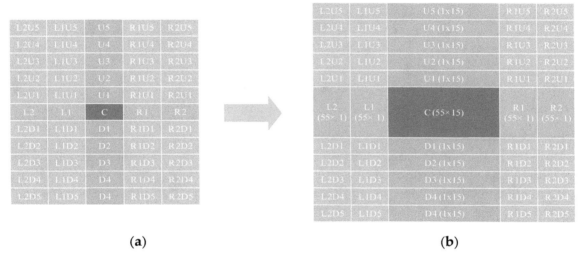

(a)   (b)

Figure 10. Schematic representation of the transformation from (**a**) small array with 11×5 sub-array elements to (**b**) a large array of 65×19 elements. In (**a**) each block represents a sub-array placed in the specific location. In (**b**) the various blocks represent sub-domains (with dimensions indicated into brackets) synthesized using the relevant embedded sub-array patterns in (**a**). The green blocks are 1×1 sub-arrays.

The transformation from the 11×5 to the 65×19 array configuration is performed as follows:

1. The radiation pattern of the central element of the 11×5 array (red element in Figure 10a is used to synthesize the central 55×15 sub-array (highlighted in red in Figure 10b).
2. The elements highlighted in orange in a are used to synthesize the corresponding domains with 55×1 sub-arrays highlighted in orange in Figure 10b.
3. Similarly to step 2, the blue elements in Figure 10a are used to synthesize the domains of 1×15 sub-arrays highlighted in blue in Figure 10b.
4. Finally, the remaining green elements in Figure 10a are used as is to complete the 65×19 array in Figure 10b.

The scan characteristics along the two principal planes of the resulting 65×19 array are illustrated in Figure 11. The achieved boresight directivity level is 46.5 dBi, which is 0.1 dBi smaller than that displayed by the 65×19 array when the embedded radiation pattern of the central sub-array is used for all the elements and, therefore, the edge effect is neglected. It is worth noting, also, that the edge effect causes faster decay of the array mainbeam with the scanning angle, as it can be observed in Figure 11. In the absence of the edge effect, the array mainbeam would follow the profile of the central sub-array pattern. This non-ideality is responsible for a reduced scan range, which can be quantified as 7.5° in the E-plane, and 24.2° in the H-plane.

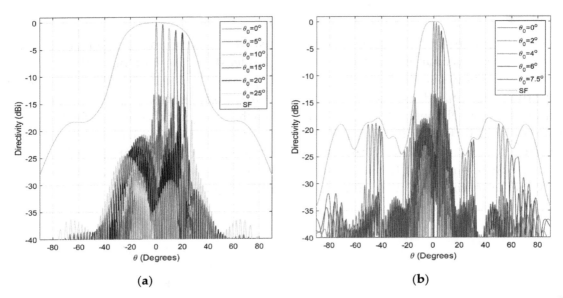

Figure 11. The radiation pattern of the 65 × 19 array for various scan angles along the (a) H-plane and (b) E-plane. The figure also includes the embedded radiation pattern of the central sub-array of the 11 × 5 array, sub-array factor (SF).

To show the impact of the edge effect on the side-lobe level characteristics, the array radiation pattern evaluated using the proposed design approach for the scan angles of 25° and 7.5° along the H- and E-plane, respectively, has been compared against the corresponding one obtained when the radiation pattern of the central sub-array is used for all elements of the array; therefore, the edge effects are neglected (see Figure 12).

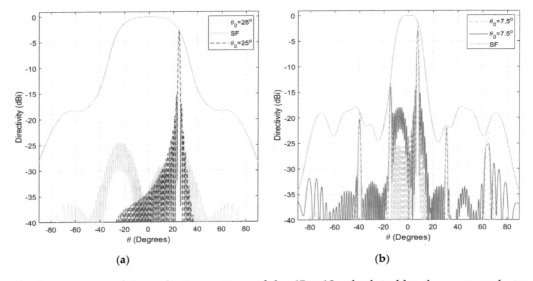

Figure 12. Comparison of the radiation pattern of the 65 × 19 calculated by the proposed approach (solid line) versus an approach that assumes identical embedded radiation (dotted line) for the (a) H-plane and (b) E-plane. The figure also includes the embedded radiation pattern of the central sub-array of the 11 × 5 array (SF).

As it can be noticed in Figure 12, neglecting the edge effect leads to an overestimation of the main-beam directivity, by 0.7 dB in the H-plane, and by 1.6 dB in the E-plane, respectively. The deviation along the E-plane is more noticeable because of the smaller number, 19, of sub-arrays in that direction, as compared to the number of elements, 65, along the H-plane. Another effect of the edge elements that can be observed in Figure 12 is relevant for the increase of the sidelobe levels, by about 14.6 dB for $\theta \simeq -25.0°$ along the H-plane, and by about 7.5 dB for $\theta \simeq -7.5°$ along the E-plane. The performed comparison clearly shows the importance of including the edge effects in the analysis of the array performance.

The filtering effect of the sub-array factor for various scan angles is visualized in Figure 13e–h. For comparison, the array factor of a 65 × 19 array is presented in Figure 13a–d. Figure 13a–d shows the array factor of a 65 × 19 array for various representative scanning angles. The array elements are excited with uniform amplitude, whereas the phase tapering is optimized so to steer the array beam along the desired direction. Because of the large distance between array elements along the u-axis, grating lobes appear. On the other hand, the array element separation along the v-axis is smaller than $0.6\lambda_0$ that is instrumental in keeping the grating lobes outside the visible region, as per Equation (3), for the targeted scanning range. Figure 13e–h shows the filtering effect on the sub-array radiation pattern that is useful in the suppression of undesired grating lobes. It is worth mentioning that even though the sidelobe level of the embedded sub-array pattern is very high in some specific directions, as is shown in Figure 8, the radiation pattern of the total array has a sidelobe level below −30 dB, at the specified directions. This performance is achieved thanks to the fact that the array scanning is restricted to the angular region defined by the contour line at −3 dB level. Therefore, it has been shown that the requirement on the sub-array side lobe level can be relaxed for certain directions which are defined by the scan range of the full array. Finally, due to the uniform amplitude excitation, the side-lobe level of the total radiation pattern is about −13.27 dB. If further sidelobe level reduction is required, a suitable amplitude tapering scheme of the array excitation has to be adopted [16].

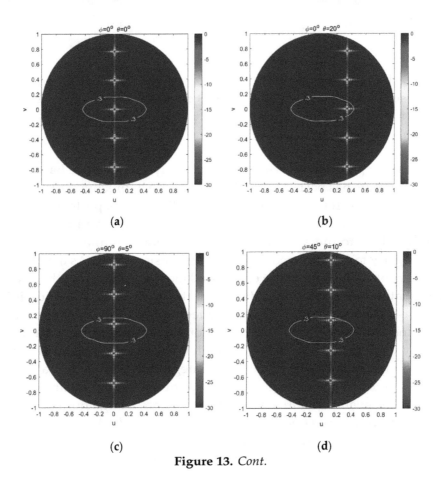

Figure 13. *Cont.*

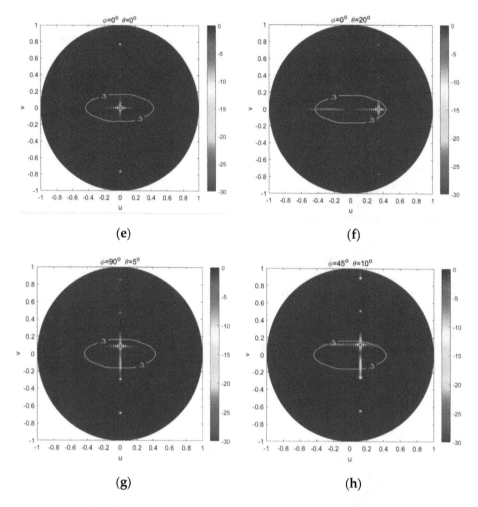

Figure 13. Array factor (**a–d**) and the radiation pattern (**e–h**) of the 65 × 19 array for various scan angles, where u = sinθcosφ and v = sinθsinφ. The −3 dB contour of the central sub-array radiation pattern of the 11 × 5 array is also shown.

4. Conclusions

In this work, we have proposed a novel antenna array architecture based on high-gain linear sub-arrays. An array structure has been designed so to achieve a directivity of 46.5 dBi, which is 2.2 dB higher than the corresponding 100% aperture efficiency limit. This performance has been obtained by exploiting the overlapping of multiple linear sub-arrays. The individual sub-array has a length of 2.653λ_0 which might lead to the appearance of grating lobes while scanning. This non-ideality, however, has been properly taken into account during the design stage. As a matter of fact, the grating lobes are filtered by a proper shaping of the embedded element pattern of the sub-array. It has been, also, demonstrated that, for an accurate prediction of the array characteristics, it is key to properly model the edge effects. The inclusion of the edge effect limited the array scan range from ±9.3° to ±7.5° along the E-plane and from ±25.7° to ±24.2° along the H-plane, respectively. We have shown that the antenna array is intrinsically matched, though the relevant impedance matching band shifts with the scan angle. To mitigate this drawback, the integration of an external impedance matching network can be considered. The proposed design supports a single linear polarization. Although this might be sufficient in several operative scenarios, there are a large number of applications where simultaneous support of two polarizations is required. To achieve dual-polarization operation, suitable radiating elements have to be utilized. One example of such an element is the square dielectric-filled waveguide operating in TE$_{01}$ and TE$_{10}$ modes simultaneously. Another area for further improvement is the extension of the bandwidth of the proposed design. One way to achieve such an extension is to investigate alternative choices for the radiating elements. Another option, especially if a wide-band operation is required, is to use active non-foster loads as waveguides terminations [17,18]. By enabling dual polarization and broadband behavior, one can develop multibeam/multicolor arrays based on the proposed reactively loading approach. In this case, the coloring can be performed on the basis of polarization and frequency separation [19].

Some general properties and implementation schemes for linear and planar arrays organized in overlapped subarrays can be found in [20].

Author Contributions: Conceptualization, R.M., D.C., G.T., and A.B.S.; methodology, R.M., D.C., G.T., and A.B.S.; software, R.M.; supervision, D.C., G.T., and A.B.S.; validation, R.M., D.C., G.T., and A.B.S.; writing—original draft preparation, R.M.; writing—review and editing, R.M., D.C., G.T., and A.B.S. All authors have read and agreed to the published version of the manuscript.

Acknowledgments: This study was conducted in the framework of a Network/Partnering Initiative between the European Space Agency, The Antenna Company, and the Eindhoven University of Technology, under the contract 4000114668/15/NL/MH.

References

1. Edstam, J.; Olsson, A.; Flodin, J.; Ohberg, M.; Henriksson, A.; Hansryd, J.; Ahlberg, J. *High-Capacity Microwave Is a Key Enabler for 5G*; Technical Report for Ericsson Microwave Outlook: Stockholm, Sweden, 2018.

2. Caratelli, D.; Toso, G.; Stukach, O.V.; Panokin, N.V. Deterministic Constrained Synthesis Technique for Conformal Aperiodic Linear Antenna Arrays—Part I: Theory. *IEEE Trans. Antennas Propag.* **2019**, *67*, 5951–5961. [CrossRef]

3. Caratelli, D.; Toso, G.; Stukach, O.V.; Panokin, N.V. Deterministic Constrained Synthesis Technique for Conformal Aperiodic Linear Antenna Arrays—Part II: Applications. *IEEE Trans. Antennas Propag.* **2019**, *67*, 5962–5973. [CrossRef]

4. Hong, W.; Jiang, Z.H.; Yu, C.; Zhou, J.; Chen, P.; Yu, Z.; Zhang, H.; Yang, B.; Pang, X.; Jiang, M.; et al. Multibeam Antenna Technologies for 5G Wireless Communications. *IEEE Trans. Antennas Propag.* **2017**, *65*, 6231–6249. [CrossRef]

5. Petrolati, D.; Angeletti, P.; Toso, G. A lossless beam-forming network for linear arrays based on overlapped sub-arrays. *IEEE Trans. Antennas Propag.* **2014**, *62*, 1769–1778. [CrossRef]

6. Maximidis, R.T.; Caratelli, D.; Toso, G.; Smolders, A.B. Design of Overlapped Subarrays Based on Aperture Reactive Loading. *IEEE Trans. Antennas Propag.* **2020**, *68*, 5322–5333. [CrossRef]

7. Angeletti, P.; Lisi, M. Multimode beamforming networks for space applications. *IEEE Antennas Propag. Mag.* **2014**, *56*, 62–78. [CrossRef]

8. Laue, H.E.A.; du Plessis, W.P. A Checkered Network for Implementing Arbitrary Overlapped Feed Networks. *IEEE Trans. Microw. Theory Tech.* **2019**, *67*, 4632–4640. [CrossRef]

9. Caille, G.; Chiniard, R.; Thevenot, M.; Chreim, H.; Arnaud, E.; Monediere, T.; De Maagt, P.; Palacin, B. Electro-magnetic band-gap feed overlapping apertures for multi-beam antennas on communication satellites. In Proceedings of the 8th European Conference on Antennas and Propagation (EuCAP 2014), The Hague, The Netherlands, 6–11 April 2014.

10. Llombart, N.; Neto, A.; Gerini, G.; Bonnedal, M.; De Maagt, P. Leaky wave enhanced feed arrays for the improvement of the edge of coverage gain in multibeam reflector antennas. *IEEE Trans. Antennas Propag.* **2008**, *56*, 1280–1291. [CrossRef]

11. Blanco, D.; Rajo-Iglesias, E.; Montesano Benito, A.; Llombart, N. Leaky-Wave Thinned Phased Array in PCB Technology for Telecommunication Applications. *IEEE Trans. Antennas Propag.* **2016**, *64*, 4288–4296. [CrossRef]

12. Smolders, A.B.; Dubok, A.; Tessema, N.M.; Chen, Z.; Al Rawi, A.; Johannsen, U.; Bressner, T.; Milosevic, D.; Gao, H.; Tangdiongga, E.; et al. Building 5G Millimeter-Wave Wireless Infrastructure: Wide-Scan Focal-Plane Arrays With Broadband Optical Beamforming. *IEEE Antennas Propag. Mag.* **2019**, *61*, 53–62. [CrossRef]

13. Mathworks Matlab. 2017. Available online: www.matlab.com (accessed on 5 July 2017).

14. Dassault Systems CST. 2019. Available online: https://www.cst.com/2019 (accessed on 10 June 2019).

15. Ugray, Z.; Lasdon, L.; Plummer, J.; Glover, F.; Kelly, J.; Marti, R. Scatter Search and Local NLP Solvers: A Multistart Framework for Global Optimization. *INFORMS J. Comput.* **2007**, *19*, 328–340. [CrossRef]

16. Mailloux, R.J. *Phased Array Antenna Handbook*; Artech House Antennas and Propagation Library: Boston, MA, USA, 2005; ISBN 9781580536899.

17. Jacob, M.M.; Long, J.; Sievenpiper, D.F. Non-foster loaded parasitic array for broadband steerable patterns. *IEEE Trans. Antennas Propag.* **2014**, *62*, 6081–6090. [CrossRef]

18. Batel, L.; Rudant, L.; Pintos, J.F.; Clemente, A.; Delaveaud, C.; Mahdjoubi, K. High directive compact antenna with non-foster elements. In Proceedings of the 2015 International Workshop on Antenna Technology (iWAT), Seoul, Korea, 4–6 March 2015.

19. Toso, G. The beauty of multibeam Antennas. In Proceedings of the 2015 9th European Conference on Antennas and Propagation (EuCAP), Lisbon, Portugal, 13–17 April 2015.

20. Angeletti, P.; Toso, G.; Petrolati, D. Beam-Forming Network for an Array Antenna and Array Antenna Comprising the Same. U.S. Patent 9,374,145, 26 November 2012.

10

Time Domain Performance of Reconfigurable Filter Antenna for IR-UWB, WLAN, and WiMAX Applications

Zhuohang Zhang * and Zhongming Pan

College of Intelligence Science and Technology, National University of Defense Technology, Changsha 410073, China
* Correspondence: zhangzhuohang23@163.com

Abstract: A novel reconfigurable filter antenna with three ports for three dependent switchable states for impulse radio-ultrawideband (IR-UWB)/wireless local area network (WLAN)/worldwide interoperability for microwave access (WiMAX) applications is presented in this paper. Three positive-intrinsic-negative diodes, controlled by direct current, are employed to realize frequency reconfiguration of one ultra-wideband state and two narrowband states (2.4 GHz and 3.5 GHz). The time domain characteristic of the proposed antenna in the ultra-wideband state is studied, because of the features of the IR-UWB system. The time domain analysis shows that the reconfigurable filtering antenna in the wideband state performs similarly to the original UWB antenna. The compact size, low cost, and expanded reconfigurable filtering features make it suitable for IR-UWB systems that are integrated with WLAN/WiMAX communications.

Keywords: time domain performance; filtering antenna; reconfigurable antenna; UWB antenna; IR-UWB system

---

1. Introduction

The need for reconfigurable antennas in various wireless communication systems has been increasing [1]. Ultrawideband (UWB) antennas are necessary for integrating other existing wireless networking technologies with impulse radio-UWB (IR-UWB) systems. Different types of radio frequency (RF) switches, such as gallium arsenide (GaAs) field effect transistor switches, micro electro-mechanical systems (MEMS) switches, positive-intrinsic-negative (PIN) diodes, and varactor diodes, have been used to enable frequency reconfiguration [2–4]. Among these components, the PIN diode is most preferred in such applications due to its fast switching time, low cost, and easy fabrication.

Many researchers have discussed using PIN diodes to enable reconfigurability, including frequency switching [2,5,6] and radiation pattern switching [7–9] with PIN diodes. Multiple reconfigurability of bandwidth switching, radiation pattern switching, and polarization switching are realized by PIN diodes in [4]. Generally, antenna radiation patterns are studied at one specific frequency in the research of reconfigurable antennas, and when referring to radiation patterns, the frequency domain is commonly considered [10].

However, in IR-UWB systems, frequency domain analysis cannot accurately and completely describe the antenna's features, as a result of adopting a narrow pulse for targeting and positioning [11]. Additional time domain analysis of the antenna is required. Some parameters of the time domain have been proposed to characterize this feature. Previous studies [12–15] have proposed a new method to describe the time domain performance by studying the forward voltage gain (S_{21}) and system fidelity factor (SFF) of the antennas. Other authors [16,17] have cited the importance of the time domain analysis of traditional UWB antennas and of the performance of their operating antennas. However, few time domain studies of the reconfigurable antenna have been conducted, which is essential to IR-UWB systems.

In this paper, a novel design of a reconfigurable filter antenna in one ultra-wideband state and two narrowband states for an IR-UWB system is presented. The time domain performance was determined by measurement compared to the original UWB antenna. The paper is organized as follows: Section 2 describes the design of the proposed antenna. Frequency and time domain analysis are studied in Section 3. Section 4 concludes the study.

2. Design of the Reconfigurable Filtering Antenna

2.1. UWB Antenna Design

We chose a round-shaped monopole antenna and a defeated ground structure, which can broaden the operating frequency band. The layout of the original UWB antenna with its parameters are illustrated in Figure 1. The detailed values of various parameters used in the proposed antenna are listed in Table 1. The original UWB antenna was printed on a 1.6-mm-thick flame retardant-4 glass epoxy (FR-4) substrate (relative permittivity (ε_r) = 4.3). The substrate had a dimension of 80 × 90 mm^2. The top layer consisted of a round-shaped patch as the radiating element, which was excited through a 50 Ω microstrip feeding line. The simulated and measured S_{11} of the original UWB antenna, shown in Figure 2, revealed that the designed antenna is able to support a wide frequency operating band of 1–4.7 GHz, which can cover the whole operating band of the IR-UWB system.

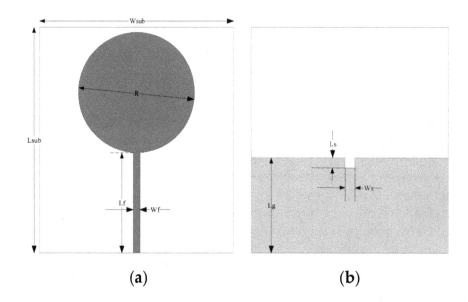

(a)   (b)

Figure 1. Layout of the original ultrawideband (UWB) antenna: (**a**) top view; (**b**) bottom view.

Table 1. Antenna parameters and their values.

Parameter	Value (mm)	Parameter	Value (mm)	Parameter	Value (mm)
W_{sub}	80	W_f	3	R	24
L_{sub}	90	L_f	41	L_g	38
W_s	4	L_s	4	-	-

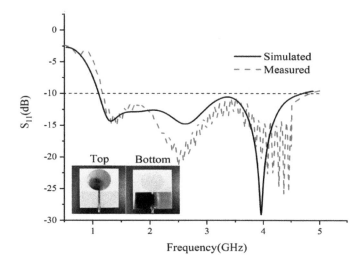

Figure 2. Simulated and measured input port voltage reflection coefficient (S_{11}) of the original UWB antenna.

2.2. Band-Pass Filter Design

Referring to [18], we adopted a half-wave-length stepped impedance resonator filter due to its compact size and high stop-band rejection range. Figure 3 shows the layout of the adopted filter; the parameters were tuned to realize the 2.4 GHz and 3.5 GHz bandpass. Figures 4 and 5 reveal that the designed band-pass filter is able to support a tuning center frequency from 2.4 GHz to 3.5 GHz with tuning parameters. Detailed values of the various parameters used in the proposed filters are listed in Tables 2 and 3. The filter structure is simple and can be easily applied in related applications.

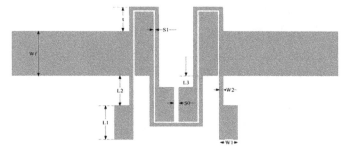

Figure 3. Layout of the band-pass filter.

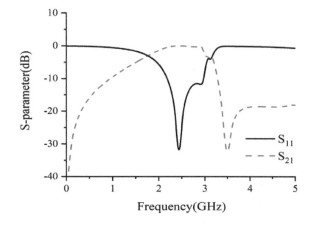

Figure 4. The simulated scattering parameters (S-parameters) of filter at 2.4 GHz.

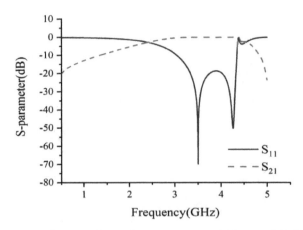

Figure 5. The simulated S-parameters of filter at 3.5 GHz.

Table 2. Different parameters of the filter and their values at 2.4 GHz.

Parameter	Value (mm)	Parameter	Value (mm)	Parameter	Value (mm)
W1	1.89	L1	3.85	S0	0.3
W2	0.38	L2	3.86	S1	0.10
t	3	L3	1.33	W_f	3

Table 3. Different parameters of the filter and their values at 3.5 GHz.

Parameter	Value (mm)	Parameter	Value (mm)	Parameter	Value (mm)
W1	1.13	L1	2.31	S0	0.3
W2	0.23	L2	2.01	S1	0.10
t	1.66	L3	0.8	W_f	3

2.3. Reconfigurable Filter Antenna Design

As shown in Figure 6a, the proposed reconfigurable filtering antenna was designed to integrate the band-pass filter and the original UWB antenna. The whole structure of the proposed antenna includes three RF ports, a monopole, two band-pass filters, 50 Ω microstrip lines, and a direct current (DC) bias circuit for PIN diodes. Frequency reconfiguration is implemented by the PIN controlling path selections. Using three paths of 50 Ω microstrip lines and two designed band-pass filters to the monopole, filtering capability is obtained. Thus, three operating states with a frequency reconfigurable antenna were achieved, with no change in size.

Three PIN diodes were welded in suitable places to control the three paths to the radiating element: diode D1 was located in the slot line of the feeding line, and diodes D2 and D3 were placed on both sides of the feeding line. In the UWB state, D1 is on, D2 and D3 have no-bias voltage, the RF signal excites the antenna through port 1 and the two filters are off. When D1 is OFF, the antenna operates in a narrowband state, depending on whether either D2 or D3 is on. When D2 is on, the antenna is excited by the RF signal through port 2 and operates in the 3.5 GHz narrowband state, and operates in the in 2.4 GHz narrowband state when D3 is on.

As shown in Figure 6b, when on, the PIN diodes can be equivalent to an inductor (L) in series with a resistor (R); when off, the PIN diodes can be equivalent to an inductor (L) in series with a capacitor (C). In this paper, MACOM MADP-000907-14020 (MACOM, Lowell, Massachusetts, USA) was used as the PIN switch due to its low capacitance, 2 ns switching speed, and up to 70 GHz operating band [19]. The circuit parameters were set to R = 7.8 Ω, C = 0.025 pF, and L= 30 nH [20].

To avoid coupling of the RF signal and the bias current, we used a proper bias in this study [21]. As shown in Figure 6c, a capacitor was adopted for DC blocking. Therefore, resistance (R) was used to restrict the max bias current. Biasing voltages (VB) of 1.67 V and 0 V were applied to the circuit for 'switch on' and 'switch off' conditions, respectively.

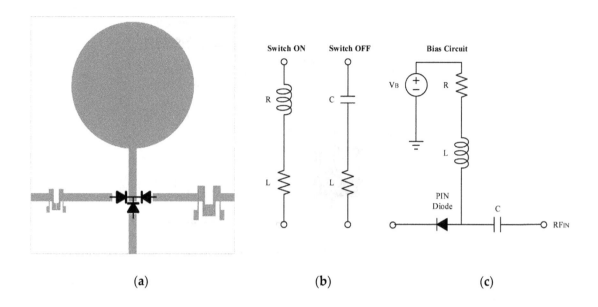

(a)   (b)   (c)

Figure 6. Proposed filter antenna and bias circuit for positive-intrinsic-negative (PIN) diode: (**a**) the proposed antenna, (**b**) equivalent circuits for the on and off states of the PIN diode, and (**c**) bias circuit for PIN diodes.

3. Results and Discussion

3.1. Frequency Domain Performance

The proposed reconfigurable filtering antenna was fabricated and measured. The simulated and measured S_{11} in three states of the antenna are shown in Figure 7a–c. In the UWB state, diode D1 is on and diodes D2 and D3 are off. Figure 7a shows that the measured S_{11} performance is basically consistent with the simulated results; the proposed antenna in the UWB state can cover from 1 GHz to more than 5 GHz, which is sufficient for IR-UWB systems [17]. In the narrowband state, diode D1 is off. Depending on the bias circuits of D2 and D3, the narrowband states are switched. Figure 7b shows the 3.5 GHz narrowband when D2 is on. The measured results show that the antenna covers the 3.5 GHz WiMAX work band. When D3 is on and D2 is off, the antenna works in the 3.5 GHz narrowband state. Figure 7c shows that the measured antenna operates in the 2.4 GHz band.

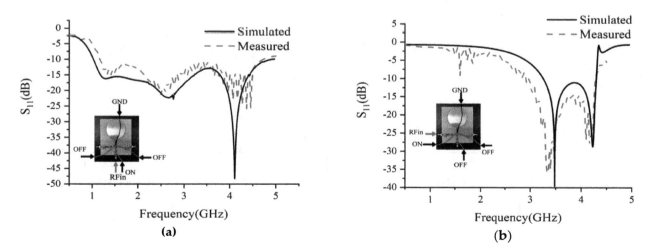

(a)   (b)

Figure 7. *Cont.*

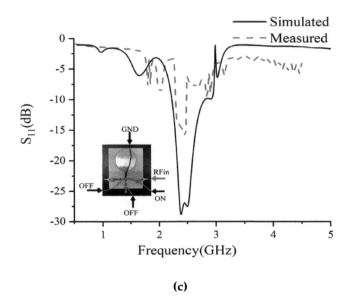

(c)

Figure 7. Summary of the S_{11} of the reconfigurable antenna: (**a**) UWB state (D1 on, D2 and D3 off); (**b**) 3.5 GHz narrowband (D2 on, D1 and D3 off); and (**c**) 2.4 GHz narrowband (D3 on, D1 and D2 off).

As illustrated in Figure 7, the measured and simulated responses were compared. The comparison showed that the measured results of the proposed antenna correspond well in the three states. However, in the wideband state and the 3.5 GHz narrowband state, S_{11} shifts at the lower frequency of the whole operating band. Nevertheless, the result is valid and acceptable. Reasonable agreements between the simulated and measured results were obtained. The error is mainly due to the inaccurate modeling of PIN diodes and the fabrication and welding processes.

The port isolations of the proposed antenna are displayed in Figure 8a–c. In the three different operating states, high isolation performance (better than 20 dB) was obtained throughout the whole operating frequency band. High isolation indicates that the three states of the proposed antenna can operate dependently. The simulated and measured results in Figure 7 prove this.

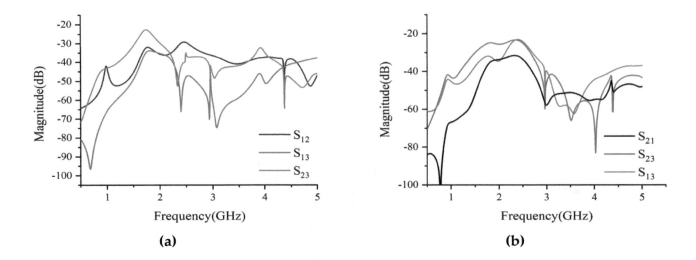

(a)   (b)

Figure 8. *Cont.*

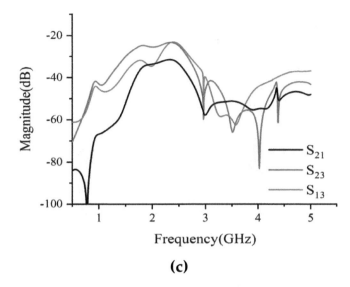

(c)

Figure 8. Summary of the simulated port isolations of the reconfigurable filtering antenna: **(a)** UWB state (D1 on, D2 and D3 off); **(b)** 2.4 GHz narrowband (D2 on, D1 and D3 off); and **(c)** 3.5 GHz narrowband (D3 on, D1 and D2 off).

The measured realized gains between the proposed reconfigurable filtering antenna and the original UWB antenna are compared in Figure 9. Figure 9 shows that the gains of the proposed filtering antenna are slightly lower than the original UWB antenna, which is mainly caused by insertion loss of additional filter structures and PIN diodes.

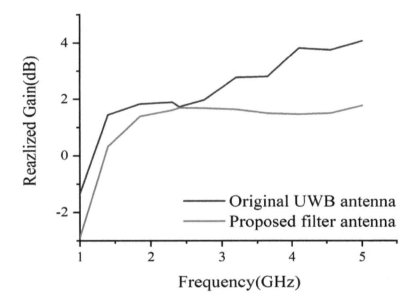

Figure 9. Comparison of the measured realized gains between the proposed antenna and the UWB antenna.

3.2. Time Domain Performance in Wideband State

As shown in Figure 10, the time domain performance of the antenna was measured in an anechoic chamber. With the transmitting antenna oriented at $\varphi = 0°$, the receiving antenna was rotated by $\varphi = 0°$, $90°$, and $180°$, which represent the typical working positions of a pair of antennas. S_{21} and group delay were measured directly from this experiment setup.

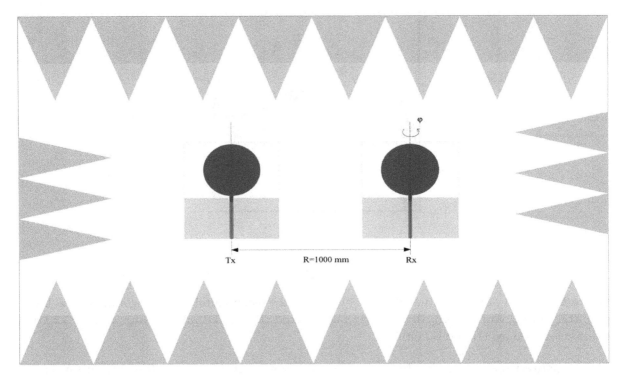

Figure 10. Measurement setup inside an anechoic chamber (Transmitted (Tx), Received (Rx)).

To characterize the time domain performance of an antenna, different parameters from the frequency domain were studied. The system fidelity factor (SFF) is the most used parameter to analyze time, frequency, and space together. The method is described in detail by the authors of [15] and in our experiment, SFF was obtained as follows. First, S_{21} was measured using the experimental setup, as shown in Figure 10. Then, the received signal was calculated by Fourier transform. The SFF was then calculated as follows (the system transform function $H(\omega)$ can be substituted by the measured S_{21}):

$$R_s(\omega) = FFT(T_S(\omega))H(\omega), \tag{1}$$

$$R_s(t) = IFFT(R_S(\omega)), \tag{2}$$

$$SFF = \max_n\left\{\frac{\int\limits_{-\infty}^{\infty} T_s(t)R_s(t+\tau)d\tau}{\left[\int\limits_{-\infty}^{\infty}|T_s(t)|^2 dt\right]^{1/2}\left[\int\limits_{-\infty}^{\infty}|R_s(t)|^2 dt\right]^{1/2}}\right\}, \tag{3}$$

where $Ts(t)$ denotes the transmitted excitation signal, $Rs(t)$ denotes the radiated signal in the received antenna, t denotes the time, $H(\omega)$ denotes the system transform function, ω denotes the frequency and τ denotes the shifted variable between $Ts(t)$ and $Rs(t)$ in convolution.

The S_{21} magnitude and group delay versus frequency characteristics for both antennas are depicted in Figure 11. For the proposed filtering antenna, the group delay is almost linear through the whole operating band. At $\varphi = 0°$, the group delay is 5 ns, and 4.5 ns at $\varphi = 90°$ and 180°. The group delay of the proposed antenna is almost consistent with the original. However, at higher frequencies, a shift away from the original antenna occurs. A similar phenomenon can be found through the comparison of S_{21}. A relative difference from 3.5 GHz to 5 GHz was observed. The difference is mainly due to the relative difference in the realized gains between the adopted filtering structure and the original UWB antenna within an acceptable range. Nevertheless, the result is valid and workable. These results indicate that the proposed filtering antenna obtains a similar S_{21} response and group delay, in comparison with the original UWB antenna.

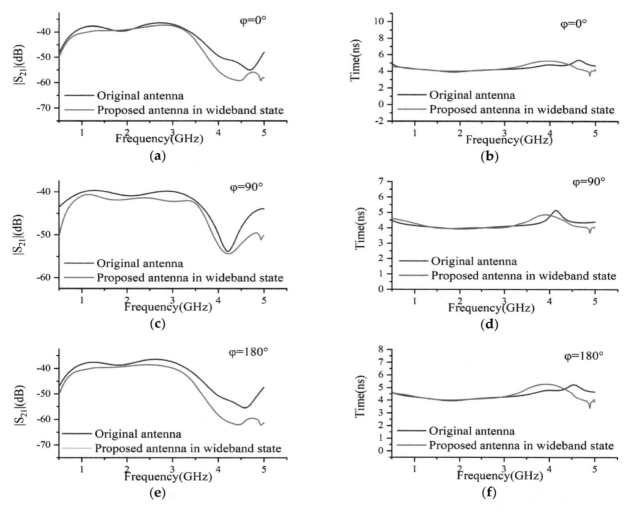

Figure 11. Measurement summary of S_{21} magnitude and group delay for the original and proposed antennas at various φ: (**a**) S_{21} magnitude and (**b**) group delay at $\varphi = 0°$, (**c**) S_{21} magnitude and (**d**) group delay at $\varphi = 90°$, and (**e**) S_{21} magnitude and (**f**) group delay at $\varphi = 180°$.

The calculated output signals for both antennas for various φ are demonstrated in Figure 12. The proposed antenna has a similar output signal in comparison with the original UWB antenna. SFF can then be calculated by the coefficient of correlation between $Ts(t)$ and $Rs(t)$. The results are provided in Table 4.

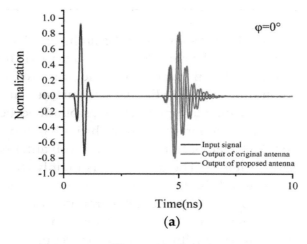

Figure 12. *Cont.*

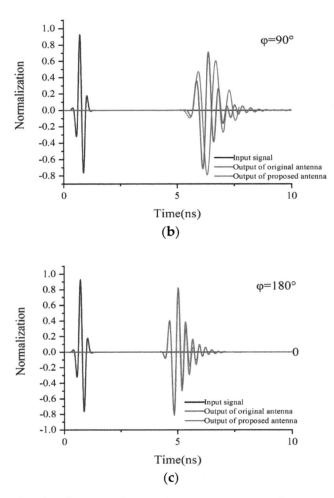

Figure 12. The calculated pulse shapes at the passive antenna output from measurement at various φ: (a) φ = 0°, (b) φ = 90°, and (c) φ = 180°.

Table 4. System fidelity factor for the UWB monopole antenna and the proposed antenna.

	System Fidelity Factor			
φ	Original UWB Antenna		Proposed Reconfigurable Filtering Antenna	
	Simulated	Measured	Simulated	Measured
0°	0.95	0.92	0.93	0.9
90°	0.91	0.86	0.87	0.83
180°	0.87	0.84	0.85	0.82

A fidelity more than 0.8 indicates that the source signal is undistorted through material propagating, such that the received pulse waveform can be completely characterized [16]. The calculated values of the fidelity factor for both the original UWB antenna and the proposed reconfigurable antenna are listed in Table 4. Table 4 shows that the highest SFF (0.93) is obtained at φ = 0° and the differences in various φ values are small. A reasonable agreement was obtained between the simulated and measured SFF results. A similar phenomenon was found through comparison of the original UWB antenna. We observed that the fidelity factor of the original UWB antenna is better than that of the proposed reconfigurable filtering antenna. The difference is mainly due to the additional filtering structures insertion loss and the use of PIN diodes. However, the measured system fidelity factor (>0.82) was still high. The calculated SFF indicates that the proposed filtering antenna obtains a similar time domain performance in comparison with the original antenna.

Table 5 compares the proposed antenna with other antennas reported in the literature. The novelty of the proposed antenna lies in its simple shape, its ability to efficiently reconfigure the frequency and extra analysis of time domain performance. It is noteworthy that the proposed antenna has expanded frequency reconfiguration, without degrading the antenna's time domain performance. The reconfigurable antenna for IR-UWB can integrate new working patterns without changing the original antenna's features and size.

Table 5. Performance comparison with other designs in the literature.

Ref.	Reconfiguration	Actuators	Mode Number	Time Domain Analysis
[2]	Frequency	5 PINs	6 (narrowband)	No
[3]	Frequency	1 FET	2	No
[7]	Frequency	4 PINs	12 (narrowband)	No
[16]	No	-	-	Yes
This work	Frequency	3 PINs	1 wideband, 2 narrowband	Yes

4. Conclusions

In this paper, a novel reconfigurable filter antenna was proposed for IR/UWB applications integrated with WLAN and WiMAX, and is suitable for many other communications. Besides the traditional frequency domain analysis, the time domain performance of the proposed antenna, which uses a PIN reconfigurable technique, was studied. The proposed antenna was fabricated and tested, and the time performance was studied in terms of measured S_{21}, group delay, and calculated SFF. A dependent operating band was obtained through frequency domain analysis, with reasonable agreement of the time domain performance between the proposed reconfigurable filtering antenna and the original UWB antenna. These results indicate that the reconfigurable filtering antenna can integrate the WLAN and WiMAX narrowband with the IR-UWB operating band without changing the size and time domain performance compared with the original UWB antenna. These features enable the proposed filtering antenna to be widely used in IR-UWB systems integrated with WLAN/WiMAX and many other communications, which is essential for the miniaturization of the RF front in WSN applications, where multiple communications are needed.

Author Contributions: Conceptualization, Z.Z.; methodology, Z.P.; validation, Z.Z.; investigation, Z.Z.; writing—original draft preparation, Z.Z.; writing—review and editing, Z.P.; visualization, Z.Z.; supervision, Z.P.; project administration, Z.P.; funding acquisition, Z.P.

Acknowledgments: The authors are grateful to the reviewers and editors for their valuable feedback on our work that helped to improve the quality of this paper.

References

1. Liang, Z.; Zhang, G.; Dong, X.; Huo, Y. Design and Analysis of Passband Transmitted Reference Pulse Cluster UWB Systems in the Presence of Phase Noise. *IEEE Access* **2018**, *6*, 14954–14965. [CrossRef]
2. Yao, C.; Longfang, Y.; Jianliang, Z.; Yanhui, L.; Liang, Z.; Miao, Z.; Qing, H.L. Frequency Reconfigurable Circular Patch Antenna with an Arc-Shaped Slot Ground Controlled by PIN Diodes. *Int. J. Antennas Propag.* **2017**, *2017*, 1–7. [CrossRef]
3. Yang, X.-L.; Lin, J.-C.; Chen, G.; Kong, F.-L. Frequency reconfigurable antenna for wireless communications using GaAs FET switch. *IEEE Antennas Wirel. Propag. Lett.* **2014**, *14*, 807–810. [CrossRef]
4. Selvam, Y.P.; Elumalai, L.; Alsath, M.G.N.; Kanagasabai, M.; Subbaraj, S.; Kingsly, S. Novel frequency-and pattern-reconfigurable rhombic patch antenna with switchable polarization. *IEEE Antennas Wirel. Propag. Lett.* **2017**, *16*, 1639–1642. [CrossRef]
5. Nguyen-Trong, N.; Piotrowski, A.; Fumeaux, C. A frequency-reconfigurable dual-band low-profile monopolar antenna. *IEEE Trans. Antennas Propag.* **2017**, *65*, 3336–3343. [CrossRef]
6. Konca, M.; Warr, P.A. A frequency-reconfigurable antenna architecture using dielectric fluids. *IEEE Trans. Antennas Propag.* **2015**, *63*, 5280–5286. [CrossRef]

7. Hossain, M.A.; Bahceci, I.; Cetiner, B.A. Parasitic layer-based radiation pattern reconfigurable antenna for 5G communications. *IEEE Trans. Antennas Propag.* **2017**, *65*, 6444–6452. [CrossRef]

8. Tran, H.H.; Nguyen-Trong, N.; Le, T.T.; Park, H.C. Wideband and multipolarization reconfigurable crossed bowtie dipole antenna. *IEEE Trans. Antennas Propag.* **2017**, *65*, 6968–6975. [CrossRef]

9. Liu, B.-J.; Qiu, J.-H.; Wang, C.-L.; Li, G.-Q. Pattern-reconfigurable cylindrical dielectric resonator antenna based on parasitic elements. *IEEE Access* **2017**, *5*, 25584–25590. [CrossRef]

10. Deng, J.; Hou, S.; Zhao, L.; Guo, L. A reconfigurable filtering antenna with integrated bandpass filters for UWB/WLAN applications. *IEEE Trans. Antennas Propag.* **2017**, *66*, 401–404. [CrossRef]

11. Huo, Y.; Dong, X.; Lu, P. Ultra-wideband transmitter design based on a new transmitted reference pulse cluster. *ICT Express.* **2017**, *3*, 142–147. [CrossRef]

12. Quintero, G.; Zurcher, J.; Skrivervik, A. Omnidirectional pulse dispersion of planar circular monopoles. In Proceedings of the 2009 IEEE International Conference on Ultra-Wideband, Vancouver, BC, Canada, 9–11 September 2009; pp. 395–399.

13. Quintero, G.; Zurcher, J.-F.; Skrivervik, A.K. System fidelity factor: A new method for comparing UWB antennas. *IEEE Trans. Antennas Propag.* **2011**, *59*, 2502–2512.

14. Koohestani, M.; Skrivervik, A.K.; Moreira, A.A. System fidelity factor evaluation of wearable ultra-wideband antennas for on-body communications. *IET Microw. Antennas Propag.* **2015**, *9*, 1054–1058. [CrossRef]

15. Koohestani, M.; Pires, N.; Moreira, A.A.; Skrivervik, A.K. Time-domain performance of patch-loaded band-reject UWB antenna. *Electron. Lett.* **2013**, *49*, 385–386. [CrossRef]

16. Singhal, S.; Singh, A.K. CPW-fed hexagonal Sierpinski super wideband fractal antenna. *LET Microw. Antennas Propag.* **2016**, *10*, 1701–1707. [CrossRef]

17. Valizade, A.; Rezaei, P.; Orouji, A.A. A design of UWB reconfigurable pulse transmitter with pulse shape modulation. *Microw. Opt. Technol. Lett.* **2016**, *58*, 2221–2227. [CrossRef]

18. Gorur, A. A novel dual-mode bandpass filter with wide stopband using the properties of microstrip open-loop resonator. *IEEE Microw. Wirel. Compon. Lett.* **2002**, *12*, 386–388. [CrossRef]

19. MACOM. Available online: https://www.macom.com/products/product-detail/MADP-000907-14020P (accessed on 15 August 2019).

20. Yang, H.H.; Yang, F.; Xu, S.H.; Li, M.K.; Cao, X.Y.; Gao, J. Design and verification of an electronically controllable ultrathin coding periodic element in Ku band. *Acta Phys. Sin.* **2016**, *65*, 54102.

21. Yeom, I.; Choi, J.; Kwoun, S.-S.; Lee, B.; Jung, C. Analysis of RF Front-End Performance of Reconfigurable Antennas with RF Switches in the Far Field. *Int. J. Antennas Propag.* **2014**, *2014*, 1–14. [CrossRef]

11

Full-Duplex NOMA Transmission with Single-Antenna Buffer-Aided Relays

Nikolaos Nomikos [1,*], Panagiotis Trakadas [2], Antonios Hatziefremidis [2] and Stamatis Voliotis [2]

[1]   Department of Information and Communication Systems Engineering, University of the Aegean, 83200 Samos, Greece
[2]   General Department, National and Kapodistrian University of Athens, 34400 Psahna, Greece; ptrakadas@uoa.gr (P.T.); ahatzie@uoa.gr (A.H.); svoliotis@uoa.gr (S.V.)
*   Correspondence: nnomikos@aegean.gr

Abstract: The efficient deployment of fifth generation and beyond networks relies upon the seamless combination of recently introduced transmission techniques. Furthermore, as multiple network nodes exist in dense wireless topologies, low-complexity implementation should be promoted. In this work, several wireless communication techniques are considered for improving the sum-rate performance of cooperative relaying non-orthogonal multiple access (NOMA) networks. For this purpose, an opportunistic relay selection algorithm is developed, employing single-antenna relays to achieve full-duplex operation by adopting the successive relaying technique. In addition, as relays are equipped with buffers, flexible half-duplex transmission can be performed when packets reside in the buffers. The proposed buffer-aided and successive single-antenna (BASSA-NOMA) algorithm is presented in detail and its operation and practical implementation aspects are thoroughly analyzed. Comparisons with other relevant algorithms illustrate significant performance gains when BASSA-NOMA is employed without incurring high implementation complexity.

Keywords: NOMA; full duplex; successive relaying; buffer-aided relaying

1. Introduction

Mobile data traffic has been increasing at a dramatic pace due to the emergence of Internet-of-Things (IoT) applications [1], resulting in coexisting human and machine traffic. So, novel techniques are required to provide improved spectral efficiency and transmission reliability. Regarding spectral-efficient communication, non-orthogonal multiple access (NOMA) has been proposed as a promising solution to overcome the inefficiency of orthogonal resource allocation in multi-user networks [2]. In contrast to orthogonal multiple access (OMA), NOMA enables multiple devices to be served on the same resources in the power domain (PD) or code domain (CD) [3]. The PD NOMA paradigm relies on combining superposition coding at the transmitters and successive interference cancellation (SIC) at the receivers and its performance depends on various parameters, such as user channel asymmetry, rate requirements and cooperation among devices [4,5]. Moreover, in order to strengthen the reliability of fifth generation (5G) communication, buffer-aided (BA) cooperative relaying has the potential to improve the wireless conditions, offering increased diversity and enhancing the outage and throughput performance, as shown in various studies [6,7]. The use of buffers enables hybrid transmission modes, due to the decoupling of the reception and transmission phases, as relays might have packets to forward even if reception was not possible in the previous phase. The survey in [7] presented various opportunistic relay selection (ORS) algorithms comprising half-duplex (HD), full-duplex (FD) and successive relaying (SuR), showing significant outage and throughput performance gains. In addition, low latency is a prerequisite for highly demanding applications, such as augmented/virtual reality, public safety applications and ultra-high definition teleconferencing among others and thus apart from FD transmission, delay-awareness must be integrated in cooperative relaying [8,9].

1.1. Background

BA relays were initially proposed in HD networks with single-antenna relays. These networks suffer from half-duplex rate losses, as relays are not able to receive and transmit at the same time. Firstly, the paper in [10] suggested to equip relays with buffers in two-hop ORS topologies. In greater detail, each time-frame was divided to two time slots, with each one allocated to the source-relay ($\{S{\to}R\}$) and relay-destination ($\{R{\to}D\}$) transmissions, forming the hybrid relay selection (HRS). Further improvements were provided by max $-$ link, an ORS algorithm that was proposed in [11], achieving increased diversity due to BA relays and flexible selection of the strongest link among all the available ones. The performance analysis showed that the use of max $-$ link offers a diversity gain of $2K$ when K relays and large buffers are available in the network. Focusing on single relay networks with multiple antennas, the authors in [12], analyzed the performance of BA relaying with adaptive link selection, concluding that the performance of HD relaying can surpass that of ideal FD relaying, given that the number of antennas at the source and the destination is greater than the number of available antennas at the relay. More recently, further diversity gains were introduced in BA relay networks through broadcast transmissions in the $\{S{\to}R\}$ links [13], resulting in beneficial packet redundancy, while the work in [14] proposed low-complexity distributed solutions to mitigate the impact of various issues, such as errors in the feedback channel and relay selection using outdated channel state information (CSI).

Without departing from single-antenna relay networks, several studies aimed at overcoming their inherent half-duplex constraint and provided throughput gains. The paper in [15] introduced hybrid BA ORS, merging HD relaying and SuR in a topology with isolated relays where the inter-relay interference (IRI), arising from the successive source and relay transmissions, was considered negligible. Both adaptive and fixed-rate transmissions were studied, revealing that space full-duplex (SFD) max–max relay selection (MMRS) can double the capacity of HD schemes in the former case, and offer coding gain, as well as diversity gain equal to the number of relays, in the latter case. Then, the degrading effect of IRI and its mitigation through proper relay-pair selection and interference cancellation were investigated in [16], while transmit power adaptation in fixed-rate communication and the integration of max $-$ link and SuR was proposed in [17]. Comparisons with standalone HD and FD solutions indicated significant resiliency against outages and increased throughput. Another improvement towards low-delay SuR was given in [18]. The proposed low-latency for the successive opportunistic relaying (LoLA4SOR) algorithm considered the buffer length in both $\{S{\to}R\}$ and $\{R{\to}D\}$ links and, if SuR failed, delay-aware (DA) HD relaying was performed, exhibiting reduced delay and robustness against outages. Furthermore, optimal scheduling towards delay-awareness and IRI mitigation based on either dirty paper coding or SIC for topologies with two relays, was examined in [19]. More recently, apart from data buffers, energy buffers storing the amount of energy that is harvested from wireless transmissions have been incorporated in BA SuR networks with single [20] or multiple relays [21], enabling network lifetime prolongation.

As spectrum scarcity threatens the operation of current wireless networks, comprising humans and machines, spectral efficient transmission is of vital importance. In this research field, NOMA has emerged as a promising technique and its integration in cooperative relaying networks has received various contributions. The work in [22] assigned users with better channel conditions to decode the messages of weak users, relaying them in the next time slot, thus promoting user fairness through homogeneous cell coverage. Then, ORS was studied in [23], introducing a two-stage relay selection, increasing the diversity gain, compared to conventional max–min ORS. Another ORS algorithm for topologies where users are employed to serve as relays was given in [24], resulting in coverage and capacity enhancement. More specifically, a range-division user relay selection scheme was devised, dividing the circular cell in continuous annular areas and performing ORS in each one. Other works, targeted the sum-rate of two-hop NOMA relay networks, jointly studying NOMA and SuR [25,26]. The proposed solutions created virtual FD relay networks by employing two dedicated relays to perform SuR and serve two users through NOMA. Likewise, a four-user topology where two users were chosen to perform SuR and serve the remaining two users using NOMA was shown in [27], achieving significant sum-rate gains.

Apart from employing SuR to overcome sum-rate losses, the use of buffers can also improve the performance of NOMA in HD relay networks. An adaptive mode selection, enabled through a single BA relay, was given in [28]. So, switching between NOMA and OMA when the former was infeasible, allowed a sum-rate increase, compared to OMA-based BA relaying and NOMA relaying without buffers. In cases when CSI at the transmitter might not be available, the authors in [29] presented adaptive link selection, deriving the outage probability expressions and showing a diversity gain of two for buffer sizes larger than two packets. In networks where multiple BA relays are present, two BA–ORS algorithms were given in [30], choosing the best relay, by considering the buffer state information (BSI) and adopting $\{R{\rightarrow}D\}$ prioritization for reduced latency. Then, the potential of BA–ORS in NOMA and hybrid NOMA/OMA networks was presented in [31,32], where hybrid NOMA/OMA selection policies were devised, further improving the sum-rate performance in the downlink and the uplink of two-hop networks, respectively.

1.2. Contributions

It is evident that significant advances have been made in NOMA-based cooperative relay networks, regarding the sum-rate performance, even in HD networks consisting of single-antenna relays and users. Still, the potential of SuR using BA relays in NOMA networks has not been examined and thus this paper aims to fill this gap. More specifically,

- A novel BA ORS algorithm is presented for two-hop NOMA networks, called buffer-aided successive single-antenna (BASSA) NOMA, targeting to provide FD relaying operation by relying only on low-complexity single antenna relays.
- Several wireless communication techniques are presented and integrated in BASSA-NOMA, such as successive broadcasting in the $\{S{\rightarrow}R\}$ links and $\{R{\rightarrow}D\}$ transmission through a selected relay, thus leading to FD operation with IRI, while guaranteeing that $\{R{\rightarrow}D\}$ transmission considers the selected transmitting relay's BSI, avoiding instances of storing packets for many time slots and thus increased delay.
- Furthermore, the steps of BASSA-NOMA are given in detail and various practical issues are discussed, such as power allocation and low-complexity network coordination for choosing the best decoding strategy at the destinations.
- Performance comparisons with other BA NOMA solutions show that even though BASSA-NOMA does not incur high implementation complexity, in terms of CSI acquisition and processing, it enhances the sum-rate and delay performance of the network, while maintaining reduced outages.

Overall, BASSA-NOMA represents an attractive low-complexity solution for two-hop NOMA networks with the potential to be adopted in heterogeneous contexts where relay nodes might not always posses high hardware and processing capabilities, such as in networks where users [33], vehicles [34] and drones [35] assume the role of relays.

1.3. Structure

The paper is organized as follows. In Section 2, the system model is given, while Section 3 presents relevant relay selection algorithms. Next, Section 4, presents BASSA-NOMA and then Section 5 includes comparisons between the proposed, as well as relevant schemes. Finally, Section 6 includes conclusions and possible future directions.

2. System Model

This work studies a two-hop cooperative relay network, comprising one source S, two destinations D_1 and D_2 and a cluster C with K decode-and-forward (DF) relays $R_k \in C$ ($1 \leq k \leq K$). DF avoids the detrimental effect of noise amplification by the relays, at the cost of higher implementation complexity, compared to amplify-and-forward relays, as it has been shown in various works on cooperative relaying in multi-user networks [36,37]. At the same time, relays are clustered in such a way, so that communication among them can be reliably established in order to ensure error-free signaling and coordination. Towards this end, the relays are positioned relatively close together, according to location-based clustering and after a long-term routing process, variations due to pathloss and shadowing effects are tracked. It should be noted that relay clustering is often assumed in the literature, as, e.g., in [16,38]. Each node is equipped with a single-antenna and relays operate in HD mode where concurrent signal transmission and reception is not possible. Furthermore, due to excessive fading conditions in both source-destination ($\{S{\rightarrow}D\}$) links, end-to-end communication is only established via the two-hop relay links. Each relay R_k has a buffer to store a maximum number L_k of successfully decoded packets in order to forward them at a later instance to the destinations. All buffers are of equal length, i.e., $L_k = L, \forall k \in \{1,2,\ldots,K\}$ while the vector including the queue sizes at the buffers of all relays is denoted by $Q \triangleq (Q_1, Q_2, \ldots, Q_K)$. Initially, each relay buffer has Q_k packets, while some buffers might be empty (i.e., $Q_k = 0$ for an arbitrary k). Figure 1 depicts an instance of this BA relay-assisted NOMA network, where the source performs broadcasting towards $K - 1$ relays, while a selected transmitting relay forwards previously received packets to the two destinations, leveraging successive source and relay transmissions for FD NOMA operation.

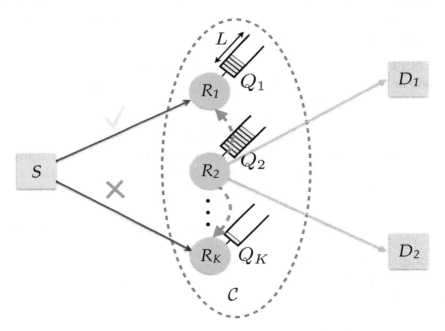

Figure 1. Source S communicates with two destinations D_j, $j \in \{1,2\}$ through a cluster of relays $R_k \in C, k \in \{1,\ldots,K\}$.

The transmission is divided in time slots and at each one simultaneous source and relay transmissions are allowed, using transmit power values P_S and P_{R_k}, respectively. A saturated source having always data for transmission, using an information rate of r_0 bps/Hz is assumed. The retransmission process is based on an acknowledgment/negative-acknowledgment (ACK/NACK) mechanism, employing the receivers (the relays and/or the destination) to broadcast short-length error-free packets over a separate narrow-band channel, notifying the network on whether or not, a specific packet was successfully received. Additionally, it is considered that the receivers are able to accurately estimate the CSI, while the relays have full CSI knowledge of the $\{R{\rightarrow}D\}$ links. As buffering at the relays is available, it is possible for the selected transmitting relay to forward a packet to the two destinations that was received in a time slot that is different from the preceding one. As a result, the destinations might have to reorder the received packets, in order to correctly decode the desired information, a necessary process in all BA relaying protocols. Still, this can be easily achieved at a low-complexity by including a sequence number in each packet.

Regarding the wireless channel conditions, it is assumed that channel quality is degraded by additive white Gaussian noise (AWGN) and for simplicity, the power of the AWGN is assumed to be normalized with zero mean and unit variance. The fading environment is modelled as frequency flat Rayleigh block fading, unless otherwise stated, following a complex Gaussian distribution with zero mean and variance σ_{ij}^2 for the i to j link. Moreover, the channel gains are $g_{ij} \triangleq |h_{ij}|^2$ and exponentially distributed.

Since successive source broadcasting and relay transmissions are performed in the network, all K relays are active at each time slot, as $K - 1$ relays attempt to receive the source signal, while a selected transmitting relay is forwarding a previously received packet to the two destinations. Thus, the HD rate loss of single-antenna relays can be alleviated through SuR. This virtual FD operation enables the destination to receive one packet per time slot, excluding the first one. Still, as a direct result of the successive transmissions, IRI arises and an efficient selection algorithm has to take into consideration its impact on the $K - 1$ relays' reception. In case SuR fails, a complete network outage can be avoided by incorporating a low-complexity mechanism, allowing at least an HD transmission towards the relays or the destinations.

3. Relevant Algorithms

In this section, relevant relay selection algorithms are given in detail, illustrating their main characteristics upon which, BASSA-NOMA is formulated.

3.1. Delay-Aware Max–Link: Adaptive Link Selection, BSI Consideration and {R→D} Prioritization

Initial attempts at providing HD BA ORS algorithms relied on fixed scheduling for the source and relay transmissions. For the first time, this limitation was relaxed in [11], enabling the flexible allocation of each time slot to either an {S→R} or {R→D} transmission, according to the instantaneous CSI and the status of the relays' buffers, activating the strongest link between all the available ones. Still, max − link did not integrate any mechanism to offer delay-awareness and so, further extensions were given in [9,39]. It was shown that the average packet delay can be reduced by prioritizing the {R→D} transmission, prompting packets to leave sooner from the buffers. The DA − max − link algorithm operates as follows:

1.   In an arbitrary time slot, from the set $\mathcal{F}_{RD}$, consisting of relays that are able to perform an {R→D} transmission, DA − max − link considers the number of packets residing in the buffer of each relay. So, the relay having the maximum number of packets is activated to perform a transmission in the {R→D} link.
2.   In case an {R→D} transmission cannot be performed, DA − max − link aims at activating the relay that can perform an {S→R} reception, having the minimum buffer length. Nonetheless, if no feasible {S→R} link exists, a network outage occurs, since packet transmission cannot be performed.

DA − max − link avoids cases of buffer overflow and underflow by integrating BSI into the relay selection process. If more than one relays have equal buffer sizes, a relay is randomly selected among them. Moreover, due to {R→D} prioritization, the average delay performance improves, thus being an attractive solution for delay-sensitive applications.

3.2. LoCo–Link: Source Broadcasting, BSI Consideration and $\{R{\to}D\}$ Prioritization

The low-complexity (LoCo − Link) selection algorithm was introduced in [14], targeting to improve the average delay performance of BA HD relaying, at a low implementation complexity, in terms of CSI requirements. LoCo − Link's operation is described below:

1. In contrast to max − link, where the strongest link was selected between the $2K$ available ones in a K relay network, LoCo − Link firstly focuses on the $\{R{\to}D\}$ links, selecting in each time slot, the $\{R{\to}D\}$ link belonging in $\mathcal{F}_{RD}$, having the maximum number of packets in its buffer. If multiple relays have the same buffer length, one of them is randomly selected.
2. If $\mathcal{F}_{RD}$ is empty, as a result of severe fading or empty buffers, the source broadcasts its packets to all the relays. Thus, there might exist more than one relays forming the set $\mathcal{F}_{SR}$, including the feasible $\{S{\to}R\}$ links, being able to successfully receive the source's packets. It is evident that in subsequent time slots, the diversity of $\{R{\to}D\}$ transmissions increases, compared to schemes with unicast $\{S{\to}R\}$ transmissions, such as DA − max − link.

LoCo − Link provides several gains to the two-hop BA networks, reducing the average end-to-end delay without degrading the diversity of the network. In addition, when a distributed implementation is adopted, the issue of relay selection with outdated CSI can be fully mitigated.

3.3. LoLa4SOR: Hybrid Successive and Half-Duplex Relaying with BSI Consideration

Various DA BA algorithms have adopted $\{R{\to}D\}$ prioritization to avoid buffers storing a large number of packets [9,39]. Still, it has been observed that although reduced latency was achieved, diversity was compromised due to instances of empty buffers. In order to safeguard the diversity of the transmission while offering increased rate through SuR, LoLa4SOR was proposed in [18], based on the hybrid combination of SuR and HD relaying. LoLa4SOR operates as follows:

1. The possibility for SuR is examined, examining possible relay pairs while considering the inter-relay CSI, at each time slot. At the same time, low-latency dictates that buffer length should be kept small without ignoring the diversity. So, relays having the largest buffer lengths and relays with the smallest ones should be prioritized as transmitting and receiving relays, respectively. Then, by denoting as $\mathcal{F}_{\mathrm{pairs}}$ the set of relay pairs that can perform SuR, at the start of each time slot, a relay-pair (R_i, R_j) in links $\{S{\to}R_i, R_j{\to}D\}$, among all the feasible ones for SuR, i.e., $(R_i, R_j) \in \mathcal{F}_{\mathrm{pairs}}$, is activated as:

$$(R_{i\star}, R_{j\star}) = \arg \max_{(R_i,R_j)\in\mathcal{F}_{\mathrm{pairs}}} \left\{ (L - Q_i)^2 + Q_j^2 \right\}, \tag{1}$$

where $(R_{i\star}, R_{j\star})$ denote the optimal relay-pair (R_i, R_j), i.e., the one providing the maximum utility in Equation (1). If two or more relay pairs have the same utility, then diversity is prioritized, and from the set of relay pairs with the maximum utility, denoted by $\mathcal{F}^{\star}_{\mathrm{pairs}}$, the one with the maximum $\{S{\to}R\}$ link utility is activated, i.e.,

$$(R_{i^o}, R_{j^o}) = \arg \max_{(R_{i\star},R_{j\star})\in\mathcal{F}^{\star}_{\mathrm{pairs}}} \left\{ (L - Q_i)^2 \right\}. \tag{2}$$

If two or more relay pairs still provide the same maximum utility, denoted by $\mathcal{F}^{o}_{\mathrm{pairs}}$, one of them is randomly selected.

2. In cases where SuR is not possible, the efficient delay- and diversity-aware max − link (DDA − max − link) HD algorithm, introduced in [9] is activated. DDA − max − link avoids the selection of relays with buffers being on the brink of underflow or overflow by setting appropriate thresholds and activating such relays only to avoid a network outage.

3.4. BA–NOMA/OMA: Flexible Multiple Access with Optimal Power Allocation

In downlink NOMA networks with a single source, multiple BA relays and two destinations, hybrid BA–NOMA/OMA was proposed in [31], integrating broadcasting in the $\{S{\rightarrow}R\}$ link and exploiting BSI in the $\{R{\rightarrow}D\}$ NOMA transmission. The steps of BA–NOMA/OMA are given in the following paragraphs:

1. In the $\{S{\rightarrow}R\}$ link, BA–NOMA/OMA enables the source to broadcast a combined packet containing the information of both destinations with rate $r_1 + r_2$, towards all the relays. Then, each relay uses the CSI of the $\{R{\rightarrow}D\}$ links to examine if PD NOMA is possible through the transmission of a superposition-coded packet to the two destinations. Still, NOMA might not be possible, i.e., due to low transmit signal-to-noise ratio (SNR), insufficient channel asymmetry between the destinations or high rate demands. So in order to avoid a complete network outage, OMA is performed, investigating if at least a single destination $D_j, j \in \{1,2\}$ can receive a packet from the relays included in $\mathcal{F}_{RD_j}$, representing the set of feasible OMA links towards D_j.
2. When OMA is activated, user fairness is guaranteed since at odd time slots, first, D_1 is examined if it can receive its packet from $R_k \in \mathcal{F}_{RD_1}$, else D_2 is examined if it can receive its packet from $R_k \in \mathcal{F}_{RD_2}$. Similarly, at even time slots this process is reversed. When relays with equal buffer lengths exist, random selection occurs in the $\{R{\rightarrow}D\}$ links.

BA–NOMA/OMA is an efficient relay selection algorithm, combining the benefits of both multiple-access schemes, thus improving the average sum-rate performance of multi-user two-hop HD relay networks.

4. The BASSA–NOMA Algorithm

4.1. Algorithmic Description

BASSA-NOMA targets to improve the sum-rate performance of two-hop NOMA cooperative relay networks, relying only on single-antenna relays with low-complexity. So, various transmission techniques are integrated, including successive source broadcasting and relay transmissions, as well as buffering at the relays. The procedure followed by each relay R_k, under the BASSA-NOMA algorithm, in each time slot, is given in Algorithm 1.

More specifically, firstly, BASSA-NOMA aims at selecting a transmitting relay R_k to simultaneously serve D_1 and D_2 using PD NOMA. Relay selection relies on the availability of both $\{R{\rightarrow}D\}$ links' CSI, determining whether or not a NOMA transmission can be successfully performed, as well as each relay's BSI, deciding which one has the maximum number of packets in its buffer (line 1). So, each candidate transmitting relay R_k appropriately chooses the power allocation coefficient, according to its respective $\{R{\rightarrow}D\}$ CSI (line 2). Simultaneously and in a distributed manner, each candidate relay R_k sets its timer to be inversely proportional to the number of packets Q_k, residing in its buffer (lines 3 and 4). In case a transmitting relay is selected (line 5), $\{S{\rightarrow}R\}$ broadcasting and $\{R{\rightarrow}D\}$ transmission are concurrently performed and thus the remaining $K - 1$ relays attempt to receive the source's broadcast signal, containing the information signals of D_1 and D_2 with rate $r = r_1 + r_2$, while subjecting to IRI (lines 6 and 7). Through broadcasting, more packets are available at the relays' buffers and more importantly, CSI at the source is not required, thus significantly reducing the implementation complexity. If, on the other hand, a transmitting relay was not selected, all K relays are available to listen to the source's broadcast transmission, as long as they have space in their buffers (line 10). Finally, when the two destinations receive packets from the transmitting relay, ACKs, containing the relevant packets IDs are broadcasted to all the relays, prompting them to discard these packets from their buffers (line 12). The adoption of BASSA-NOMA enables the network to operate in FD mode without depending on relays with multiple antennas, nor on advanced self-interference cancellation schemes, as it is usually the case in two-hop networks, where simultaneous transmission and reception take place through the same relay.

Algorithm 1 BASSA-NOMA

1: **input** Q_k, CSI for $\{R_k \rightarrow D_j\}$, $j \in \{1, 2\}$ links
2: Each candidate transmitting relay R_k selects α_k for NOMA
3: **if** $R_k \in \mathcal{F}_{RD}$ **then**
4:    R_k's timer is set to be inversely proportional to Q_k
5:    **if** R_k is selected for transmission **then**
6:       R_k performs a NOMA transmission towards D_1 and D_2.
7:       Each R_i, $i \neq k$ with a non-full buffer attempts to receive the source signal with IRI.
8:    **end if**
9: **else**
10:   Each R_i with a non-full buffer attempts to receive the source signal without IRI.
11: **end if**
12: Discard packets from the buffers based on the ACKs from destinations, containing packet IDs
13: **output** Links $\{R_k \rightarrow D_j\}$, $j \in \{1, 2\}$ for transmission and $R_i \in \mathcal{F}_{SR}$ for reception.

As at an arbitrary time slot, BASSA-NOMA initially checks the $\{R \rightarrow D\}$ links on whether or not a NOMA transmission is possible, firstly, details on the $\{R \rightarrow D\}$ transmission are given.

4.2. Transmission in the $\{R \rightarrow D\}$ Link

When a transmitting relay R_k is selected, the information symbols of D_1 and D_2 are superimposed following the PD NOMA paradigm. In greater detail, the transmitted superimposed information symbol, containing the information symbols x_1 and x_2 of each destination, is described as:

$$x = \sqrt{\alpha_k} x_1 + \sqrt{1 - \alpha_k} x_2 \tag{3}$$

with $\mathbb{E}[|x_1^2|] = \mathbb{E}[|x_2^2|] = 1$ and $0 \leq \alpha_k \leq 1$.

Then, D_1 will receive an information symbol y_1 containing the desired symbol, as well as the symbol of D_2, i.e.,

$$y_1 = h_{R_k D_1} \sqrt{\alpha_k P_{R_k}} x_1 + h_{R_k D_1} \sqrt{(1 - \alpha_k) P_{R_k}} x_2 + \eta_1; \tag{4}$$

correspondingly, the received information symbol y_2 at D_2 is given by:

$$y_2 = h_{R_k D_2} \sqrt{\alpha_k P_{R_k}} x_1 + h_{R_k D_2} \sqrt{(1 - \alpha_k) P_{R_k}} x_2 + \eta_2, \tag{5}$$

where η_1 and η_2 denote the AWGN at each destination.

An important parameter that must be optimally determined is α_k representing the power allocation of the transmitting relay R_k to the superimposed signals. Here, the relays poses full CSI of the $\{R \rightarrow D\}$ links and power allocation can be dynamically adjusted in each time slot. The value of α_k is determined in order to increase the chances of successful SIC to decode x_1 and x_2. Thus, in order to achieve SIC, the signal-to-interference-plus-noise ratio (SINR), for at least one of the symbols, should be greater than or equal to a threshold at both destinations. This process is followed by all the relays except from the possible transmitting one, leading to the formation of the set of relays that can perform $\{R \rightarrow D\}$ NOMA transmissions.

As an illustrative example, the decoding of x_2 at both destinations is performed as follows:

$$\Gamma_{R_k D_j}(P_{R_k}) = \frac{(1 - \alpha_k) P_{R_k} g_{R_k D_j}}{\alpha_k P_{R_k} g_{R_k D_j} + \sigma_{D_j}^2} \geq \gamma_j, \quad j \in \{1, 2\}. \tag{6}$$

Note that $\gamma_j \equiv 2^{r_j} - 1$. Then, once x_2 is successfully decoded and subtracted, x_1 can be decoded interference-free at D_1 according to:

$$\Gamma_{R_k D_1}(P_{R_k}) = \frac{\alpha_k P_{R_k} g_{R_k D_1}}{\sigma_{D_1}^2} \geq \gamma_1. \tag{7}$$

There are different methods of determining the power allocation coefficient α_k in NOMA networks. In this work, power allocation follows the method that was developed in [31] which considered a two-hop relay network with two destinations with possibly heterogeneous rate requirements, as it is the case of coexisting IoT devices and users. Since a similar $\{R \rightarrow D\}$ link setting is studied here, readers are referred to that work for more details on calculating α_k.

The outage probability of NOMA in the $\{R \rightarrow D\}$ link is equal to:

$$P_{\text{out}\{R \rightarrow D\}} = \mathbb{P}\left[\alpha_{k,\min} > \min\{1, \alpha_{k,\max}\}\right]. \tag{8}$$

where $\alpha_{k,\min}$ and $\alpha_{k,\max}$ are expressed as [31]:

$$\alpha_{k,\min} \triangleq \frac{\gamma_1 \sigma_{D_1}^2}{P_{R_k} g_{R_k D_1}},$$

$$\alpha_{k,\max} \triangleq \min\left\{\frac{P_{R_k} g_{R_k D_1} - \gamma_2 \sigma_{D_1}^2}{P_{R_k} g_{R_k D_1}(1 + \gamma_2)}, \frac{P_{R_k} g_{R_k D_2} - \gamma_2 \sigma_{D_2}^2}{P_{R_k} g_{R_k D_2}(1 + \gamma_2)}\right\}.$$

Likewise, for decoding x_1 first at both destinations, we have:

$$\alpha_{k,\min} \triangleq \max\left\{\frac{\gamma_1(P_{R_k} g_{R_k D_1} + \sigma_{D_1}^2)}{P_{R_k} g_{R_k D_1}(1 + \gamma_1)}, \frac{\gamma_1(P_{R_k} g_{R_k D_2} + \sigma_{D_2}^2)}{P_{R_k} g_{R_k D_2}(1 + \gamma_1)}\right\},$$

$$\alpha_{k,\max} \triangleq \frac{P_{R_k} g_{R_k D_2} - \gamma_2 \sigma_{D_2}^2}{P_{R_k} g_{R_k D_2}}.$$

Let vector $b_{RD} \triangleq (b_{R_1 D}, b_{R_2 D}, \dots, b_{R_K D})$ be the binary representation of the $\{R \rightarrow D\}$ links satisfying Equations (6) and (7) and so, if a NOMA transmission on the set of links $\{R_k \rightarrow D_1\}$, $\{R_k \rightarrow D_2\}$ is possible, then $b_{R_k D} = 1$. Correspondingly, let vector $q_{RD} \triangleq (q_{R_1 D}, q_{R_2 D}, \dots, q_{R_K D})$ be the binary representation of the feasible links due to the fulfillment of the buffer conditions, i.e., $\{R \rightarrow D\}$ links where relays have non-empty buffers. By $\mathcal{F}_{RD}$, we denote the sets $\{R \rightarrow D\}$ links that are feasible with cardinality F_{RD}.

4.3. Transmission in the $\{S \rightarrow R\}$ Link

In this topology, a time slot is dedicated to concurrent $\{S \rightarrow R\}$ and $\{R \rightarrow D\}$ transmissions. As a result, if a transmitting relay R_k has been selected, the reception of the other $K - 1$ relays will experience the degrading effect of IRI. Regarding the source broadcast signal, as each destination might demand a different rate r_j, $j \in \{1, 2\}$, and in order to avoid instances of buffer overflowing or underflowing, the source transmits with rate $r_1 + r_2$ [29]. Therefore link SR_i, $i \neq k$ is not in outage when:

$$\Gamma_{SR_i}(P_S) \triangleq \frac{g_{SR_i} P_S}{g_{R_k R_i} P_{R_k} + \sigma_i^2} \geq 2^{r_1 + r_2} - 1. \tag{9}$$

On the contrary, link SR_i is in outage if $\gamma_{R_i} < 2^{r_1 + r_2} - 1$, and its outage probability is given by:

$$P_{\text{out}\{S \rightarrow R\}} \triangleq \mathbb{P}\left[g_{SR_i} < \frac{(2^{r_1 + r_2} - 1)(g_{R_k R_i} P_{R_k} + \sigma_i^2)}{P_S}\right]. \tag{10}$$

The vector $b_{SR} \triangleq (b_{SR_1}, b_{SR_2}, \ldots, b_{SR_K})$ with binary elements includes the $\{S \rightarrow R\}$ links that are not in outage. So, if transmission on link SR_i is feasible, then $b_{SR_i} = 1$. Similarly, $q_{SR} \triangleq (q_{SR_1}, q_{SR_2}, \ldots, q_{SR_K})$ represents in a binary form, the feasible $\{S \rightarrow R\}$ links due to the fulfillment of the queue conditions. More specifically, buffer conditions are not violated in case of non-full buffers in the $\{S \rightarrow R\}$ links. Set $\mathcal{F}_{SR}$ comprises the feasible $\{S \rightarrow R\}$ links. If $b_{SR_i} = 0$ or $q_{SR_i} = 0$, the reception of the source broadcast signal is not possible on link SR_i and thus it is considered to be in outage.

4.4. Discussion and Summary

Regarding the practical implementation of BASSA-NOMA, each relay R_k with packets in its buffer must compute the values $\alpha_{k,min}$ and $\alpha_{k,max}$, examining if a NOMA transmission is feasible. Considering the availability of $\{R \rightarrow D\}$ CSI at each R_k, the range of the α_k values can be precisely derived. After this process has been completed, if $\mathcal{F}_{RD} \neq \emptyset$, $R_k \in \mathcal{F}_{RD}$ is chosen as the transmitting relay, having the maximum number of packets in its buffer. In order to avoid relay selection with outdated CSI, as a result of centralized CSI acquisition and processing, distributed solutions should be prioritized [14]. A fully distributed method for network coordination relying on BSI has been presented in [9], based on the use of timers at the relays, setting their values to be inversely proportional to the buffer size of each relay belonging to $\mathcal{F}_{RD}$.

In addition, the use of $\{S \rightarrow R\}$ broadcasting might introduce duplicate packets among the K relays. As a countermeasure to avoid the possibility of duplicate packets, the two destinations are employed to broadcast ACKs, including the identity of the successfully received packets, thus enabling the relays, storing the same packets in their buffers to discard them. If $\mathcal{F}_{RD} = \emptyset$, no relay exists to be activated for transmission and the source broadcast transmission is performed without any IRI degrading the reception of the relays.

Furthermore, the power allocation process in NOMA requires the order with which, the packets will be decoded at each destination. Thus, the selected transmitting relay R_k must notify both destinations on the decoding strategy that they must follow. This process can be performed with an additional bit to the packet's header. In case the bit value is "0", D_1 will be directed to adopt SIC and decode its packet interference-free after decoding and subtracting the packet of D_2. Simultaneously, D_2 will attempt to decode its packet by treating the signal of D_1, as interference. The reverse strategies are followed when the bit value is "1". It must be noted that each candidate transmitting relay R_k evaluates all the possible decoding strategies and when the first one satisfying the rate requirements is identified, the decoding strategy is broadcasted to D_1 and D_2.

5. Performance Evaluation

Following the description of the system model in Section 2, BASSA-NOMA's performance is evaluated for different network parameters, in terms of outage probability, average sum-rate and average delay for the transmitted packets. In the simulations, the noise level is assumed to be equal to 1 mW while the wireless channels are independent non identically distributed (i.n.i.d). Moreover, two destinations exist, requiring equal rates $r_1 = r_2 = r_D = 2$ bps/Hz and their channel asymmetry is defined as $\sigma_{SR_k}^2 = \sigma_{R_k D_1}^2 = 4\sigma_{R_k D_2}^2$. In addition, BASSA-NOMA is evaluated under different channel conditions, in terms of line-of-sight (LoS). When LoS conditions occur, the Rician factor K_{Rice} is assumed to be equal to 10 dB.

In the following paragraphs, comparisons are presented among BASSA-NOMA and other NOMA and OMA schemes. In order to ensure fairness while comparing NOMA and OMA algorithms, the OMA transmission in each hop is performed with twice the rate, as for end-to-end communication, twice the time slots of NOMA are required, with D_1 receiving in odd time slots and D_2 in even time slots. Moreover, outages in OMA occur if at two consecutive time slots, all the relays fail to successfully transmit/receive a packet, while in NOMA, outages are experienced if at an arbitrary time slot, no relay exists to perform a successful transmission/reception.

5.1. Comparisons with Other Schemes

Here, comparisons include the OMA schemes of [14] (LoCo − Link) and [39] (OMA RD), as well as equivalent NOMA-based versions (LoCo − Link–NOMA and NOMA RD). Finally, the hybrid BA–HD–NOMA/OMA of [31] is compared, where if NOMA fails, a single user D_i is served with rate r_i, $i \in \{1, 2\}$. However, BA–HD–NOMA/OMA is excluded from the outage results, as it cannot be directly compared to standalone NOMA and OMA [29].

5.1.1. Outage Probability

The first comparison focuses on the outage probability performance, in a network with $K = 3$ relays and $L = 8$. The results are included in Figure 2. Firstly, it can be seen that among the HD schemes, the NOMA-based versions provide improved outage probability compared to their OMA equivalents, as optimal power allocation is performed, exploiting the availability of $\{R \rightarrow D\}$ CSI. Moreover, for high transmit SNR values, BASSA-NOMA outperforms the outage performance of both HD NOMA schemes. This behaviour stems from the successive $\{S \rightarrow R\}$ and $\{R \rightarrow D\}$ transmissions at each time slot, as SuR allows more packets to be distributed at the relays due to source broadcasting, while instances of full buffers and reduced diversity are avoided, as a result of the transmitting relay's NOMA transmissions at each time slot.

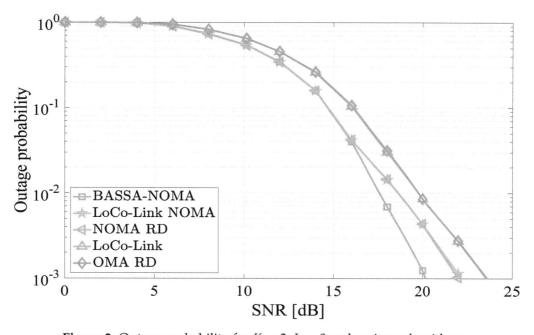

Figure 2. Outage probability for $K = 3$, $L = 8$ and various algorithms.

5.1.2. Average Sum-Rate

Regarding the average sum-rate performance, Figure 3 presents the results for different relay selection schemes, in a network with $K = 3$ relays and $L = 8$. For low transmit SNR, BA–HD–NOMA/OMA provides the best performance, as it can maintain a higher sum-rate by switching to OMA when NOMA power allocation is insufficient to improve the SIC performance. However, after 12 dB, BASSA-NOMA provides clear gains, as it is the only scheme that can achieve FD transmissions, capitalizing on the SuR principle, even if it introduces IRI to the network. More importantly, the source's broadcast transmissions, enabling $K - 1$ relays to attempt a reception improves the chances of successful $\{S \rightarrow R\}$ communication, without requiring CSI at the source, thus entailing reduced coordination overheads.

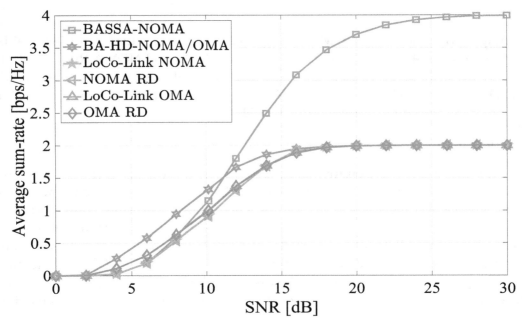

Figure 3. Average sum-rate for $K = 3$, $L = 8$ and various algorithms.

Next, the effect of increasing the number of available relays K is evaluated in Figure 4. It is evident that by increasing K, the effect of IRI in the network can be mitigated, as with each additional relay, the available degrees of freedom are increased. Moreover, it can be seen that having the minimum number of relays for SuR, i.e., $K = 2$ results in significantly worse performance, compared to $K = 3$ and $K = 4$. In this case, the network is much more vulnerable to both IRI and instances of reduced diversity when buffers are empty or full.

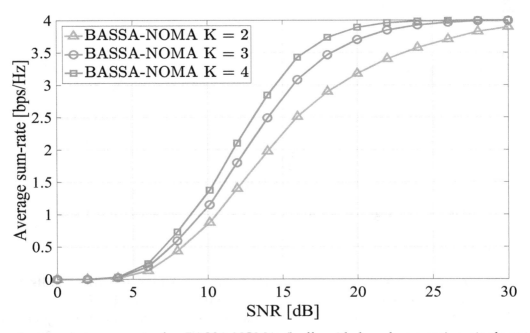

Figure 4. Average sum-rate for BASSA-NOMA (buffer-aided and successive single-antenna non-orthogonal multiple access)for varying K and $L = 8$.

Another sum-rate comparison is given in Figure 5. Here, $K = 3$ relays and $L = 8$ and each time different rate requirements are considered. From the results, it can be observed that as the transmit SNR increases, the sum-rate that can be supported by the network increases as well. Thus, it should be noted that, in highly dynamic topologies where the average SNR conditions abruptly change, there is significant potential in adopting BASSA-NOMA and adjusting its operation for adaptive rate transmissions by the source and the transmitting relay.

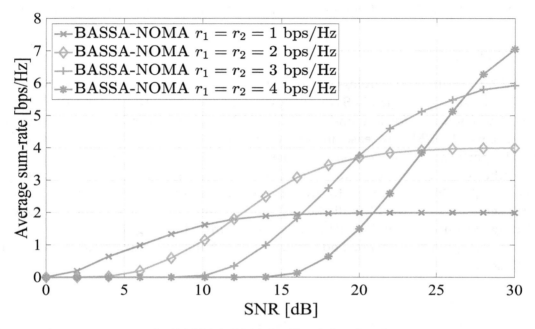

Figure 5. Average sum-rate for BASSA-NOMA for $K = 3$, $L = 8$ and varying rate requirements.

5.1.3. Average Delay

In BA relay networks, the average delay performance is of high importance, as inefficient selection schemes might lead to packets residing in the buffers for a large number of time slots. So, the final comparison in Figure 6 deals with the average delay performance of the transmitted packets for various selection schemes. In this comparison, it can be seen that all the schemes converge to an average delay of one time slot. This is the result of integrating DA mechanisms in every scheme. Nonetheless, as it was shown in the sum-rate comparisons, BASSA-NOMA transmits twice the number of packets for high transmit SNR and thus the average delay performance corresponds to an increased number of transmitted packets, compared to the HD schemes.

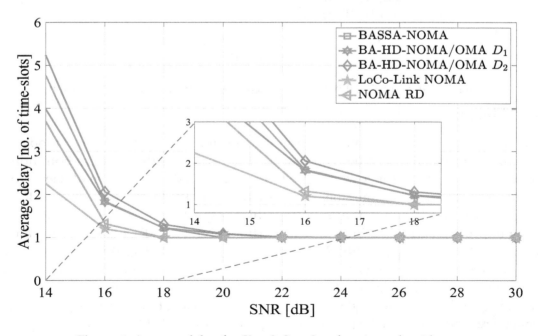

Figure 6. Average delay for $K = 3$, $L = 8$ and various algorithms.

5.2. Evaluation for Varying Channel Conditions

Departing from comparing different relay selection schemes under Rayleigh fading, we now focus on evaluating BASSA-NOMA for different wireless channel conditions, in terms of LoS connectivity in each hop.

5.2.1. Outage Probability

In Figure 7, four different wireless settings are compared, depending on the LoS conditions in each hop. It can be seen that when LoS conditions occur in the $\{R \rightarrow D_1\}$ link while the $\{S \rightarrow R\}$ links experience NLoS channels, for low and medium SNR, improved performance is observed as the chances for successful NOMA increase due to the channel asymmetry, allowing more efficient power allocation to be performed. Still, as SNR increases beyond 18 dB and thus NOMA can be efficiently performed without LoS conditions, the same performance is experienced as in the case of Rayleigh fading in both hops. On the contrary, for the case of LoS conditions only in the $\{S \rightarrow R\}$, improved diversity is observed after 18 dB. Overall, the best performance is exhibited in the topology where LoS and Rice fading exist in both the $\{S \rightarrow R\}$ and $\{R \rightarrow D_1\}$ links, limiting the instances of unsuccessful reception by the relays in the first hop and increasing the chances for successful NOMA transmission to both destinations.

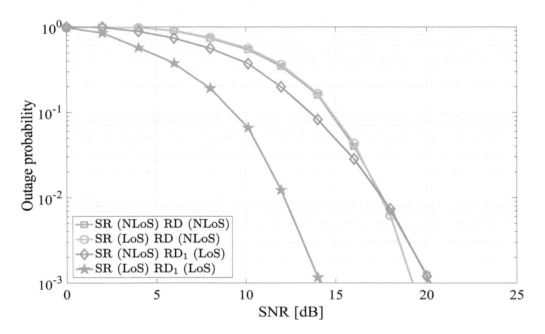

Figure 7. Outage probability for $K = 3$, $L = 8$ and varying channel conditions.

5.2.2. Average Sum-Rate

Then, Figure 8 depicts the average sum-rate performance of the four topologies where BASSA-NOMA is adopted. It can be seen that the case of Rayleigh fading in both hops provides the worst performance, as in the low and medium SNR, NOMA cannot be efficiently performed, while the source's broadcast transmissions reaches a diversity threshold in the fow SNR regime. The mixed cases of Rice and Rayleigh fading follow their respective outage probability performance, each offering improved performance for low to medium SNR ($\{S \rightarrow R\}$ with Rayleigh and $\{R \rightarrow D_1\}$ with Rice) and high SNR ($\{S \rightarrow R\}$ with Rice and $\{R \rightarrow D_1\}$ with Rayleigh). Again, the best sum-rate results are achieved in the topology with Rice fading in both the $\{S \rightarrow R\}$ and $\{R \rightarrow D_1\}$ links, thus highlighting the importance of using relays, having optimized positions in relation to the source and the destinations. This behavior strengthens the importance of using mobile relays in wireless networks, where dynamic repositioning can be performed, exploiting the CSI at the relays.

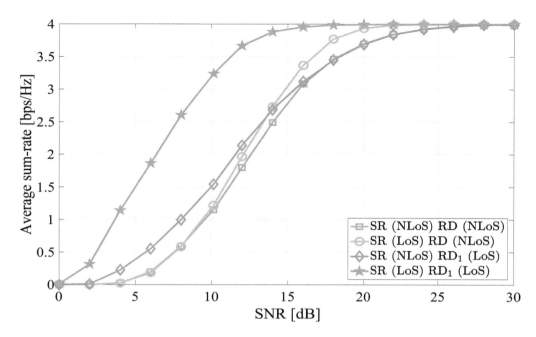

Figure 8. Average sum-rate for $K = 3$, $L = 8$ and varying channel conditions.

6. Conclusions and Future Directions

The ever-increasing desire of users to enjoy highly demanding services, in terms of rate and delay has put excessive stress on wireless networks. At the same time, recent breakthroughs have lead to the development of novel transmission techniques, improving the performance of wireless deployments. Moreover, the heterogeneity of wireless networks, comprising nodes with different hardware and processing capabilities cannot be ignored and thus efficient algorithms must be devised, offering high performance with low-complexity implementation, exploiting apart from fixed infrastructure-based relays, users and mobile relay nodes. This work focused on leveraging the capabilities provided by low-complexity single-antenna cooperative relays with buffers and non-orthogonal multiple access. More specifically, an opportunistic relay selection algorithm was developed, allowing a cluster of single-antenna relays to operate in full-duplex mode through the successive relaying principle. In addition, buffering facilitated the seamless combination of half-duplex relaying with successive source broadcasting and relay transmissions, enhancing the reliability of the two-hop NOMA relay network, and lowering the delay of the transmission, while increasing the number of transmitted packets to both destinations.

The findings in this study provide further opportunities to investigate the applicability of BASSA-NOMA in different scenarios. An important process towards improving the performance of NOMA networks is to form user clusters, in order to exploit channel asymmetries and different rate requirements. So, the integration of efficient clustering algorithms [40], improving the chances for SIC and user fairness, is of high importance. Moreover, in this work, we focused on a two-user topology and leveraged efficient power allocation for improved NOMA performance. Nonetheless, aiming to support massive connectivity requirements of 5G and beyond networks, the generalization of the BASSA-NOMA operation to topologies with arbitrary numbers of users is necessary. Moreover, the advantage of BASSA-NOMA, in terms of reliability can be beneficial to aerial networks where connectivity conditions are subject to abrupt changes. Finally, adjusting the operation of BASSA-NOMA with adaptive rate transmissions represents another interesting direction to further improve the performance of NOMA relay networks.

Author Contributions: Conceptualization, N.N.; methodology, N.N. and P.T.; validation, P.T., A.H. and S.V.; formal analysis, N.N. and P.T.; investigation, N.N., P.T. and A.H.; writing–original draft preparation, N.N. and P.T.; writing–review and editing, A.H. and S.V.

Abbreviations

The following abbreviations are used in this manuscript:

5G	Fifth Generation
ACK	Acknowledgment
AWGN	Additive White Gaussian Noise
BA	Buffer-Aided
BASSA	Buffer-Aided Successive Single-Antenna
BSI	Buffer State Information
CD	Code Domain
CSI	Channel State Information
DA	Delay-Aware
DDA	Delay- and Diversity-Aware
DF	Decode-and-Forward
FD	Full-Duplex
HD	Half-Duplex
HRS	Hybrid Relay Selection
IoT	Internet-of-Things
IRI	Inter-Relay Interference
LoCo	Low-Complexity
LoLA4SOR	Low-Latency for Successive Opportunistic Relaying
LoS	Line-of-Sight
MMRS	Max-Max Relay Selection
NACK	Negative-Acknowledgment
NLoS	Non-Line-of-Sight
NOMA	Non-Orthogonal Multiple Access
OMA	Orthogonal Multiple Access
ORS	Opportunistic Relay Selection
PD	Power Domain
RD	Relay-Destination
SD	Source-Destination
SFD	Space Full-Duplex
SIC	Successive Interference Cancellation
SINR	Signal-to-Interference-plus-Noise Ratio
SNR	Signal-to-Noise Ratio
SR	Source-Relay
SuR	Successive Relaying

References

1. Ericsson Mobility Report. Available online: https://www.ericsson.com/assets/local/mobility-report/documents/2019/ericsson-mobility-report-june-2019.pdf (accessed on 11 November 2019).
2. Ding, Z.; Liu, Y.; Choi, J.; Sun, Q.; Elkashlann, M.; Chih-Lin, I.; Poor, H.V. Application of Non-Orthogonal Multiple Access in LTE and 5G Networks. *IEEE Commun. Mag.* **2017**, *55*, 185–191. [CrossRef]
3. Dai, L.; Wang, B.; Yuan, Y.; Han, S.; Chih-Lin, I.; Wang, Z. Non-Orthogonal Multiple Access for 5G: Solutions, Challenges, Opportunities, and Future Research Trends. *IEEE Commun. Mag.* **2015**, *53*, 74–81. [CrossRef]
4. Islam, S.M.R.; Avazov, N.; Dobre, O.A.; Kwak, K.S. Power-Domain Non-Orthogonal Multiple Access (NOMA) in 5G Systems: Potentials and Challenges. *IEEE Commun. Surv. Tutor.* **2017**, *19*, 721–742. [CrossRef]
5. Ding, Z.; Lei, X.; Karagiannidis, G.K.; Schober, R.; Yuan, J.; Bhargava, V.K. A Survey on Non-Orthogonal Multiple Access for 5G Networks: Research Challenges and Future Trends. *IEEE J. Sel. Areas Commun.* **2017**, *35*, 2181–2195. [CrossRef]

6. Zlatanov, N.; Ikhlef, A.; Islam, T.; Schober, R. Buffer-Aided Cooperative Communications: Opportunities and Challenges. *IEEE Commun. Mag.* **2014**, *52*, 146–153. [CrossRef]

7. Nomikos, N.; Charalambous, T.; Krikidis, I.; Skoutas, D.N.; Vouyioukas, D.; Johansson, M.; Skianis, C. A Survey on Buffer-Aided Relay Selection. *IEEE Commun. Surv. Tutor.* **2016**, *18*, 1073–1097. [CrossRef]

8. Qiao, D.Q.; Gursoy, M.C. Buffer-Aided Relay Systems Under Delay Constraints: Potentials and Challenges. *IEEE Commun. Mag.* **2017**, *55*, 168–174. [CrossRef]

9. Nomikos, N.; Poulimeneas, D.; Charalambous, T.; Krikidis, I.; Vouyioukas, D.; Johansson, M. Delay- and Diversity-Aware Buffer-Aided Relay Selection Policies in Cooperative Networks. *IEEE Access* **2018**, *6*, 73531–73547. [CrossRef]

10. Ikhlef, A.; Michalopoulos, D.S.; Schober, R. Max-Max Relay Selection for Relays with Buffers. *IEEE Trans. Wirel. Commun.* **2012**, *11*, 1124–1135. [CrossRef]

11. Krikidis, I.; Charalambous, T.; Thompson, J.S. Buffer-Aided Relay Selection for Cooperative Diversity Systems without Delay Constraints. *IEEE Trans. Wirel. Commun.* **2012**, *11*, 1957–1967. [CrossRef]

12. Zlatanov, N.; Schober, R. Buffer-Aided Half-Duplex Relaying Can Outperform Ideal Full-Duplex Relaying. *IEEE Commun. Lett.* **2013**, *17*, 479–482. [CrossRef]

13. Oiwa, M.; Sugiura, S. Reduced-Packet-Delay Generalized Buffer-Aided Relaying Protocol: Simultaneous Activation of Multiple Source-to-Relay Links. *IEEE Access* **2016**, *4*, 3632–3646. [CrossRef]

14. Nomikos, N.; Charalambous, T.; Vouyioukas, D.; Karagiannidis, G.K. Low-Complexity Buffer-Aided Link Selection with Outdated CSI and Feedback Errors. *IEEE Trans. Commun.* **2018**, *66*, 3694–3706. [CrossRef]

15. Ikhlef, A.; Junsu, K.; Schober, R. Mimicking Full-Duplex Relaying Using Half-Duplex Relays with Buffers. *IEEE Trans. Veh. Technol.* **2012**, *61*, 3025–3037. [CrossRef]

16. Nomikos, N.; Vouyioukas, D.; Charalambous, T.; Krikidis, I.; Makris, P.; Skoutas, D.N.; Johansson, M.; Skianis, C. Joint Relay-Pair Selection for Buffer-Aided Successive Opportunistic Relaying. *Trans. Emerg. Telecommun. Technol.* **2014**, *25*, 823–834. [CrossRef]

17. Nomikos, N.; Charalambous, T.; Krikidis, I.; Skoutas, D.N.; Vouyioukas, D.; Johansson, M. A Buffer-Aided Successive Opportunistic Relaying Selection Scheme with Power Adaptation and Inter-Relay Interference Cancellation for Cooperative Diversity Systems. *IEEE Trans. Commun.* **2015**, *63*, 1623–1634. [CrossRef]

18. Nomikos, N.; Charalambous, T.; Pappas, N.; Vouyioukas, D.; Wichman, R. LoLA4SOR: A Low-Latency Algorithm for Successive Opportunistic Relaying. In Proceedings of the IEEE International Conference on Computer Communications: Workshop on Ultra-Low Latency in Wireless Networks (ULLWN), Paris, France, 29 April–2 May 2019.

19. Simoni, R.; Jamali, V.; Zlatanov, N.; Schober, R.; Pierucci, L.; Fantacci, R. Buffer-Aided Diamond Relay Network with Block Fading and Inter-Relay Interference. *IEEE Trans. Wirel. Commun.* **2016**, *15*, 7357–7372. [CrossRef]

20. Gupta, S.; Zhang, R.; Hanzo, L. Throughput Maximization for a Buffer-Aided Successive Relaying Network Employing Energy Harvesting. *IEEE Trans. Veh. Technol.* **2016**, *65*, 6758–6765. [CrossRef]

21. Shabbir, G.; Ahmad, J.; Raza, W.; Amin, Y.; Akram, A.; Loo, J.; Tenhunen, H. Buffer-Aided Successive Relay Selection Scheme for Energy Harvesting IoT Networks. *IEEE Access* **2019**, *7*, 36246–36258. [CrossRef]

22. Ding, Z.; Peng, M.; Poor, H.V. Cooperative Non-Orthogonal Multiple Access in 5G Systems. *IEEE Commun. Lett.* **2015**, *19*, 1462–1465. [CrossRef]

23. Ding, Z.; Dai, H.; Poor, H.V. Relay Selection for Cooperative NOMA. *IEEE Wirel. Commun. Lett.* **2016**, *5*, 416–419. [CrossRef]

24. Zhang, Y.; Wang, X.; Wang, D.; Zhao, Q.; Zhang, Y. A Range-Division User Relay Selection Scheme and Performance Analysis in NOMA-based Cooperative Opportunistic Multicast Systems. *Electronics* **2019**, *8*, 544. [CrossRef]

25. Liau, Q.Y.; Leow, C.Y.; Ding, Z. Amplify-and-Forward Virtual Full-Duplex Relaying-Based Cooperative NOMA. *IEEE Wirel. Commun. Lett.* **2018**, *7*, 464–467. [CrossRef]

26. Wang, S.; Cao, S.; Ruby, R. Optimal Power Allocation in NOMA-Based Two-Path Successive AF Relay Systems. *EURASIP J. Wirel. Commun. Netw.* **2018**, *2018*, 273. [CrossRef]

27. Liau, Q.Y.; Leow, C.Y. Successive User Relaying in Cooperative NOMA System. *IEEE Wirel. Commun. Lett.* **2019**, *8*, 921–924. [CrossRef]

28. Luo, S.; Teh, K.C. Adaptive Transmission for Cooperative NOMA System with Buffer-Aided Relaying. *IEEE Commun. Lett.* **2017**, *21*, 937–940. [CrossRef]

29. Zhang, Q.; Liang, Z.; Li, Q.; Qin, J. Buffer-Aided Non-Orthogonal Multiple Access Relaying Systems in Rayleigh Fading Channels. *IEEE Trans. Commun.* **2017**, *65*, 95–106. [CrossRef]

30. Nomikos, N.; Charalambous, T.; Vouyioukas, D.; Karagiannidis, G.K.; Wichman, R. Relay Selection for Buffer-Aided Non-Orthogonal Multiple Access Networks. In Proceedings of the IEEE GLOBECOM: Workshop on Non-Orthogonal Multiple Access Techniques for 5G, Singapore, 4–8 December 2017.

31. Nomikos, N.; Charalambous, T.; Vouyioukas, D.; Karagiannidis, G.K.; Wichman, R. Hybrid NOMA/OMA with Buffer-Aided Relay Selection in Cooperative Networks. *IEEE J. Sel. Top. Signal Process.* **2019**, *13*, 524–537. [CrossRef]

32. Nomikos, N.; Michailidis, E.T.; Trakadas, P.; Vouyioukas, D.; Zahariadis, T.; Krikidis, I. Flex-NOMA: Exploiting Buffer-Aided Relay Selection for Massive Connectivity in the 5G Uplink. *IEEE Access* **2019**, *7*, 88743–88755. [CrossRef]

33. Dun, H.; Ye, F.; Jiao, S.; Liu, D. Power Control for Device-to-Device Communication with a Hybrid Relay Mode in Unequal Transmission Slots. *Electronics* **2018**, *7*, 17. [CrossRef]

34. Khan, U.A.; Lee, S.S. Three-Dimensional Resource Allocation in D2D-Based V2V Communication. *Electronics* **2019**, *8*, 962. [CrossRef]

35. Michailidis, E.T.; Nomikos, N.; Bithas, P.S.; Vouyioukas, D.; Kanatas, A.G. Optimal 3-D Aerial Relay Placement for Multi-User MIMO Communications. *IEEE Trans. Aerosp. Electron. Syst.* **2019**. [CrossRef]

36. Singh, K.; Gupta, A.; Ratnarajah, T.; Ku, M. A General Approach Toward Green Resource Allocation in Relay-Assisted Multiuser Communication Networks. *IEEE Trans. Wirel. Commun.* **2018**, *17*, 848–862. [CrossRef]

37. Singh, K.; Gupta, A.; Ratnarajah, T. A Utility-Based Joint Subcarrier and Power Allocation for Green Communications in Multi-User Two-Way Regenerative Relay Networks. *IEEE Trans. Commun.* **2017**, *65*, 3705–3722. [CrossRef]

38. Krikidis, I.; Thompson, J.; McLaughlin, S.; Goertz, N. Amplify-and-Forward with Partial Relay Selection. *IEEE Commun. Lett.* **2008**, *12*, 235–237. [CrossRef]

39. Tian, Z.; Gong, Y.; Chen, G.; Chambers, J.A. Buffer-Aided Relay Selection with Reduced Packet Delay in Cooperative Networks. *IEEE Trans. Veh. Technol.* **2017**, *66*, 2567–2575. [CrossRef]

40. Singh, K.; Wang, K.; Biswas, S.; Ding, Z.; Khan, F.A.; Ratnarajah, T. Resource Optimization in Full Duplex Non-Orthogonal Multiple Access Systems. *IEEE Trans. Wirel. Commun.* **2019**, *18*, 4312–4325. [CrossRef]

12

A Novel High Gain Wideband MIMO Antenna for 5G Millimeter Wave Applications

Daniyal Ali Sehrai [1], Mujeeb Abdullah [2], Ahsan Altaf [3], Saad Hassan Kiani [1],
Fazal Muhammad [1,*], Muhammad Tufail [4], Muhammad Irfan [5,*], Adam Glowacz [6]
and Saifur Rahman [5]

[1] Department of Electrical Engineering, City University of Science and Information Technology, Peshawar 25000, Pakistan; danyalkhan134@gmail.com (D.A.S.); saad.kiani@cusit.edu.pk (S.H.K.)

[2] Department of Computer Science, Bacha Khan University, Charsadda 24420, Pakistan; mujeeb.abdullah@gmail.com

[3] Department of Electrical Engineering, Istanbul Medipol University, Istanbul 34083, Turkey; aaltaf@st.medipol.edu.tr

[4] Department of Mechatronics Engineering, University of Engineering and Technology, Peshawar 25000, Pakistan; tufail@uetpeshawar.edu.pk

[5] Electrical Engineering Department, College of Engineering, Najran University Saudi Arabia, Najran 61441, Saudi Arabia; srrahman@nu.edu.sa

[6] Department of Automatic, Control and Robotics, AGH University of Science and Technology, 30-059 Krakow, Poland; adglow@agh.edu.pl

* Correspondence: fazal.muhammad@cusit.edu.pk (F.M.); miditta@nu.edu.sa (M.I.)

Abstract: A compact tree shape planar quad element Multiple Input Multiple Output (MIMO) antenna bearing a wide bandwidth for 5G communication operating in the millimeter-wave spectrum is proposed. The radiating element of the proposed design contains four different arcs to achieve the wide bandwidth response. Each radiating element is backed by a 1.57 mm thicker Rogers-5880 substrate material, having a loss tangent and relative dielectric constant of 0.0009 and 2.2, respectively. The measured impedance bandwidth of the proposed quad element MIMO antenna system based on 10 dB criterion is from 23 GHz to 40 GHz with a port isolation of greater than 20 dB. The measured radiation patterns are presented at 28 GHz, 33 GHz and 38 GHz with a maximum total gain of 10.58, 8.87 and 11.45 dB, respectively. The high gain of the proposed antenna further helps to overcome the atmospheric attenuations faced by the higher frequencies. In addition, the measured total efficiency of the proposed MIMO antenna is observed above 70% for the millimeter wave frequencies. Furthermore, the MIMO key performance metrics such as Mean Effective Gain (MEG) and Envelope Correlation Coefficient (ECC) are analyzed and found to conform to the required standard of MEG < 3 dB and ECC < 0.5. A prototype of the proposed quad element MIMO antenna system is fabricated and measured. The experimental results validate the simulation design process conducted with Computer Simulation Technology (CST) software.

Keywords: 5G; MIMO; wideband; high isolation; envelope correlation coefficient

1. Introduction

In recent years, Fifth Generation (5G) has acquired a lot of attention in the field of wireless communication. The reason behind the great interest towards the development of the 5G technology is the rapid increase in mobile phone traffic, demanding a higher data rate and bandwidth [1]. So far, the implementation up to 4G technology has been achieved. But these advancements in the technology couldn't fulfil the demand of higher data rate and bandwidth of the modern period [2]. The mobile data traffic generated from video streaming, social applications and cloud services etc., will probably go beyond the potentials of the current 4G infrastructure before 2020 [3]. Therefore, research has started on the Fifth Generation (5G) technology. After much efforts by researchers, 5G is now being standardized by most countries to accomplish the need for the higher bandwidth and data rates, which is itself a challenging task [4]. To solve this problem, the Multiple Input Multiple Output (MIMO) technologies with a wide bandwidth characteristic are crucial to improve the spectrum efficiency and channel capacity by utilizing the multipath property with no need for increasing the input power [5,6]. Furthermore, the characteristics of high element isolation and broadband should be possessed by the MIMO system to contribute promising performance [7,8]. The higher mutual coupling between the MIMO antenna elements would affect the throughput of the MIMO antenna system [9,10]. Thus, to design a MIMO antenna system with a high element isolation is also a challenge.

The centimeter and millimeter wave spectrum (3–300 GHz) have been mostly targeted by the 5G technology, which can further help to achieve the higher bandwidth with a data rate up to several Gigabit-per-second (Gbps) [11,12]. Another reason for choosing this spectrum range is that the lower spectrum's portion is already under the use of several wireless networks and applications like wireless fidelity (Wi-Fi), Worldwide Interoperability for Microwave Access (WiMAX), Bluetooth, Industrial, Scientific and Medical (ISM), and mobile communication, etc., while most of the higher portion of the spectrum is still not utilized, and can be exercised for 5G technology [13].

However, focusing on the higher portion of the spectrum has also raised some challenges for this 5G technology. One of the challenges is the free space propagation of these frequencies, as the signals at the lower frequency propagate for more than tens of miles and can easily penetrate through high tall buildings and trees. On the other hand, the signals at higher frequency bands can travel only a few miles and cannot penetrate through dense materials very well, resulting a lower coverage area. Nevertheless, these properties are not essentially being disadvantaged [14]. These propagation losses can be exploited to increase the frequency reuse, by introducing the small cell base stations known as pico-cells and femtocells. Another challenge in the wireless communication at the higher frequencies is the rain and the atmosphere which make these higher frequencies impractical [15,16]. In other words, these frequencies are badly affected by rain, snowfall and fog, etc. Thus, the electromagnetic (EM) waves experience higher losses in terms of signal quality and strength, etc., by these atmospheric attenuations [17]. However, this problem can be resolved by designing the antenna, which is highly directive and possesses a high gain [18,19].

In the literature [20–28], different MIMO antennas have been reported; covering the wide range of frequencies of the 5G targeted spectrum. Some of them have proposed the MIMO configuration for lower portion of spectrum i.e., below 6 GHz while above 20 GHz frequency, the MIMO configuration has also been presented. In [20], the four-antenna structure is implemented, covering the 5G mm-wave frequency band from 25.5–29.6 GHz with a peak gain of 8.3 dB. Similarly, a MIMO array with an effective bandwidth of 3.4 to 3.6 GHz at −6 dB reference is proposed in [21] for 5G applications. The total volume of the proposed antenna is $145 \times 75 \times 6$ mm^3. Moreover, the MIMO antenna system with a total volume of $90 \times 90 \times 1.6$ mm^3 is proposed in [22]. The proposed antenna possesses the impedance bandwidth of 3 to 9 GHz. In [23], a dual band MIMO antenna array having dimensions of $150 \times 75 \times 7$ mm^3 is presented, with a bandwidth of 3.4 to 3.6 GHz and 4.8 to 5.1 GHz at −6 dB reference. Likewise, a MIMO antenna array for millimeter wave communication with a SIW fed slotted is presented in [24]. The proposed antenna covers the 27.5–28.35 GHz and 24.25–27.5 GHz frequency bands for 5G, whereas the gain alters from 8.2 to 9.6 dB over the desired wideband region. In [25], a high gain MIMO antenna for 5G applications is presented, which covers the frequency band ranging from 26–29.5 GHz. The peak gain achieved is 14 dB. Similarly, in [26], a 5G metamaterial-based antenna for MIMO systems with a maximum gain value of 7.4 dB at the 26 GHz frequency band is presented. A broadband MIMO antenna with an impedance bandwidth of 2.6 to 13 GHz is reported in [27], while the overall size of the proposed antenna is $66.8 \times 40 \times 0.8$ mm^3. Likewise, in [28], the four element MIMO antenna system covering the 5G frequency band 27.5 to 40 GHz with an overall size of 158×77.8 mm^2 is proposed. It is observed from the above literature review that the

MIMO configuration presented is either large or complex in structure. Furthermore, the reported antennas [20–28] possess poor bandwidth due to which the number of frequency channels gets limited, while some of them also achieve low gain at the desired frequency band, which is an important factor at the mm-wave spectrum.

In this paper, a quad element MIMO antenna for 5G mm-wave applications is demonstrated. The proposed antenna possesses a wideband and high gain with a good MIMO characteristic for 5G millimeter wave applications. The wideband of the antenna is further helpful to achieve a high data rate transmission. The radiating elements of the proposed MIMO antenna design consist of four different arcs which mainly contribute in achieving a wideband performance. Moreover, the proposed MIMO configuration antenna elements are orthogonally assembled to each other, while the elements in diagonal position are assembled in the anti-parallel mode to lower the mutual interference between the MIMO antenna elements. Furthermore, to ensure that the proposed MIMO antenna system contains the same voltage; the antenna elements ground surfaces are connected. The remaining research work is sequenced in the following way. Section 2 presents the geometry of the proposed MIMO antenna and discussion is also made on the design evolution steps of the antenna element. Results are discussed in Section 3, while Section 4 concludes the paper.

2. Antenna Geometric Configuration and Design

The proposed quad element MIMO antenna is printed on RT-5880 substrate with a loss tangent and relative permittivity of 0.0009 and 2.2, respectively, whereas its thickness is 1.57 mm. The geometrical layout of the proposed MIMO configuration is illustrated in Figure 1, while Table 1 provides the dimensions of different design parameters. The proposed MIMO configuration consists of four antenna elements. A finite ground plane made up of copper with a size of 80 mm × 80 mm is used to back the substrate. Copper with a very stable conductivity of 5.8×10^7 S/m is used for the radiating element. Due to the very stable conductivity of copper, its effect on the impedance matching is very low. All the simulations and modeling of the proposed MIMO configuration are carried out in the CST Microwave Studio software.

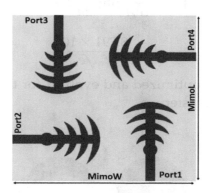

Figure 1. Proposed wideband multiple-input multiple-output (MIMO) antenna geometric layout.

Table 1. Proposed wideband antenna element dimensions.

Parameter	Value (mm)	Parameter	Value (mm)
SubW	40	Arc4	6.0
SubL	40	FeedL	12.76
FeedW	4.0	FW	2.15
Arc1	15	FL	2.54
Arc2	15	Rad	3.0
Arc3	9.0	MimoL	80
MimoW	80	–	–

2.1. Single Antenna Element Design

The schematic of the unit antenna element with a size of 40×40 mm^2, as illustrated in Figure 2, serves as a building block for the four port or quad element MIMO antenna. The substrate of the antenna element is backed by a full ground plane to reduce to the flow of the antenna radiated waves in the backward direction and to achieve a maximum gain.

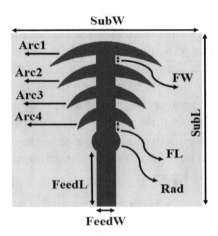

Figure 2. Layout of the proposed wideband antenna element.

The number of arcs and their dimensions must be carefully chosen to achieve the proper wideband response for the desired frequency range. The arc-shaped stripe is evolved from a circular patch as shown in Figure 3. First of all, a simple circular patch geometry is approximated by using the standard theory of circular patch antenna [29], i.e.,

$$rad_1 = \frac{F}{\{1 + \frac{2h}{\pi \epsilon_r F}[\ln(\frac{\pi F}{2h}) + 1.7726]\}^{1/2}} \tag{1}$$

$$F = \frac{8.791 \times 10^9}{f_r \sqrt{\epsilon_r}} \tag{2}$$

Then, the arc-shaped stripe is optimized and evolved for the targeted mm-wave spectrum by subtracting the circular patch step by step.

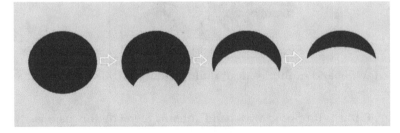

Figure 3. Arc-shaped stripe design evolution from a circular patch.

The evolution of the antenna element for the proposed MIMO system is discussed, as depicted in Figure 4. The design process consists of four antennas, i.e., Antenna-A, Antenna-B, Antenna-C and Antenna-D. Figure 4a illustrates the geometry of Antenna-A which is the basic monopole structure widely used in variant shapes as reported in [30–34], while Figure 4d depicts the geometry of Antenna-D; the proposed unit antenna which is used for the quad element MIMO antenna. The reflection coefficient (S_{11}) at -10 dB reference for these different design stages is depicted in Figure 5.

At the first stage, Antenna-A with just single arc-shaped stripe is designed on the top of the substrate as shown in Figure 4a, which shows an approximately wideband response above 31 GHz, whereas, below that frequency, a single resonance is obtained at the 28.5 GHz frequency band. Thus, at the second stage, Antenna-B with two arc-shaped stripes is produced to improve the response of the antenna element as illustrated in Figure 4b. This time a wideband response is observed for the frequency band of 23 to 34.8 GHz and 36.5 GHz to onwards. To improve the response further, Antenna-C with three arc-shaped stripes is introduced at stage three as depicted in Figure 4c, which gives a satisfactory wideband ranging from 23.2 to 39.2 GHz. Finally, at stage four, Antenna-D with four arc-shaped stripes is designed as shown in Figure 4d, to observe the effect on wideband response obtained at stage three. However, it is worth mentioning that the arc-shaped stripe introduced at stage four does not have a major effect on the wideband response achieved at stage three. Thus, a final geometry containing four different arc-shaped stripes is achieved, shown in Figure 4d for the MIMO configuration. Figure 6a–c illustrate the surface current distribution at the 28 GHz, 33 GHz and 38 GHz frequency bands, respectively. It is observed that the entire effective resonant length of the proposed antenna element is responsible for achieving the wideband characteristics.

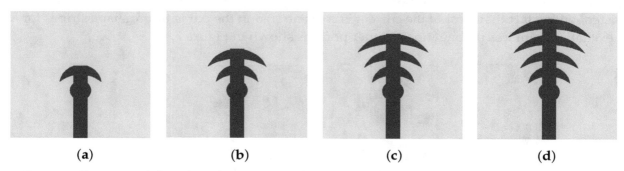

(a)   **(b)**   **(c)**   **(d)**

Figure 4. Geometrical design evolution steps of the proposed wideband antenna element. (**a**) step 1 Antenna-A). (**b**) step 2 (Antenna-B). (**c**) step 3 (Antenna-C). (**d**) step 4 (Antenna-D).

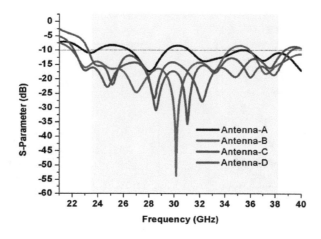

Figure 5. Reflection coefficient comparison of the different geometrical design evolution steps.

It is evident from the comparison of the simulated scattering-parameter (reflection coefficient) or (S_{11}) that the unit antenna exhibits improvement in the impedance bandwidth by adding additional arch-shaped stripes. It is evident that bending, meandering and tapering the monopole antenna design gives excellent properties such as compactness and less complex structure having multiple frequency bands at a reasonable cost of production [30–34]. However, the reported structure can be further improved to accommodate the higher frequency bands for 5G communication systems. Thus, inspired by the related work [30–34], a new tree-shaped quad element MIMO antenna is conceived.

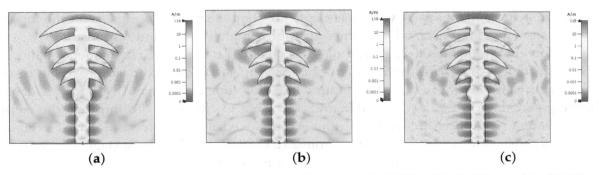

Figure 6. Wideband antenna element current distribution at (**a**) 28 GHz, (**b**) 33 GHz, and (**c**) 38 GHz.

2.2. Integeration of Four Unit Antenna for MIMO Configuration

In this section, a quad element MIMO antenna is built as depicted in Figure 1, by using the unit antenna element design obtained in the previous section. The individual antenna element occupies an area of 40×40 mm^2, are placed symmetrical and rotational in the 90-degree interval, forming a square shape. The ports isolation of greater than 20 dB is achieved for the proposed MIMO antenna. This is consistent with a fact that most of the current is concentrated in the parasitic arc shape stripes for each port excitation with less propagation to other ports as shown in Figure 7.

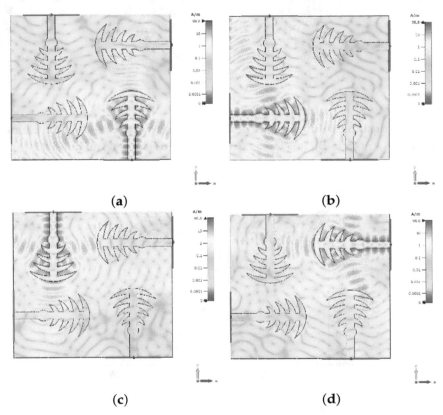

Figure 7. MIMO antenna system current distribution at 28 GHz for (**a**) port 1, (**b**) port 2, (**c**) port 3, and (**d**) port 4.

The mutual interference among the radiating elements makes the designing of the quad port diversity antenna complex task. The presence of manifold similar elements in the MIMO configuration leads to multiple increases in the mutual interference and Envelope Correlation Coefficient (ECC) among different antenna elements. That is why the placement of diagonal elements in the anti-parallel mode for the quad element MIMO antenna is chosen, while the four elements are assembled orthogonally to each other. To ensure that the proposed MIMO wideband antenna ground plane contains the same voltage; the four monopole resonating elements ground surfaces are allied.

The prototype of the fabricated MIMO wideband 5G antenna is revealed in Figure 8. The surface current distribution of the four-port MIMO antenna on the excitation of all ports sequentially at the frequency 28 GHz is depicted in Figure 7. It is observed that the current flow is mainly concentrated along the edges of arcs and around the feedline of the proposed MIMO antenna elements. Moreover, the concentration of the coupling current among the elements of MIMO antenna is insignificant.

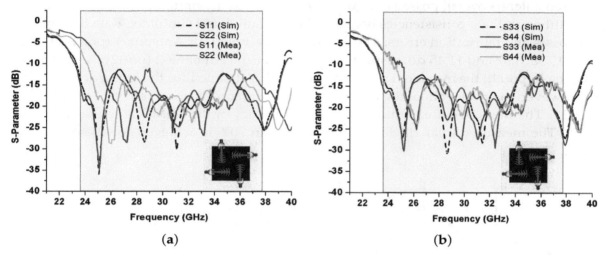

Figure 8. Proposed wideband MIMO antenna reflection coefficients: (a) port 1 and 2; (b) port 3 and 4.

3. Measured Results

The detailed analysis of the measured and simulated results is performed in this section. The southwest RF connector is used to feed the MIMO antenna elements. The antenna measurements are carried out using Agilent 8722ES vector network analyzer. While measuring transmission and reflection coefficient for the fabricated quad element MIMO antenna, the idle ports are terminated in 50 Ω load. The same procedure is adopted for the radiated far-field measurement in the anechoic chamber.

3.1. S-Parameters

A good coherence in the simulated and measured reflection coefficients is observed with a slight shift in the frequency bands due to the use of cables during the measurement or fabrication losses as shown in Figure 8. It is noted that the impedance bandwidth based on −10 dB criterion of the proposed MIMO antenna is 23 to 40 GHz. The mutual interference among the MIMO antenna elements is well above 20 dB for the entire desired wideband region, as depicted in Figure 9.

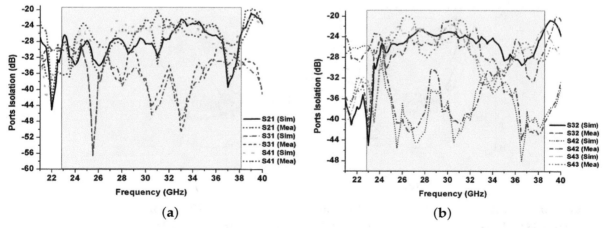

Figure 9. Proposed wideband MIMO antenna S-parameters at (a) port 1 and (b) other port.

3.2. Far Field Measurment

The measured radiation patterns for two principal planes, namely E-plane and H-plane at 28 GHz, 33 GHz and 38 GHz are shown in Figure 10. The radiating elements of the proposed MIMO antenna by virtue of its orthogonal placement have pattern diversity which is helpful to mitigate the multipath effect for communication systems. The MIMO antenna system exhibits directional radiation patterns. The proposed antenna overall possesses a good resemblance in the simulated and measured radiation patterns. Although the inconsistencies between the measured and simulated data are noted due to the cable losses and fabrication errors. The measured gain noted for Antenna-1 (port1) at 28, 33 and 38 GHz is 10.58, 8.87 and 11.45 dB, respectively while for the Antenna-2 (port2), Antenna-3 (port3) and Antenna-4 (port4); the measured gain of above 9 dB is obtained for the entire desired wideband region. However, the peak gain of the proposed MIMO antenna is observed at 12 dB, as illustrated in Figure 11. The measured and simulated totally efficiency of the MIMO antenna is presented in Figure 12. The measured total efficiency of greater than 70% is achieved for the overall desired wideband.

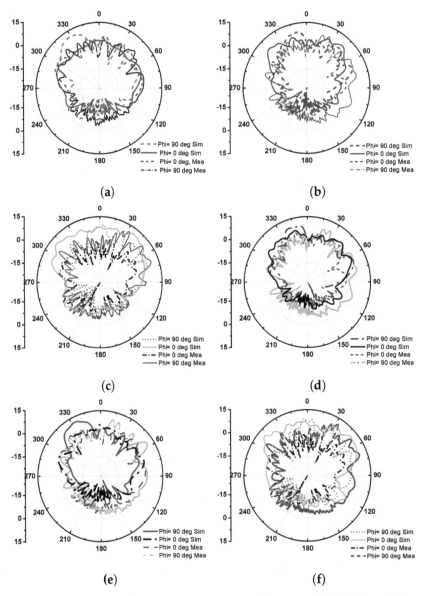

Figure 10. Radiation patterns of MIMO antenna: (**a**) port-1/28 GHz; (**b**) port-1/33 GHz; (**c**) port-1/38 GHz; (**d**) port-2/28 GHz; (**e**) port-2/33 GHz; (**f**) port-2/38 GHz.

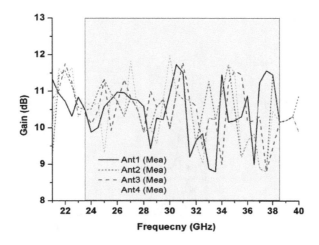

Figure 11. Maximum gain over frequency of the MIMO antenna elements.

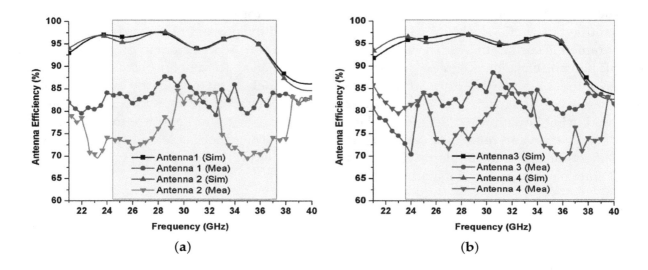

Figure 12. Efficiencies of the proposed MIMO antenna system: (**a**) antenna 1 and 2; (**b**) antenna 3 and 4.

3.3. MIMO Performance Metrics for the Quad Antenna System

The essential MIMO performance metrics such as the Mean Effective Gain (MEG) and Envelope Correlation Coefficient (ECC) for the proposed MIMO antenna system are discussed in this section. Figure 13 shows the ECC between the different ports of the MIMO antenna. The following expression [35] can be employed to compute the ECC between port 1 and port 2 of the quad element MIMO antenna:

$$\rho_{eij} = \frac{|S_{ii} \times S_{ij} + S_{ji} \times S_{jj}|^2}{(1 - |S_{ii}|^2 - S_{ij}^2)(1 - |S_{ji}|^2 - S_{jj}^2)} \tag{3}$$

Likewise, the ECC between the MIMO antenna other ports can be computed as well. It is observed that the ECC is below 0.0014 for the entire wideband region, obeying the practical standard of <0.5 required for the optimal diversity performance and ensuring independent channel operation.

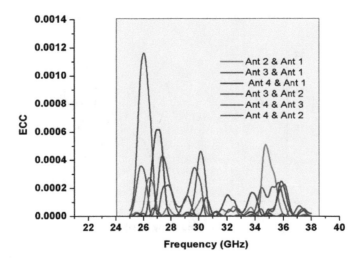

Figure 13. Envelope Correlation Coefficients (ECCs) between some adjacent antenna-elements.

The antenna elements mutual interaction and statistical properties of the propagation environment are quantified by MEG. The MEG is helpful to understand the power imbalance by considering the vital parameters such as total efficiency, gain and propagation environment in multiple branches or multiple antenna elements, degrading the diversity performance. To fulfil the balance power standard and for optimal diversity performance with a good channel characteristic; the difference between MEGs of any two antennas equal should be less than 3 dB. The numerically estimated values of MEG calculated using Equation (4), [20] are tabulated in Table 2. The terms i and k represent the antenna under observation and number of antennas, respectively. The MEG calculated based on measured results meet the required standard with the ratio of any two MEGs of the proposed quad element MIMO antenna being nearly equal to 1.

$$MEG_i = 0.5(1 - \sum_{j=1}^{k} S_{ij}) \tag{4}$$

Table 2. MEG for Antenna-1 to Antenna-4.

Frequency (GHz)	XPR	MEG Ant.1 (dB)	MEG Ant.2 (dB)	MEG Ant.3 (dB)	MEG Ant.4 (dB)
28	1	−5.24	−5.36	−5.45	−5.33
	6	−6.13	−6.34	−6.53	−6.64
33	1	−5.32	−5.43	−5.34	−5.37
	6	−6.63	−6.43	−6.33	−6.67
38	1	−5.88	−5.83	−5.65	−5.56
	6	−6.87	−6.31	−6.26	−6.34

The comparison of the proposed wideband MIMO antenna for 5G applications and other antennas reported in the literature is presented in Table 3. It is observed from the comparison that the proposed MIMO antenna owns numerous advantages over the previously reported MIMO antennas [20–28], in terms of total efficiency, impedance bandwidth, isolation among MIMO antenna elements, number of radiating elements, and gain. Furthermore, the radiating elements are assembled in the anti-parallel mode and orthogonally to provide better isolation among wireless elements, while the proposed quad element MIMO antenna possesses a common ground plane to achieve a stable operation.

Table 3. Performance comparison with previous published literatures.

Ref.	Ports	Bandwidth (GHz)	Peak Gain (dB)	Size (mm^3)	Isolation (dB)	Total Eff. (%)	ECC
[20]	4	25.5–29.6	8.3	30 × 35 × 0.76	>10	80–85	<0.01
[21]	8	3.4–3.6	1.6–4.5	145 × 75 × 6	>15	42–73	<0.16
[22]	4	3-9	11–12	90 × 90 × 1.6	>13	–	–
[23]	8	3.4–3.6, 4.8–5.1	–	150 × 75 × 7	>11.5	48–85	<0.08
[24]	–	21–34	9	–	–	–	–
[25]	4	26–29.5	14	19 × 19 × 7.608	>20	–	<0.015
[26]	–	24–28	7.4	30 × 30.5 × 0.508	–	–	–
[27]	2	2.6–13	0.76–6.02	66.8 × 40 × 0.8	>15	>75	<0.02
[28]	4	27.5–40	5.8–7.2	158 × 77.8 × 0.381	>17	<75	<0.001
Prop.	4	23–40	12	80 × 80 × 1.57	>20	>70	<0.0014

4. Conclusions

In this paper, a quad element MIMO antenna for 5G mm-wave applications is demonstrated. The proposed antenna possesses a wideband and high gain with a good MIMO characteristic for 5G mm-wave applications. The operation band of the proposed antenna covers from 23 GHz to 40 GHz. Each radiating element of the proposed MIMO configuration contains four different arc-shaped stripes which help to achieve the wide bandwidth and high element isolation of more than 20 dB. The peak gain of 12 dB is achieved for the proposed MIMO antenna. The high gain of the proposed antenna can be further helpful to overcome the atmospheric attenuations face by the higher frequencies. The mean effective gain and envelope correlation coefficient are also provided for the proposed MIMO configuration. In addition, the measured total efficiency of the proposed MIMO antenna is observed to be above 70% for the desired millimeter wave frequencies. Apart from this, the prototype of the MIMO wideband antenna is fabricated and tested. A good coherence between the experimental and simulated results is achieved. The proposed MIMO antenna operates efficiently with a significant return loss, wide bandwidth, high gain and high element isolation, which make it a potential candidate for 5G mm-wave applications.

Author Contributions: Conceptualization, D.A.S. and S.H.K.; methodology, D.A.S.; Software, D.A.S., S.H.K. and A.A.; Validation, M.A., F.M.; Formal Analysis, F.M., M.A.; Investigation, F.M., M.I., A.G., S.R.; Resources, A.A., M.T., M.I., S.R., and A.G.; Data Curation, D.A.S.; writing—original draft preparation, D.A.S.; writing—review and editing, D.A.S., M.A., S.H.K.; F.M., and M.T., Visualization, D.A.S; supervision, M.A.; project administration, M.A., M.I., and A.G. All authors have read and agreed to the published version of the manuscript.

References

1. Sharaf, M.H.; Zaki, A.I.; Hamad, R.K.; Omar, M.M.M. A novel dual-band (38/60 GHz) patch antenna for 5G mobile handsets. *Sensors* **2015**, *20*, 2541. [CrossRef]
2. Haroon, M. S.; Muhammad, F.; Abbas, G.; Abbas, Z.H.; Kamal, A.; Waqas, M.; Kim, S. Interference Management in Ultra-Dense 5G Networks with Excessive Drone Usage. *IEEE Access* **2020**, 1–10. [CrossRef]
3. Khan, J.; Sehrai, D.A.; Ali, U. Design of dual band 5G antenna array with SAR analysis for future mobile handsets. *J. Electr. Eng. Technol.* **2019**, *14*, 809–816. [CrossRef]
4. Pervez, M.M.; Abbas, Z.H.; Muhammad, F.; Jiao, L. Location-based coverage and capacity analysis of a two tier HetNet. *IET Commun.* **2017**, *11*, 1067–1073 [CrossRef]
5. Sun, L.; Li, Y.; Zhang, Z.; Feng, Z. Wideband 5G MIMO antenna with integrated orthogonal-mode dual-antenna pairs for metal-rimmed smartphones. *IEEE Trans. Antennas Propag.* **2020**, *68*, 2494–2503. [CrossRef]
6. Abdullah, M.; Kiani, S.H.; Iqbal, A. Eight element multiple-input multiple-output (MIMO) antenna for 5G mobile applications. *IEEE Access* **2019**, *7*, 134488–134495. [CrossRef]

7.   Yuan, X.; He, W.; Hong, K.; Han, C.; Chen, Z.; Yuan, T. Ultra-wideband MIMO antenna system with high element-isolation for 5G smartphone application. *IEEE Access* **2020**, *8*, 56281–56289. [CrossRef]

8.   Altaf, A.; Alsunaidi, M.A.; Arvas, E. A novel EBG structure to improve isolation in MIMO antenna. In Proceedings of the IEEE USNC-URSI Radio Science Meeting (Joint with AP-S Symposium), San Diego, CA, USA, 9–14 July 2017; pp. 105–106.

9.   Wang, F.; Duan, Z.; Wang, X.; Zhou, Q.; Gong, Y. High isolation millimeter-wave wideband MIMO antenna for 5G communication. *Int. J. Antennas Propag.* **2019**, *2019*, 4283010. [CrossRef]

10. Abdullah, M.; Kiani, S.H.; Abdulrazak, L.F.; Iqbal, A.; Bashir, M.A.; Khan, S.; Kim, S. High-performance multiple-input multiple-output antenna system for 5G mobile terminals. *Electronics* **2019**, *8*, 1090. [CrossRef]

11. Haroon, M.S.; Abbas, Z.H.; Abbas, G.; Muhammad, F. SIR analysis for non-uniform HetNets with Joint decoupled association and interference management. *Comput. Commun.* **2020**, *155*, 48–57.

12. Haroon, M.S.; Abbas, Z.H.; Muhammad, F.; Abbas, G. Coverage Analysis of Cell Edge Users in Heterogeneous Wireless Networks using Stienen's Model and RFA scheme. *Int. J. Commun. Syst.* **2019**, *33*, e4147. [CrossRef]

13. Khan, J.; Sehrai, D.A.; Khan, M.A.; Khan, H.A.; Ahmad, S.; Ali, A.; Arif, A.; Memon, A.A.; Khan, S. Design and performance comparison of rotated Y-shaped antenna using different metamaterial surfaces for 5G mobile devices. *Comput. Mater. Contin.* **2019**, *60*, 409–420. [CrossRef]

14. Wang, P.; Li, Y.; Song, L.; Vucetic, B. Multi-gigabit millimeter waves wireless communications for 5G: From fixed access to cellular networks. *IEEE Commun. Mag.* **2015**, *53*, 168–178. [CrossRef]

15. Sulyman, A.I.; Alwarafy, A.; MacCartney, G.R.; Rappaport, T.S.; Alsanie, A. Directional radio propagation path loss models for millimeter-wave wireless networks in the 28-, 60-, and 73-GHz bands. *IEEE Trans. Wirel. Commun.* **2016**, *15*, 6939–6947. [CrossRef]

16. Shayea, I.; Rahman, T.A.; Azmi, M.H.; Islam, M.R. Real measurement study for rain rate and rain attenuation conducted over 26 GHz microwave 5G link system in malaysia. *IEEE Access* **2018**, *6*, 19044–19064. [CrossRef]

17. Zhang, J.; Ge, X.; Li, Q.; Guizani, M.; Zhang, Y. 5G millimeter-wave antenna array: design and challenges. *IEEE Wirel. Commun.* **2017**, *24*, 106–112. [CrossRef]

18. Roh, W.; Seol, J.Y.; Park, J.; Lee, B.; Lee, J.; Kim, Y.; Cho, J.; Cheun, K.; Aryanfar, F. Millimeter-wave beamforming as an enabling technology for 5G cellular communications: Theoretical feasibility and prototype results. *IEEE Commun. Mag.* **2014**, *52*, 106–113. [CrossRef]

19. Khalily, M.; Tafazolli, R.; Xiao, P.; Kishk, A.A. Broadband mm-Wave microstrip array antenna with improved radiation characteristics for different 5G applications. *IEEE Trans. Antennas Propag.* **2018**, *66*, 4641–4647. [CrossRef]

20. Khalid, M.; Iffat Naqvi, S.; Hussain, N.; Rahman, M.; Mirjavadi, S.S.; Khan, M. J.; Amin, Y. 4-port MIMO antenna with defected ground structure for 5G millimeter wave applications. *Electronics* **2020**, *9*, 71. [CrossRef]

21. Liu, Y.; Ren, A.; Liu, H.; Wang, H.; Sim, C. Eight-port MIMO array using characteristic mode theory for 5G smartphone applications. *IEEE Access* **2019**, *7*, 45679–45692. [CrossRef]

22. Haq, M.A.U.; Khan M.A.; Islam, M.R. MIMO antenna design for future 5G wireless communication systems. In *Software Engineering, Artificial Intelligence, Networking and Parallel/Distributed Computing*; Springer: Cham, Switzerland 2016; p. 653.

23. Guo, J.; Cui, L.; Li, C.; Sun, B. Side-edge frame printed eight-port dual-band antenna array for 5G smartphone applications. *IEEE Trans. Antennas Propag.* **2018**, *66*, 7412–7417. [CrossRef]

24. Yang, B.; Yu, Z.; Dong, Y.; Zhou, J.; Hong, W. Compact tapered slot antenna array for 5G millimeter-wave massive MIMO systems. *IEEE Trans. Antennas Propag.* **2017**, *65*, 6721–6727. [CrossRef]

25. Hussain, N.; Jeong, M.; Park, J.; Kim, N. A broadband circularly polarized fabry-perot resonant antenna using a single-layered PRS for 5G MIMO applications. *IEEE Access* **2019**, *7*, 42897–42907. [CrossRef]

26. Jiang, H.; Si, L.; Hu, W.; Lv, X. A symmetrical dual-beam bowtie antenna with gain enhancement using metamaterial for 5G MIMO applications. *IEEE Photonics J.* **2019**, *11*, 1–9. [CrossRef]

27. Patre, S.R.; Singh, S.P. Broadband multiple-input–multiple-output antenna using castor leaf-shaped quasi-self-complementary elements. In *IET Microwaves, Antennas & Propagation*; IET: Hertford, UK, 2016; Volume 10, pp. 1673–1681.

28. Abbas, E.A.; Ikram, M.; Mobashsher, A.T.; Abbosh, A. MIMO antenna system for multi-band millimeter-wave 5G and wideband 4G mobile communications. *IEEE Access* **2019**, *7*, 181916–181923. [CrossRef]

29. Balanis, C.A. *Antenna Theory: Analysis and Design*, 4th ed.; John Wiley and Sons: Hoboken, NJ, USA, 2016.

30. Deng, J.; Hou, S.; Zhao, L.; Guo, L. Wideband-to-narrowband tunable monopole antenna with integrated bandpass filters for UWB/WLAN applications. *IEEE Antennas Wirel. Propag. Lett.* **2017**, *16*, 2734–2737. [CrossRef]

31. Ding, K.; Gao, C.; Wu, Y.; Qu, D.; Zhang, B. A broadband circularly polarized printed monopole antenna with parasitic strips. *IEEE Antennas Wirel. Propag. Lett.* **2017**, *16*, 2509–2512. [CrossRef]

32. Alsariera, H.; Zakaria, Z.; Awang Md Isa, A. A broadband p-shaped circularly polarized monopole antenna with a single parasitic strip. *IEEE Antennas Wirel. Propag. Lett.* **2019**, *18*, 2194–2198. [CrossRef]

33. Wong, K.; Chang, H.; Chen, J.; Wang, K. Three wideband monopolar patch antennas in a Y-shape structure for 5G multi-input–multi-output access points. *IEEE Antennas Wirel. Propag. Lett.* **2020**, *19*, 393–397. [CrossRef]

34. Fang, X.; Wen, G.; Inserra, D.; Huang, Y.; Li, J. Compact wideband CPW-fed meandered-slot antenna with slotted Y-shaped central element for Wi-Fi, WiMAX, and 5G applications. *IEEE Trans. Antennas Propag.* **2018**, *66*, 7395–7399. [CrossRef]

35. Sharawi, M.S. Printed multi-band MIMO antenna systems and their performance metrics [wireless corner]. *IEEE Antennas Propag. Mag.* **2013**, *55*, 218–232. [CrossRef]

13

Different Antenna Designs for Non-Contact Vital Signs Measurement

Carolina Gouveia [1,2,†], **Caroline Loss** [3,†], **Pedro Pinho** [1,4,*,†] **and José Vieira** [1,2,†]

[1] Instituto de Telecomunicações, 3810-193 Aveiro, Portugal; carolina.gouveia@ua.pt (C.G.); jnvieira@ua.pt (J.V.)

[2] Departamento de Eletrónica, Telecomunicações e Informática, Universidade de Aveiro, 3810-193 Aveiro, Portugal

[3] FibEnTech Research Unit, Universidade da Beira Interior, 6200-001 Covilhã, Portugal; carol@ubi.pt

[4] Departamento de Engenharia Eletrónica, Telecomunicações e de Computadores, Instituto Superior de Engenharia de Lisboa, 1959-007 Lisboa, Portugal

* Correspondence: ptpinho@av.it.pt

† These authors contributed equally to this work.

Abstract: Cardiopulmonary activity measured through contactless means is a hot topic within the research community. The Doppler radar is an approach often used to acquire vital signs in real time and to further estimate their rates, in a remote way and without requiring direct contact with subjects. Many solutions have been proposed in the literature, using different transceivers and operation modes. Nonetheless, all different strategies have a common goal: enhance the system efficiency, reduce the manufacturing cost, and minimize the overall size of the system. Antennas are a key component for these systems since they can influence the radar robustness directly. Therefore, antennas must be designed with care, facing several trade-offs to meet all the system requirements. In this sense, it is necessary to define the proper guidelines that need to be followed in the antenna design. In this manuscript, an extensive review on different antenna designs for non-contact vital signals measurements is presented. It is intended to point out and quantify which parameters are crucial for the optimal radar operation, for non-contact vital signs' acquisition.

Keywords: antennas; radar; vital-signs; CW; UWB

---

1. Introduction

The ability to measure physiological signals accurately has several applications in different areas, from health care to the full medicine procedures. Here, the main vital signals include breathing and heart rate. By combining this concept with the consumer demand for flexible and wireless sensors, it is possible to have a positive impact on society and open a new door in the health care industry.

Until now, the conventional measuring equipments are directly in contact with the subject, which requires some wires' usage. For this reason, the research in this area is focused on the development of solutions with a high degree of freedom, robustness, and obviously maintaining the same accuracy. One possible solution is to use wearable sensors in contact with the subject body, although this solution is more evasive and not comfortable. In this sense, the concept of bio-radar has emerged. The bio-radar uses the Doppler radar principle to evaluate the breathing and heart rate of a subject in a convenient contactless way. It uses an antenna for transmission (TX), which focuses the energy towards the subject chest-wall, and another antenna for reception (RX), to acquire its reflection. This approach stands out as being more advantageous when compared to the traditional devices using contact sensors on the human body, since the subject can be remotely monitored.

The antenna design plays a crucial role in the performance of bio-radar system, in order to obtain the best signal quality and to maintain a signal-to-noise ratio (SNR) at a superior level. Several types of antenna have been used in literature, covering a wide range of frequencies, polarization modes, and half-power beamwidths (HPBW), according to the application at hand. However, most of the papers related to non-contact vital signal (NCVS) measurement using radar-based systems, do not have the antenna information clearly detailed. Mostly, they focus on the global system performance, the main challenges that had to be overcome and new algorithms to extract the vital signs successfully. The majority of papers do not mention the design of the antennas and its importance in the global behavior of the system. Even though, articles that refer to antennas are focused on specific aspects that can enhance the main goal of those works.

Antennas can be designed as single elements or in array configuration, using conventional or alternative materials, depending on the final application. Moreover, in the industry framework, selection of the proper parameters should be done with care, aiming to search for portable solutions (decrease the system size) and using low cost materials that are also able to keep the optimal antenna performance. Therefore, there is a need to review the antenna side, in order to seek which are the best antenna characteristics to properly apply in radar systems for vital signs' acquisition.

In [1], a preliminary investigation on the best antenna features for bio-radar applications was made. This paper was the only review work related with the antennas impact that we have found so far, and the authors also mentioned the lack of reviewing on this matter and in which way this is important. Their study encompasses different antenna designs, which have features for different applications. The authors conclude that increasing frequency has many advantageous aspects, since shorter wavelength helps on decreasing the antennas size, and are more sensitive to low amplitude motions (for example, it could improve on the cardiac signal detection). They have also presented a comparative table that shows a higher gain for these designs.

In our work, it is intended to review the state-of-the-art on antennas for NCVS monitoring, namely for respiratory and cardiac signals. Different frequencies and radar operation modes will be explored, as well as the different designs, and several implementation examples are also presented and discussed. By gathering this information, it is intended to determine which are the antenna features that are more suitable for NCVS acquisition, such as directivity, gain, or polarization. Moreover, it is also intended to present a panoply of low cost substrates and efficient antennas integration that can contribute to system portability and low-profile. It is important to note that different applications and environments require different characteristics, but the goal of this manuscript is to define common guidelines that are required to be maintained for any optimal system performance.

This manuscript is divided as the following: first of all, bio-radar theory is briefly presented in Section 2. In this section, different radar operation modes are also introduced. Then, Section 3 presents some antenna examples developed mainly for Continuous Wave (CW) radar. This section is sub-divided according to the different goals of the authors. Similarly, Section 4 presents more antenna solutions but focused on Ultra-Wideband (UWB) radars. Section 5 presents a briefly summary and discussion about the characteristics of the antennas developed for bio-radar applications. Moreover, a group of trade-off decisions to design the most suitable antenna for bio-radar applications is presented, based on the review made. Finally, conclusions and guidelines are presented in Section 6.

2. Bio-Radar Theory

The bio-radar system can measure vital signs using electromagnetic waves. The operation principle of this system is based on the Doppler effect, which relates the received signal properties with the distance change between the radar antennas and the person's chest-wall [2]. The frequency change on the received signal, caused by the motion of the chest-wall, depends on the total number of wavelengths in the two-way path, between the radar and the target that can be determined through (1):

$$N_\lambda = \frac{2R}{\lambda},$$

(1)

which is given in waves per second (Wave/s) and where R is the radar range and λ is the wavelength of the transmitted signal.

Considering that the radar is located in a fixed position, the same frequency is received over the time if the target is also stationary because the same number of N_λ is received. However, if the target moves toward or away from the radar, different frequencies are received: if the motion is toward to the radar, higher frequency is received due to the bigger number of received N_λ. On the other hand, if the target is moving backwards, a lower frequency is perceived, due to the fewer number of received N_λ [3].

The frequency shift effect can also be perceived as phase change, once each wavelength corresponds to a phase change equal to 2π. Thus, if the path travelled by the wave changes due to the target's motion, also the number of N_λ changes, hence the total phase change relative to range R is given by the Equation (2):

$$\phi = 2\pi N_\lambda. \tag{2}$$

The Doppler radar is the basis of some radar operation modes, namely the Continuous Wave (CW) radar or the Ultra-Wideband (UWB) radar, as depicted in Figure 1. Their operation principle influences the antennas specification directly. In the next sections, a brief introduction about radar operation modes is presented for contextualization purposes.

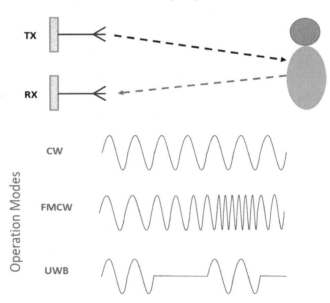

Figure 1. Doppler radar applied to vital-signs acquisition, using different operation modes.

2.1. Continuous Wave Radar

CW radar transmits and receives a radio-frequency (RF) signal continuously. Usually, the transmitted signal is defined as a single-tone. A radar system that uses this operation mode is composed of a signal generator, to generate the signal for TX and also to down-convert RX signal on the receiver side. Once the system usually handles with narrow band signals, it is possible to perceive the frequency shift due to the Doppler effect when the target is moving. Thus, CW radar can measure the velocity of the target's motion, and this feature allows for distinguishing between a moving target from stationary objects, independently of their distance (within the radar's range) [4].

CW radar presents a significant advantage regarding the implementation complexity. It uses a single oscillator for both TX and RX, and the filters used in the receiver chain can be quite simple because TX and RX signals have a narrow bandwidth.

Nonetheless, some disadvantages can be pointed out about this type of Doppler radars. Once TX and RX are continuous operations, due to circuitry or antenna coupling, a portion of the transmitted signal can directly affect the signal at the receiver causing leakage [3]. Moreover, the clutter from multiple reflections of stationary objects located within the range can contribute to the signal power, generating low frequency noise and DC offsets.

Frequency-Modulated Continuous Wave Radar

Frequency-Modulated Continuous Wave (FMCW) allow the computation of the target's velocity and distance between the target and the radar, once it has more range resolution that does not exist for a single frequency waveform [4].

In this operation mode, frequency modulation is performed, usually with a triangular modulation in order to increase and decrease the frequency linearly over time.

Then, another triangular signal is the received signal, which is the delayed version of the transmitted one, and the delay can be quantified by $T = 2R/c$ s, with R being the radar range and c the speed of light. The modulated signal bandwidth determines the accuracy of the range measurement, and the modulation rate determines the maximum detectable range without ambiguity. The mix between the transmitted and the received signals result in a frequency difference f_r that changes according to the target motion, and from where the vital signs are extracted.

2.2. Pulsed Radar

Pulse radar transmits pulsed bursts and then listens to the resultant echoes. In contrast with the previously mentioned radars, TX and RX operations are not done simultaneously. Therefore, the pulse repetition period should be longer than the round-trip path length of the transmitted wave, to receive the echoes between transmissions [5].

Since the reflections are not acquired at the same time as the signal's transmission, the leakage from the transmitter and the parasitic reflections are separated temporally from the long-range targets. This means that, as we have to wait for the full transmission, the reflections that occur immediately from short-range objects are not detected, which represents a clear advantage when comparing to the CW radar. Similarly to the FMCW radar, it is possible to measure the target range.

On the other hand, the pulsed radar is more complex to implement [5]. Moreover, there is a lack of velocity resolution, which implies a limit in target's velocity.

Ultra-Wideband Radar

UWB radar is a special application of the pulsed radar since very short duration pulses are generated, performing wide bandwidth signals. The Federal Communications Commission (FCC) has established that a signal can be categorized as a UWB signal if it has a bandwidth equal to or higher than 500 MHz [3].

The operation of UWB radars starts by generating short pulses and transmits them through the antenna. Then, the target will reflect a portion of the transmitted signal. The total range ΔR of this radar is given by the Equation (3) [6]:

$$\Delta R = \frac{c}{2 * BW} = \frac{\tau c}{2},$$ (3)

where BW is the bandwidth of the radar pulse in frequency domain and τ the bandwidth in time domain. With this type of radar, it is possible to compute the distance to target d_T (Equation (4)):

$$d_T = \frac{\Delta t * c}{2},$$ (4)

where Δt is the delay between the transmitted and the received signal.

Applications of the UWB radars encompass motion detection beyond different materials, due to the vast panoply of different wavelengths that can be used. With this type of technology, it is possible to perceive periodic or quasi-periodic motions because they cause periodic changes in the received signal.

This periodic change is reflected across multiple scans, which are then compared with a reference scan in order to locate the target.

3. Antennas for CW Operation Mode

Generally, designing an antenna is a challenging task, since it is made by following a guideline based on a group of trade-off decisions. The optimal performance of the antenna can be dictated by an infinite gain, lack of side or back lobes, infinite S_{11}, among many other characteristics. However, these features should also be achieved having in mind specific restrictions and requirements regarding the application at hand. In the framework of NCVS applications, this care must also be accounted due to the system sensitivity to environmental clutter, to the very limited radar cross-section (RCS) and to the lowpass signal characteristics.

Although there are multiple antenna types, shapes, and layouts that can be combined in order to fulfill those trade-off decisions, there is not a single solution once the carrier frequency of the transceiver and its operation mode largely influence the antenna behavior. The transceiver operation mode (such as CW, FMCW, or UWB) also depends on the application of the bio-radar. For example, applications that aim to distinguish between different individuals, i.e., where there are multiple subjects to monitor, can use FMCW technology, since the usage of chirps can help to indicate the subject location. Furthermore, rescue applications imply that electromagnetic waves cross obstacles. For this purpose, ultra-wideband front-ends are more indicated. Finally, if the goal is to monitor bedridden patients, CW is enough and the hardware and signal processing are less complex.

In this section, some examples of different antenna designs are presented, considering the CW operation mode. The examples are subdivided in different antenna characteristics according to the improvement goal of authors.

3.1. Directivity

Taking into account the CW operation mode, a system performance evaluation was made in [7]. Considering four different antennas, all operating at 2.4 GHz, several tests were performed inside an anechoic chamber, where the vital signs of a subject were monitored using the following antenna types and their respective beamwidths:

1. Single patch antenna $\Rightarrow$ 92°;
2. Yagi antenna $\Rightarrow$ 47°;
3. Log-periodic antenna $\Rightarrow$ 62°;
4. Helical antenna $\Rightarrow$ 49°.

The radiation pattern of each antenna was verified individually before the experiments. Wide beams are susceptible to acquire more clutter and noise, hence narrower beams are preferable. Taking this into account, among the four tested antennas, Yagi and Helical were the ones that have narrower radiation patterns, and they also have identical values of HPBW. Moreover, they are low cost alternatives. However, a Yagi antenna does not have a symmetrical beam across E- and H-planes unlike the Helical antenna.

The influence that the pair TX/RX can have on each other is also studied, by checking the radiation pattern with another antenna by its side (antennas' feeding points were 24 cm apart). The authors concluded that the influence is minimal, as long as TX and RX antennas are at least a wavelength apart.

The experiment results showed that the error in signal rate computation was smaller for helical antennas. The authors believe that the optimal performance is directly related to the directivity and symmetry in E- and H- planes. An additional test was performed by equalizing the gain for all antennas and repeating the vital signs measurement. Helical antennas stand out again, the beat rate was computed accurately, and the Signal-to-Noise-plus-Interference Ratio (SNIR) increased.

One of the first usages of helical antennas for NCVS acquisition was proposed in [8]. In this work, two antennas with 4-turns and, therefore, directive beams ($\approx$40° of HPBW) were used to acquire vital signs in non-stationary environments, such as vehicles. Compensation of motion artifacts was implemented through a differential measurement, using these antennas in separate. Antennas were designed to operate at different but near frequencies, namely, 2.46 GHz and 2.51 GHz, respectively. Both antennas were implemented with different linear polarization orientation to isolate each signal.

Helical antennas operating at 2.4 GHz are also proposed in [9]. More specifically, the axial-mode helical antenna design was implemented, with an 8-turn design and a simple matching process to feed the antenna without increasing its total size. Helical antennas operating at axial mode are known for having high directivity and hence high gain, which also enables more detection range. Nonetheless, to build a helical antenna for NCVS acquisition, a trade-off should be respected, since the radiation pattern is as directive as the number of turns. Therefore, for a highly directive antenna, a larger size antenna is required. The work presented in [9] resulted in a 8-turn antenna with 20 cm length, HPBW equal to 44.6° and gain equal to 9.80 dBi. Moreover, these antennas also have circular polarization to achieve better SNIR and thus better accuracy on detecting vital signs.

Finally, in [9], the performance of a 1-turn helical antenna was compared with a single patch antenna. Both antennas were operating at 2.4 GHz and had the same HPBW (around 60°). Comparison results showed that helical antennas have more gain, up to 2 dB above the one from a single patch antenna.

Besides helical antennas, any antenna design that can provide directivity can be advantageous for bio-radar systems. In [2], an antenna evaluation is done, by comparing the performance of the antenna array operating at 5.8 GHz with the performance of a single patch operating at 2.5 GHz. As mentioned previously, array antennas have a directive beam, which focuses the radiated energy on the subject chest-wall. On the other hand, a single patch has a wider beamwidth and radiates in different directions, hence the received signal has more noise from the clutter and parasitic reflections. By taking into account these issues, the comparison mentioned above was made by estimating the DC component of each signal acquired with each antenna. The authors concluded that the signal acquired with a single patch has a higher DC component rather than the one acquired with a narrower beam.

Customized Directivity

Until now, we have seen that narrow radiation patterns are preferable to focus all the energy on the desired target. However, some applications require different beamwidth customization to cover the area of interest. For example, in [10], a one-dimensional patch array is developed to measure vital signs from elderly people located in a room. The antenna has to be designed so its beamwidth can cover the full room. The radar front-end was located in the middle of the room, using a microstrip patch antenna array, operating at 24 GHz. The radiation pattern is defined as a triangle-shaped with ±33.7° opening angle, in order to illuminate the corners of the room. On the final prototype, it is intended to use a standard patch array, beamshiped on the H-plane used as a TX antenna and a linear-array, beamshiped on the E-plane used as an RX antenna, with 10-dB gain higher for the ±33.7° angle than for the 0° angle. In this way, multiplication of both radiation patterns will rectify power losses in corner spots.

The antenna array proposed in [10] corresponds to the RX antenna beamshiped on the E-plane. In summary, the designed antenna is an array with three microstrip patch resonators fed in series through a microstrip line, where the last patch is a single-port standard patch. Patches are optimized to operate at 24.125 GHz and are built using an RO4835 substrate, with a relative permittivity of $\varepsilon_r = 3.66$. The total array length is equal to 15.694 mm and a width equal to 6.339 mm. In the simulation, it was possible to achieve the required features respecting to radiation pattern, with 8 dB higher gain at 33.7°. The measurements meet the specifications, but the authors state that there are some improvements that need to be done.

3.2. Different Carrier Frequency Applications

The usage of high carrier frequencies increases the sensibility to detect imperceptible target motions. Planar microstrip antennas tuned at these frequencies enable the on-chip integration for portable applications. For example, in [11], a circularly polarized (CP) 2×2 patch array antenna is presented, to be further integrated with a 60-GHz Doppler radar. The low-temperature co-fired ceramic (LTCC) substrate is used to increase bandwidth, which is also crucial for single-tone applications to guarantee good performance in the case of manufacturing issues.

Another millimeter-wave microstrip array antenna was presented in [12], but with a lower operating frequency, 24 GHz. The antenna was developed to evaluate the organism adaption to physical and mental stress remotely. Each antenna element from the transmitter and receiver array is single-patch antenna U-slot shaped. The TX array included 16 elements (8×2 array) for a narrow beam, and the RX was composed by 48 elements (8×6 array) for a large aperture area. Each single-patch is fed using a power divider network to adopt an unequal current amplitude distribution instead using an equal amplitude. The antenna module was designed using FR-4 substrate, with $\varepsilon_r = 3.55$, and copper cladding for the conductive pats. The final dimensions of the antenna module were 73×64 mm^2. This paper presents only the simulation of these antennas and the numerical results obtained were: bandwidth 24–25 GHz (RX) and 23.3–23.7 GHz (TX); S_{11} less than -10 dB for both TX and RX antennas, and a gain equal to 17.15 dBi for the TX antenna, while the RX antenna exceeds 17 dBi. The mutual coupling was below -35 dB at 24 GHz, providing good isolation between the TX and RX antennas.

Later, in 2016, the same research group has proposed a new microstrip array antenna, operating at 77 GHz, to be used as TX and RX antennas, in a bio-radar module [13]. The array is composed by 160 single-patch elements (16×10) for a narrow beam, which are fed using a power divider network. The antenna module was designed using an RO3003 substrate, with $\varepsilon_r = 3.0$, and copper cladding for the conductive parts. The final dimensions of the antenna module were 43.79×25.04 mm^2. The simulated S_{11} showed that the antenna is capable of operating between 74.6 GHz and 79.7 GHz, having a bandwidth equal to 5.1 GHz. At 76.5 GHz, the calculated gains on the H- and E-plane are higher than 25 dBi, and the HPBW on both planes is 12°. In addition, in this frequency, the side lobe level is -12.5 dB and -11.5 dB for the H- and E-plane, respectively. The numerical results prove that this array module can be used for bio-radar applications.

In addition, a continuous wave harmonic radar system is presented in [14]. The system is composed of two slot array antennas designed using a Substrate Integrated Waveguide (SIW) technique, to operate at 12 GHz and 24 GHz. The SIW slot array antennas had 13 dBi and 24 dBi of gain, for the 12 GHz and 24 GHz, respectively. The experimental results show that, by using harmonic radar, the received signal power can be increased while the noise level decreased. Regarding the sensitivity of the system, as expected, the detection of vital signs was possible when the noise level was below the received signal power.

On the other hand, low carrier frequencies can also be interested to be explored, since they allow high EM penetration in the human body and thus increase accuracy in vital sign detection. In [15], a fractal-slot patch antenna operating at 915 MHz and combined with an ultra-wideband Low Noise Amplifier (LNA) was used for this purpose. In order to evaluate the performance of this CW Doppler radar operating with such low frequency, the path losses were computed and compared for 915 MHz and 2.45 GHz signals, when propagating in the subject body that was located 80 cm apart from the radar. The authors could conclude that, with the 915 MHz radar, it was possible to achieve signals with more information. For example, the cardiac signal achieved a 7 dB superior level for 915 MHz radar rather for 2.45 GHz.

A fractal-slot patch antenna design, using FR-4 substrate $\varepsilon_r = 4.4$ and 1 mm thickness, helped to decrease the antenna size and also increase the path length of EM wave, keeping the antenna performance. The final patch size was equal to 70.4 mm, which is 7.9% smaller than a conventional patch. Moreover, the total physical size reduced 15.2% when compared with a conventional patch antenna operating at the same frequency. The fractal slot in the center of the patch was implemented using Koch snowflake patterns to increase the surface current on the patch. The patch was designed with truncated edges to achieve circular polarization. The obtained performance characteristics were

26 dB of return loss, 17 MHz of bandwidth, HPBW equal to 108°, 12% of radiation efficiency, and 5.8 dBi of gain. The total antenna size was 140.8×140.8 mm^2.

As seen in the examples presented in this sub-section, the carrier frequency also influences the antenna size. Although, due to some hardware limitations, using high frequency carriers can not be always an option; therefore, other solutions will be explored to optimize the size of the radar system and are discussed in the next sub-section.

3.3. Techniques to Reduce the Antenna/System Size

Usually, two separate antennas are used to perform TX and RX. However, they have to be separated at least a half-wavelength to avoid mutual coupling, which increases the total size and hampers the antenna integration in more compact systems [16]. One alternative to decrease size is using a single antenna, but this should be done with care. The most obvious way to separate TX and RX signals is using circulators, although they are composed by ferrite materials, have a fixed size and are costly components [16,17]. The other option is to use couplers, but they divide signal power, and at least 6 dB of power loss is induced. Nonetheless, a dual-antenna is still an option if antenna design techniques are applied specifically to reduce the antenna size. Below, we describe some works that present techniques to reduce the size of the antennas for bio-radar.

3.3.1. Single Antenna for TX/RX

Although the authors do not mention the radar operating mode, in [18,19], they proposed a single antenna for TX/RX. In [18], a low cost system using a Yagi patch antenna as a single antenna for the RF front-end was planned. The Yagi patch antenna has a simple structure, is discreet, and has a highly directive lobe. The antenna was developed using FR-4 substrate, with relative permittivity $\varepsilon_r = 4.4$ and a total size of 120×80 mm^2. The achieved performance parameters were $S_{11} = -22.87$ dB at 2.45 GHz and gain equal to 8.69 dBi and a minimal back lobe (-10.50 dB). Since a single antenna was used for both TX and RX, the separation of the incoming and outcoming signals was performed by a circulator.

Similarly, a transceiver with a single antenna was also implemented in [19]. In this case, a horn antenna operating in the V-band (50–75 GHz) was used, with 25 dBi gain and beamwidth equal to 7°. A circulator with 18 dB isolation was used for TX and RX signals separation. This amount of isolation was not sufficient to avoid leakage at the receiver stage, hence a clutter canceller was also implemented and it is explained in detail in [19].

As seen so far, when the transceiver has a single antenna for TX and RX, a circulator is used to separate signals. Although this hardware component is costly, has large dimensions, and sometimes it is not as effective as demonstrated in [19]. Therefore, alternative solutions for single antenna implementation can be explored. One possible solution is sharing the same radiating aperture using a CP antenna and using opposite polarizations for TX and RX functions [16,17,20,21]. For example, in [17], a single antenna with CP and operating at 24 GHz, was shared by TX and RX. Different polarization directions were used, more specifically left-hand circular polarization (LHCP) for TX and right-hand circular polarization (RHCP) for RX. A quadrature coupler (Langer coupler) was also used to avoid the circulator usage, and TX and RX were located in opposite ports. The full radar system size was 4×4 cm. It was tested by acquiring vital signs at a distance of 50 cm. An average of 21 beats/min and 69 beats/min was detected for respiration and heartbeat rate, respectively.

A similar solution was presented in [16], this time using a radar front-end operating at 2.4 GHz. The antenna is composed by three metallic layers wrapped in FR-4 substrate with 50 mm of diameter. The bottom layer is a ring-shaped quadrature hybrid coupler (QHC) for the vias' excitation. The middle layer is the ground plane and the top layer is a circular patch fed with two vias, for TX and RX, respectively. This layout was selected to help reduce the total system size. With this sharing method using opposite polarization directions, it was possible to obtain 30 dB isolation between TX and RX.

Then, this antenna was used to monitor both heartbeat and respiratory signal accurately within a range of 60 cm.

The same approach was used in [20], where a bio-radar system is presented with a phase-locked loop and uses a single antenna for TX and RX. In this work, the authors have developed a circular-polarized annular ring microstrip antenna, with RHCP at transmitting mode and LHCP at receiving mode. This antenna was designed to operate at 2.4 GHz and presented an HPBW of 132°, and measured return loss of 20 dB. In further measurements, the radar was able to detect the respiratory signal of a human seated at 30 and 50 cm away from the radar. Later, in [21], the authors have made a study comparing this system with a 10 GHz bio-radar with separated antennas for TX and RX. The 10 GHz super-heterodyne bio-radar system is fully described in [22]. The TX and RX array presented a linear polarization, 8 dBi of gain, HPBW of 30°, and measured return loss of 30 dB. The vital signs of a subject were acquired from different distances, starting with 0.3 m until 2.9 m away from the radar. This measurements were performed with both systems [21]. Comparing the obtained results, the authors have concluded that, despite the advantages of the 10 GHz bio-radar (higher antenna gain and RCS), the system had not shown significant improvement compared to the 2.4 GHz bio-radar system. In addition, the 10 GHz bio-radar is more expensive and complex to implement.

Single antennas are indeed an alternative to decrease the front-end size. However, some drawbacks were identified. In [23], a test was performed to evaluate the system performance, using a single antenna or two separate antennas for TX and RX. Results showed that a single antenna has good accuracy in vital signs' detection, considering a short-range. The distance between the subject and the system antennas was equal to 5 cm, although two separated antennas could cover a wider range with the same level of signal quality. The subject was monitored successfully within the range of 25 to 200 cm. Antennas were located 20 cm apart from each other to avoid cross-talk effect.

In [23], the authors carefully selected the microstrip type of antenna for their experiments. Apart from having an acceptable gain, microstrip antennas can also be embedded on chip devices, which reduces the total size of the system and use a low-cost fabrication process. Moreover, microstrip antennas can be arranged in arrays to increase gain, decrease the side-lobe level, narrow the beamwidth, and thus focus the energy only on the target of interest. In this sense, three different microstrip antenna arrays of ultra-wide elements were designed and tested, and the following gains were achieved:

1. 2×1 elements $\Rightarrow \approx 13$ dBi;
2. 3×3 elements $\Rightarrow \approx 16$ dBi;
3. 6×2 elements $\Rightarrow \approx 18$ dBi.

All of the antennas were designed to operate on a 60 GHz band, using Duroid substrate with $\varepsilon_r = 2.2$. The main purpose was to integrate these antennas in a CW front-end. However, antennas were designed with an ultra-wide band (57.24 GHz–65.88 GHz) to enhance the fabrication tolerance when embedding in a low cost printed circuit board (PCB).

In practice, it was possible to observe that the achieved bandwidth did not have an increased behavior in relation to the number of elements like the gain. In fact, it decreased with the number of elements in the array.

3.3.2. Size Reduction Techniques Applied with Dual-Antenna for TX and RX

Now, considering the dual-antenna option, other techniques were explored to reduce the antenna size as much as possible.

In [24], a 3D design was chosen to use as a portable life-detecting system. In this paper, the authors presented a 3D-orthogonal patch antenna that was designed to operate from 900 MHz to 12 GHz and was developed using FR-4 substrate, with $\varepsilon_r = 4.4$ and 1.2 mm of thickness. The antenna module is composed by the TX and the RX arrays, with 15 and 20 patch elements, respectively. The TX and RX arrays are connected together by a power splitter and power combiner. Comparing the simulated

(900 MHz to 12 GHz) and measured performances of both antennas for all bandwidth frequencies, the best results were from 800 MHz to 4.4 GHz for the TX and 9.4 GHz to 12 GHz for the RX antenna. As the authors have used only one average permittivity value ($\varepsilon_r = 4.4$) during the simulation, and the antenna covers an extra wideband; they concluded that the variation of the substrate permittivity at different frequencies has influenced the results.

The authors also tested the applicability of this extra wideband antenna for a Doppler radar system. The test was performed using a PNA-X in a non-controlled environment, where the subject was placed 80 cm away from the antenna. The measurements were carried out using five different frequencies (0.953 GHz, 2.198 GHz, 3.362 GHz, 5.556 GHz, and 10 GHz) and during 3 min. The system was more sensitive for higher frequencies. As for the go-through materials applications—such as finding survivors—lower frequencies are required, and similar measurements were made with a wood barrier (3 cm of thickness) between the subject and the antenna. In this case, in higher frequencies, the system was negatively affected because the amplitude of the signal at 5.556 GHz was 0.7 mm while at 3.637 GHz was 2 mm.

In [25], the authors combined two developed arrays with a commercial 24 GHz radar system chip in Silicon Germanium technology, reducing the size of the radar module. The TX and RX are 4×5 patch arrays, operating at 24 GHz, designed and fabricated using RO4350 with 0.254 mm and $\varepsilon_r = 3.48$, as a substrate. The simulated gain was about 17.1 dBi and the beamwidth was around 24° and 20° for the E- and H-plane, respectively. This compact radar module was tested by measuring the breathing and heartbeat of an adult seated 1.5 m away from the radar. In these conditions, the radar was able to measure 15 breath/min and 76 beats/min, which agrees with the results obtained with a pulse sensor (76 beats/min).

Furthermore, portable and low-power radar was designed and built on a palm-size PCB, and it is presented in [26]. In this work, the authors proposed two 2×2 printed patch antenna arrays, for 5 GHz, to be used as TX and RX antennas. The arrays had 9.7 dBi (at 5.8 GHz) of gain and HPBW equal to 20°. To compare the results, the measurements using a UFI1010 wired fingertip pulse sensor were also performed. In comparison, the heart rate accuracy obtained with the non-contact radar was 98.82%, 92.40% and 81.35% at 0.5 m, 1.5 m, 2.8 m away from the radar, respectively.

Techniques to integrate the radar in a complementary metal-oxide-semiconductor (CMOS) chip were explored in [27,28]. In [27], the authors have proposed a 5.8 GHz radar in a 0.18 μm CMOS chip. Generally, this on-chip integration is done for higher operating frequencies. However, harmonic interference can occur on those frequencies, and, due to this issue, the authors opted to use a lower frequency. The antenna attached on this chip was a 2×2 patch antenna array, fabricated with a RO4003 substrate with $\varepsilon_r = 3.38$ and thickness equal to 0.73 mm. Adjacent patch elements were located with a distance of $0.74\lambda_0$, where λ_0 is the wavelength in free space, in order to decrease side lobes. The developed antenna had a gain equal to 10.77 dBi and HPBW equal to 14.5°.

Later, in [28], a 100 GHz low-IF CW radar transceiver was presented in a 65 nm CMOS chip. In this work, the radar was applied for mechanical vibration and biological vital sign detections. Two horn antennas, operating at 100 GHz, with 20 dBi of gain, were used as TX and RX antennas. Vital signs were acquired, with the subject being seated 2 meters away from the radar. In this case, 65 beats/min were obtained, which agrees with the pulse rate measured using a finger oximeter. In addition, the heart rate of a bullfrog was measured, at 0.6 m away from the radar. These measures were performed under different temperatures—10 °C, 20 °C, 30 °C, and the obtained results were about 22, 41, and 80 beats/min, respectively. In this way, the authors could conclude that, at 100 GHz, the radar was able to detect much smaller chest displacements.

In the same framework of the on-chip embedded alternatives, a microstrip dipole array antenna was developed in [29], to be later integrated along with the radar front-end on the same chip. Dipole antennas have a low profile, are low cost, and they can be arranged in arrays. The authors used a structure with three dipoles disposed symmetrically to obtain a higher gain and used a reflector to achieve directivity. The proposed antenna operates at 2.4 GHz, within a bandwidth from 2.35 to 2.5 GHz. The main lobe has 10.8 dBi gain, with a beamwidth equal to 49°.

3.4. Mutual Coupling Reduction

Mutual coupling effect occurs when one antenna receives part of the energy radiated from a second antenna located nearby. This can happen due to three main reasons: the radiation pattern of each antenna, the separation between both antennas (which should be at least equal to half wavelength), and the main lobe orientation of both antennas [30]. Furthermore, mutual coupling can alter the radiation pattern of each radiating element, since it shifts the maximum and nulls location, filling the nulls when it was not supposed to [30].

As seen in the previous subsection, when two antennas are used to perform TX and RX, they should be separated sufficiently enough to avoid mutual coupling (generally half wavelength), but the distance between antennas should be enough to guarantee the monostatic radar function. In this sense, the mutual coupling effect must be taken into account, and strategies to decrease its effect should be adopted.

The literature about bio-radar systems that specifies which antennas are being used and their design constraints does not focus on the mutual coupling effect and does not explore different strategies to reduce it. Even though we have seen so far that some authors care about cross-talk and present measures to prove that the TX/RX pair has acceptable levels of coupling. In addition, it is mentioned that the usage of CP approach can be one possible solution for this issue, and it can also enhance other problems inherent to the bio-radar systems.

As we will see in the next subsection, it is possible to use orthogonal polarization in TX and RX because an incident CP wave flips its polarization propagation when reflecting on a surface. Furthermore, if the system uses different rotation directions, such as RHCP for TX and LHCP for RX, there is no power reduction due to the signal rotation when reflecting at the target surface, and there is no mutual interference because, at the front-end stage, antennas have crossed polarization. Thus, a system using CP antennas has low mutual coupling [16].

3.5. Circular Polarization

Linear polarization (LP) is the less complex approach. However, some problems arise with its usage for NCVS applications. After transmitting an LP signal, during propagation, the reflected signal can rotate θ degrees in total, hence the signal at the receiver input has its power decreased by a factor of $\cos\theta$, and the radar sensitivity is largely reduced [17]. Thus, CP is generally the best alternative, since CP antennas are not affected by polarization mismatch. Furthermore, in [31], the authors have pointed out other emerging problems due to the usage of LP antennas. There is fading RCS due to the scattering reflection on the target. The human body is composed of different materials, shapes, sizes, or thickness. Hence, different surfaces cause electric vector rotation, which lead to a misalignment with the receiver antenna. Moreover, the target at hand is moving; consequently, a time-varying RCS arises. CP antennas stable the RCS over time and keep the alignment between scatter signals and the receiver.

In this sense, it is possible to conclude that CP is the best strategy to achieve better SNR and to isolate the system from other radar based systems nearby. In [11], TX and RX antennas with opposite polarization were used, (LHCP for TX and RHCP for RX), in order to isolate the signal reflected by the target from signals derived from other transmitting systems.

The impact of circular polarization in the bio-radar performance was tested in [9]. A set-up was settled with three antennas, where two are for RX and one for TX. Antennas were disposed according to the following scheme: RX1 with RHCP, TX with RHCP and RX2 with LHCP. Antenna TX transmitted a signal towards a metallic reflector, and the signal strength received in RX1 and RX2 were evaluated. The authors observed that the signal at RX2 was 1.2 dB stronger than the signal at RX1 input. This effect can be explained with the isolation of each signal. Thus, it is possible to conclude that the antenna pair TX/RX with opposite polarization is a better option to enhance the SNIR.

More tests were conducted to analyze the effect of the antenna polarization along with beamwidth in the performance of Doppler radar, in [32,33]. Four microstrip patch antennas for bio-radar, operating at 2.4 GHz, were designed and manufactured:

1. LP single patch antenna;
2. CP single patch antenna;
3. 2×2 LP array;
4. 2×2 CP array.

To achieve the same gain for all antennas, different substrates were used to manufacture them, being 1 and 2 printed on RT-Duroid 5880, with $\varepsilon_r = 2.2$ and 1.6 mm thickness, and 3 and 4 made using FR-4, with $\varepsilon_r = 4.4$ and 1.6 mm thickness. In this way, the authors were able to compare only the beamwidth effects on radar performance, excluding the influence of the gain. Sixteen combinations were presented, e.g., CP array used as TX and LP single patch antenna as RX antennas. Five measurements per combination were performed inside an anechoic chamber, during 30 s, using a linear programmable actuator as a moving target. The measured gains of all antennas are within 5.8 dBi. The HPBW is 37° for the array and 81° for the single antennas. All combinations were capable of estimating the motion frequency, and, in summary, the authors concluded that the best radar performance is obtained when they used the antennas 1 and 4 as TX and RX antennas, respectively. In this case, the received signal showed the strongest fundamental frequency amplitude, −6.41 dB. On the other hand, the combination of 4 and 2, used as TX and RX, respectively, showed the worse performance, the fundamental frequency amplitude being equal to −22.77 dB.

By taking advantage of different antenna polarization, in [34], a solution is proposed to mitigate the random body motion using crossed-polarized antennas. Two transceivers are used, one at the back of the subject and other in the front. Since vital signs have the same waveform pattern and periodicity, this system combines both signals and mitigate any waveform that has different pattern and frequency. Since two antennas face each other, their design should be done carefully. In this sense, a 2×2 patch array antenna was used, with vertical polarization for front and horizontal polarization for back transceivers, respectively. The other antennas' parameters are not specified in the paper.

3.6. Customized Antennas for Commercial Transceivers

Some commercially available bio-radars have also been tested. In [35], the authors proposed a new method of blind source separation for signal from multiple subjects. In this work, the authors used a stepped frequency CW bio-radar BioRASCAN-4 (RSLab, Moscow, Russia) [36] that is composed by two horn antennas, operating at 3.6–4.0 GHz, with a constant gain equal to 20 dBi. The combined size of these antennas are $370 \times 150 \times 150$ mm, and it has a 60 dB dynamic range to detect signals. In this experiment, the authors were capable of identifying three different respiratory patterns of the subjects located at the same distance from the radar.

In the same framework of commercial products, a K-LC5 transceiver from RFbeam Microwave GmbH (St. Gallen, Switzerland) [37], operating with 24 GHz carrier, was used in [38] to evaluate if bio-radar technology can be used to access the psychophysiological state of 35 healthy volunteers. In this sense, vital signs were acquired and classified using data-mining techniques and thus determine if there is any mental or physical stress among the subjects. TX and RX antennas of the K-LC5 are patch arrays, which can be used in two different configurations: 3 array, with 80° of HPBW, or 1 array, with 34° of HPBW. The transceiver has a compact size $25 \times 25 \times 6$ mm, and it was designed for short range applications (within 1 m range). The authors on this paper could identify correctly whether the subject was calm or stressed with 80% of accuracy.

In addition, from RFbeam, a K-LC2 transceiver [39] was tested in [40]. The K-LC2 also works at 24 GHz and uses for TX and RX a 2×4 array. The antenna gain is 8.6 dBi and the beamwidth is around 80° and 34° for the E- and H-plane, respectively. In this study, measurements of the vital signals from a man seated 50 cm away from the radar were done. The developed radar using the K-LC2 transceiver was able to measure 19 beats/min and 64 beats/min, for breathing and heartbeat, respectively, which agrees with the results obtained using a Healthcare HM10 pulse sensor.

3.7. Other Radar Transceivers

In this sub-section, non conventional radar transceivers are approached and the antennas used are described. First, different radar front-ends are seen and front-ends designed to perform beam-steering are mentioned afterwards.

3.7.1. Customized Front-Ends

The research community suggests a varied panoply of solutions to mitigate the inherent problems in the proper detection of vital signs. Solutions vary not only in the antenna design but also in the transceiver hardware. The majority of works presented until now use single hardware components, interconnect and/or integrated on chips which implied analog signal processing. On the other hand, Software Defined Radios (SDR) perform part of the signal processing digitally, which confer more flexibility to the system. In [41], a radar for vital signs measurement using SDR is proposed. Two microstrip patch antenna arrays are used for TX and RX, respectively. Each antenna has eight elements to achieve directivity (8×2 array). The authors underline the advantage of using patch antennas instead of horn antennas which would require a bigger footprint. Antennas are optimized to operate at 5.77 GHz and have a 16 dBi gain. Vital signs were successfully acquired at a distance of 0.5 m.

Some other solutions were presented to adapt the transceiver operational mode to environment issues. For example, the motions of the proper radar handling were considered in [42] as a noise source and a possible solution was presented. The system proposed by the authors uses a Doppler radar that transmits a fundamental signal component at 2.4 GHz towards the target, and simultaneously transmits a harmonic signal component at 4.8 GHz towards a stationary reflector, located on the opposite side. This system receives the reflected signal from the main target, as well as the signal reflected by the reflector. Both signals are phase modulated due to the chest-wall motion and the radar handling motion, respectively. Thus, it is possible to remove the noise caused by radar handling, through self-cancellation implementation.

The antennas used in this set-up had a gain equal to 8 dBi approximately and a HPBW equal to 80° for the fundamental antennas and 70° for the harmonic antenna. Since two pairs of antennas are located back to each other, a high front-to-back ratio was required and it was accomplished, being better than 21 dB. The coupling between antennas was -32 dB for the fundamental antenna pair and -40 dB for the harmonic antenna pair.

In [43], a radar system for NCVS was prepared using standard equipment that can be found in any RF laboratory. In this way, some inherent problems are immediately solved, such as the DC component offset calibration or the In-phase and Quadrature (IQ) imbalance, and thus a bio-radar prototype can be rapidly built for research purposes, without caring with possible hardware problems that can come along with portable transceivers. With this set-up, it was possible to retune the system for different carrier frequencies and monitor multiple targets. The authors have captured vital signs using two different antennas, in order to evaluate the system's performance for single-frequency operation mode, for frequency-tuning experiments, and for capturing vital signs with objects in front of the target. The first antenna was a 2×2 microstrip patch array, operating at 2.4 GHz. Two equal antennas were fabricated together, where one was used for TX and the other for RX. The FR-4 substrate was used, with $\varepsilon_r = 4.6$ and thickness equal to 1.5 mm. The final size of dual-incorporated antennas was 459×222 mm, and it was measured in an anechoic chamber. It had as a maximum gain 6.5 dBi and as HPBW 40°. The second antenna, was a commercial horn antenna (HD-10180DRHA, Hengda Microwave, Xi'an, China), with broadband between 1 GHz to 18 GHz. The gain of this antenna varied according to the frequency, between 10 and 16.5 dBi.

The experiments were carried out successfully for all three test scenarios. With single-frequency operation mode, vital signs were acquired accurately, as compared with reference measurements. On the other hand, for variable frequency carriers, the results were not always accurate. With fixed transmitted power, the amplitude of vital signs decreased for the higher frequency component (18 GHz), due to cable attenuation, antenna coupling, and free-space attenuation. Finally, the obstacles experiment was conducted with the fixed 2.4 GHz frequency and vital signs were measured accurately.

3.7.2. Beam-Steering Systems

Many problems can arise when detecting vital signs in moving subjects. As seen previously, the human chest-wall has a small RCS, which has a limited area around 0.5 m^2, and the proper alignment should be guaranteed to acquire signals with optimal SNR. Fixed-beam antennas require this alignment for every different subject, since humans have different heights and body structures. In this framework, beam-steering technology can be advantageous since it can re-direct the beam seeking for the best SNR.

In [44], an adaptive beam-steering antenna is proposed, in order to increase the detectable range without increase the system size. A 2 × 2 microstrip patch antenna array and two phase shifters are embedded on the same board. The developed antenna has 200 MHz of bandwidth, centered in 5.8 GHz and it can steer from −22° to 22° within the H-plane. Phase array antennas can steer to different angles if the adjacent antennas are fed with different phased signals, by using phase shifters [44]. Antennas are built using a Rogers4350B substrate. The S_{11} values were far below −10 dB for all steering angles, from −22° to 22°. The HPBW was equal to 41°, which leads to total coverage of 85°.

To test the performance of a system with a beam-steering approach, a subject located at the angle of 21.8° was monitored using both fixed-beam and beam-steering antennas. By using beam-steering antennas, several angles were scanned. The cardiac signal was detected with the beam-steering antenna at the angle where the subject was located and the remain angles, as well as the fixed-beam antenna, presented only noise.

A NCVS system was designed for multi-target monitoring in [45]. For this purpose, a phased-array CW radar, operating at 2.4 GHz was developed to generate two different beams concurrently. Thus, it was possible to capture respiratory signals from two subjects at the same time using the same carrier. To perform the phased-array radar, two linear arrays were used with four elements each, for both TX and RX. Single elements were rectangular patches, spaced 8 cm between each other. Measurements of antennas were performed inside and outside of the anechoic chamber, to verify the impact of propagation environment on its radiation pattern. For both cases, it was possible to obtain good agreement between simulated and measured radiation patterns. Furthermore, as expected, the side-lobe level increased on measurements outside the anechoic chamber since there are other systems operating at the same frequency. Dual-beam mode was also measured under the same conditions. Two situations were considered: in the first case, one beam was directed to −15° and the other directed to 25°, while the second situation had one beam directed to −25° and the other directed to 30°. Results were similar for main lobes, for both simulation and measurement cases, but had slightly differences respecting the sidelobes that were justified due to some minor errors in circuits and cables.

Metamaterials were explored in the NCVS radar systems, in [46]. In this work, metamaterial-based scanning leaky-wave antenna is developed for beam-scanning Doppler radar, with the steering angles between −33° and 26°. This system can track the human subject and measure vital signs. The main beam is frequency controlled, which means that the system should be retuned within the frequency range of 5.1–6.5 GHz in order to be able to scan all steering angles.

The antenna was fabricated using two layers of FR-4 substrate, with $\varepsilon_r = 4.4$, loss tangent of 0.02 and with thickness equal to 0.2 mm and 1.23 mm, respectively. The antenna is composed by 30-cells (where one cell corresponds to a single element), disposed linearly. Antenna gains varied with the frequency, being 5 dBi for 5.1 GHz, 7.1 dBi for 5.8 GHz and 6.8 dBi for 6.5 GHz.

With this set-up, it was possible to detect two subjects located at $0°$ and $26°$ angles, respectively; thus, frequencies of 5.8 GHz and 6.5 GHz were the ones being used. Their vital signs were successfully measured.

4. Antennas for UWB Operation Mode

Similarly to the previous section, herein it is presented antenna design examples, considering the UWB operation mode for NCVS applications. UWB radars have a different working principal, which requires high operation bandwidth. Therefore, the antenna design should be suitable for this purpose.

4.1. General Antenna Considerations for UWB Systems

Different antennas were evaluated and compared in [47]. An experiment was made, where the respiratory signal was acquired using three antennas with different designs, as described in Table 1. Thus, different characteristics concerning the radiation pattern, gain and cross-polarization were tested. Their set-up was composed by a UWB radar operating within the bandwidth 3.1 GHz–5.3 GHz.

Table 1. Antennas tested and compared in [47].

Antenna Design	Radiation Pattern	Gain	Bandwidth (GHz)
Broadspec UWB antenna	Omnidirectional	low gain	3.4–10.4
Horn antenna	Directive	high gain	2–18
Doubled layered Vivaldi antenna	Directive	medium gain	3.3–10

The best results were achieved with the Vivaldi antenna, which has a medium gain, a directive radiation pattern, and the best co- and cross-polarization ratio. The remaining antennas had the worst performance, due to different aspects. First, the omnidirectional antenna transmits the same power in every direction, which means that unwanted reflections are equally received as the wanted ones. This antenna also had the worst co- and cross-polarization ratio, which means high cross-polarization components. Then, the horn antenna has bigger dimensions and its radiation pattern is highly directive. This obliges a perfect alignment between the target and both TX and RX antennas that have be separated. This balance is difficult to achieve.

Moreover, UWB antennas are known to have dispersive behavior, i.e., radiate in different frequency components. This occurs due to a phase center instability, as stated in [48]. The phase center deviation can cause signal distortion and can contribute to the error on respiratory rate computation. Phase center deviations were highly perceived for the Broadspec UWB and horn antenna and were less evident for Vivaldi antenna.

The performance was evaluated after the respiratory rate error computation, considering a reference rate. Several experiments were performed, where different set-up parameters varied, such as the target motion frequency (to imitate the respiratory rate), the motion amplitude, and the nominal distance between the radar and the target. It is important to note that all the error rates had increased with this latter parameter, but the error was more significant for the Broadspec UWB and Horn antennas rather than Vivaldi antenna.

4.2. Antenna Performance Improvement

The characteristics for an acceptable UWB antenna were identified in the previous sub-section: directional radiation pattern, high gain, compact size, good co- and cross-polarization ratio, and non-existence of phase center deviations. Nevertheless, antenna features should be optimized for each application, and some techniques applied for this purpose are described in this sub-section.

4.2.1. Gain and Bandwidth

Different techniques to improve the gain and broadband characteristics were developed in [49]. They used a partial ground plane patch to achieve wideband coverage (4.3 GHz to 7 GHz), combined with parasitic patches emulating the Yagi concept, to improve gain. In the end, the authors also compared the performance of the final antenna with a single-patch with a partial ground plane as well. It was possible to conclude that single-patch is worse than the proposed one since it does not have sufficient gain, has a wide beam (hence, has lack of directivity) and it does not confer enough isolation when two antennas are side by side.

Furthermore, the antenna developed in [49] was also focused on indoor applications considering a specific location inside a room. The presented layout encompassed a ground plane working as a reflector and parasitic patches working as directors. The beam angle was designed to allow the antenna to be located on the ceiling, in the corner of a room. For this purpose, the main lobe was directed to a specific angle (35°), through a mutual coupling effect. Regarding the radiation pattern, the HPBW was equal to 43°.

In [50], another antenna for indoor applications is proposed. A Ground Surround Antenna (GSA) array was designed, considering the framework of NCVS acquisition. The development of antennas for indoor applications should meet some beamwidth and range requirements, in order to mitigate multipath interferences that can occur due to reflections on walls or furniture. In this sense, the authors state that the beamwidth should be no larger than 30° and the gain can be improved by applying a superstrate layer. The GSA also has higher bandwidth rather than a conventional patch. Thus, the antenna array developed in [50] had a maximum gain equal to 12.2 dBi, an HPBW of 26° and bandwidth within 2.28 GHz–2.49 GHz, centered at 2.45 GHz.

4.2.2. Polarization

Within general performance characteristics, the advantages of CP were explored for the UWB radar in [31]. The Axial Ratio (AR) of the developed antenna is preserved in all bandwidths, by using an array of LP antennas arranged in a sequential rotation manner, with 90° rotations. The selected antenna design is a dual elliptically tapered antipodal slot and two different sequential arrangements were considered: a 'box' and a 'cross' arrangement. Antennas were fabricated using a Rogers 4003C substrate and with a total size of 9×9 cm^2. The LHCP was the selected polarization since it has less interference from the existent wireless equipment, which is operating in the same band. They proved the robustness of the system using CP antennas by comparing the amplitude of the received signal with four different polarization angles and the wave amplitude remained approximately the same. Furthermore, the accuracy on detecting the heartbeat was 2 beats/min superior rather than using LP antennas.

4.3. Antennas Customization for Different UWB Radar Applications

Many radar applications were presented in the literature with a UWB front-end. For example, in [51], the authors use a UWB bio-radar, with impulse radios, to monitor multiple human targets. The proposed system consists of one TX and three RX antennas, to form three independent channels. All antennas are bow-tie dipoles working at centered frequency 500 MHz and with equal bandwidth. The antennas were tested in four different scenarios: with no target, single target, and two and three targets behind a brick wall with 28 cm of thickness. The results have shown that, besides the acquired respiration waveform being quite different, depending on the channel, in all cases, the UWB radar was able to detect the respiration patterns of the human targets.

Then, a non-contact system to acquire vital signs were presented in [52], with a different operation mode. Instead of using a Doppler radar, the respiratory rate is evaluated through phase differences in the S_{11} coefficient. For this purpose, they combine the UWB with CW principles to take advantage of both techniques. Nonetheless, their concern about antenna characteristics is the same. The goal of their work is to develop an antenna with small size, low cost, with directional radiation pattern, large bandwidth, and good impedance matching over all bands. The front-to-back lobe ratio was also a concern to reduce the possible interference from parasitic reflections in objects within the antenna's range.

To fulfill all these requirements, a microstrip slot antenna was developed to operate within 3 GHz to 5 GHz bandwidth. A circular patch was used to cover this bandwidth with an acceptable match impedance, and a reflector plane was added behind the ground plane, to provide a directive beam. Finally, a metallic box was used in the antenna and reflector surroundings, to isolate the antenna from adjacent reflections, and thus improve the front-to-back lobe ratio. The metallic box also helps to narrow the radiation pattern. In the end, the final antenna presented the following characteristics, at the central frequency of 4 GHz: HPBW equal to $46°$, gain equal to 9.3 dBi, and S_{11} coefficient between -20 and -10 dB within all bandwidths.

5. Discussion

As several antennas have been reported throughout this manuscript, for a better overview, Table 2 summarizes some characteristics of the antennas that were developed for bio-radar applications.

Table 2. Summary of the parameters of the reviewed antennas.

Radar Mode	Ref.	Antenna Type	Central Frequency (GHz)	Bandwidth (GHz)	Gain (dBi)	HPBW
CW	[9]	Axial-mode helical antenna	2.4	-	9.80	44.6°
		Patch antenna	2.4	-	-	60°
	[10]	One-dimensional patch array	24	-	8	33.7°
	[12]	Microstrip patch array (8 × 2)	24	24–25	17.15	-
		Microstrip patch array (8 × 6)	24	23.3–23.7	17	-
	[13]	Microstrip patch array (10 × 16)	77	74.6–79.7	25	12°
	[14]	SIW slot array	12	-	13	-
		SIW slot array	24	-	24	-
	[15]	Fractal-slot patch antenna	0.915	0.907–0.924	5.8	108°
	[18]	Yagi patch antenna	2.45	-	8.69	-
	[19]	Horn antenna	-	50–75	25	7°
	[20]	Annular ring patch antenna	2.4	-	-	132°
	[22]	Microstrip patch array	10	-	8	30°
	[23]	Microstrip patch array (2 × 1)	60	57.24–65.88	13	-
		Microstrip patch array (3 × 3)	60	57.24–65.88	16	-
		Microstrip patch array (6 × 2)	60	57.24–65.88	16	-
	[25]	Microstrip patch array (4 × 5)	24	-	17	24°
	[26]	Printed patch array (2 × 2)	5	-	9.7	20°
	[27]	Patch array attached to CMOS chip	5.8	-	10.77	14.5°
	[29]	Dipole array	2.4	2.53–2.4	10.8	49°
	[32]	Microstrip patch antenna	2.4	-	5.8	81°
		Microstrip patch array	2.4	-	5.8	37°
	[35]	Horn antenna	-	3.6–4	20	-
	[38]	Three Patch array	24	-	8.6	80°
		Single patch array	24	-	8.6	34°
	[40]	Microstrip patch array (2 × 4)	24	-	8.6	80°
	[41]	Microstrip patch array	5.77	-	16	-
	[42]	Aperture-coupled patch antenna	2.4	-	8	80°
		Aperture-coupled patch antenna	4.8	-	8	70°
	[43]	Microstrip patch array (2 × 2)	2.4	-	6.5	40°
	[44]	Adaptive beam-steering antenna	5.8	5.7–5.9	-	41°
UWB	[49]	Partial ground plane combined with parasitic patches emulating Yagi concept	4.37	4.3–7	-	43°
	[50]	GSA array	2.45	2.28–2.49	12.2	26°
	[52]	Microstrip slot antenna	4	3–5	9.4	46°

As one can see in Table 2, several designs have been proposed for different bio-radar application scenarios. Planar patch antennas were the type of antennas most used in the reviewed papers. This preference can be explained by the fact that this type of antenna has a low cost, low profile, is easy to be fabricated, and can potentially be integrated with the radar circuitry on the same printed circuit board.

Observing the characteristics of the revised antennas, the most used frequencies are 2.4 GHz, 5.8 GHz, 24 GHz, and 60 GHz, and typically the antennas presented a narrow bandwidth. In addition, the gains can vary between 5.8–25 dBi. Generally, the arrays show higher gain (10.8–25 dBi), when compared with a single antenna (5.8–9.80 dBi). As an exception, the single horn antennas presented in [19,35] had gain equal to 25 dBi and 20 dBi, respectively.

Regarding the materials used for the development of antennas for bio-radar, it seems to be a preference to commercial available low-cost materials, with well-known electromagnetic characteristics. Table 3 summarizes the materials used as a dielectric substrate in the reviewed antennas.

Table 3. Summary of the materials used as a dielectric substrate for bio-radar antennas.

Ref.	Antenna Frequency (GHz)	Substrate	ε_r
[18]	2.45		
[34]	2.4	RF-4	4.4
[24]	0.9–12		
[15]	0.9–2.5		
[12]	24	FR-4	3.55
[23]	60	Duroid	2.2
[25]	24	RO4350	3.48
[32]	2.4	RT-Duroid5880	2.2
[27]	5.8	RO4030	3.38
[10]	24.125	RO4835	3.66
[13]	77	RO3003C	3.0

Although the commercially available substrate materials had the permittivity values very well characterized by the manufacturer, some attention has to be taken during the design of ultra-wideband antennas. In this case, as reported before, the use of a single permittivity value during the simulation process for all bands can influence the results of the antenna in different frequencies.

After analyzing all the examples and the most common choices, some trade-offs can be established for the optimal antenna design, considering NCVS applications.

Starting with the signal quality matter, larger antennas confer directive beams, but antennas with larger sizes also have a larger near-field region [5], where the electromagnetic propagation can not be modelled linearly. In these cases, there is more area where the antenna behavior can not be foreseen. Nonetheless, directivity is seen by many authors as a crucial characteristic to accomplish high SNR and accuracy in NCVS signals, since it reduces the parasitic reflections and decreases clutter interference. In parallel, the main lobe beamwidth should not be too narrow, since, subsequently, a perfect alignment with the subject's chest-wall would be required. These also hamper the generalization of system operation by extending it to multiple subjects, since humans have different body structures and heights and prior calibrations would be needed.

The system size is also a concern, to enable its portability and to facilitate its usage. Using a single antenna for both TX and RX operations could be an immediate solution. However, proper isolation must be assured where the selection of low cost and simpler hardware components is preferable. Opting by two separate antennas implies that antennas should be closed together, so the radar operation is approximately as a monostatic, and the RCS is preserved [5]. Either way, i.e., using a single antenna or two separate antennas, the cross-talk must be minimized.

Regarding the frequency of operation, it is directly related to the radar operation mode, even though there are some studies that aim to determine the optimal frequency band for NCVS applications. In [53], a mathematical model and its simulation are presented to seek for the best operating frequency that allows both respiratory signal and cardiac detection. They conclude that the signal strength increases when using frequencies above 5 GHz, and it stabilizes until the lower region of the K-band. Above 20 GHz, the signal strength decreases slightly. Moreover, nonlinear phase modulation causes harmonic intermodulation, which is more evident for frequencies above 27 GHz than for frequencies around 5 GHz.

In [53], the authors also studied the relation between frequency and beamwidth. They considered large beamwidth above 20° of HPBW and narrow beamwidth around 7°. It was possible to conclude that signal strength from heartbeat and respiration varies with the frequency, if a large beamwidth is used. On the other hand, the signal strength is stable for narrow beamwidths. Hereinafter, the gain range for bio-radar applications was proposed in [29], where the authors state that a gain under 5 dBi is not enough, and the typical gain value for antennas applied in NCVS acquisition is around 9 dBi.

In summary, some key features directly influence the quality of the signal acquisition and take action in the system manufacturing. Figures 2 and 3 present the main characteristics that the authors took into account, when developing antennas for bio-radar applications.

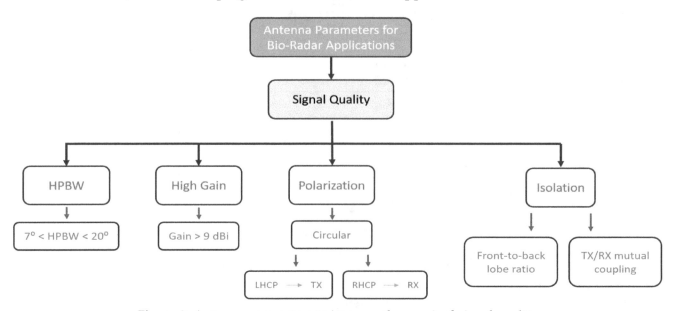

Figure 2. Antenna parameters to improve the acquired signal quality.

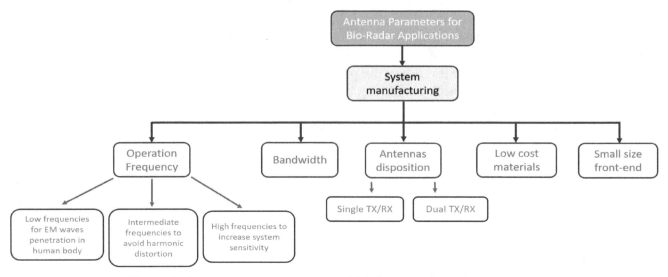

Figure 3. Antenna parameters to consider for manufacturing purposes.

6. Conclusions

With the advance of wireless technologies, remote health monitoring has become an emerging solution to promote well-being and to detect emergencies at indoor and outdoor scenarios.

In this way, several studies using Doppler radar for NCVS have been presented in the past years. However, most of them have described the development and/or improvement of the radar circuit based on demodulation methods and algorithms. As the antenna is a crucial part of the NCVS systems, a gap of information regarding the antennas parameters suitable for bio-radar was identified.

The knowledge of the effects of the antenna characteristics can help to improve the performance of radar systems, such as detection accuracy and sensitivity. Throughout this work, several solutions to improve signal acquisition were described. In summary, there is not a single solution to enhance antenna performance. Despite that, based on the presented survey, some guidelines for the design of antennas for vital signs acquisition using radar technology, are pointed out:

- Frequencies—High frequencies increase the sensibility of the radar system and intermediate frequencies can avoid harmonic distortion. Furthermore, low frequencies can be applied for surveillance and finding people applications due to the good penetration capacity of the EM wave through the materials.

- Circular polarization, with different rotation direction in TX and RX, is preferable—using different rotation direction increases the isolation between the transmitter and the receiver, minimizing negative influences of second order reflections and multi patch effects on the system;

- Avoid circulator usage if a single antenna is used. Instead, use alternative hardware components for signals division that do not decrease signal power;

- Directivity—directive beams can achieve high SNR and accuracy in NCVS signals, since its reduces parasitic reflections and decreases clutter interference;

- High gain—above 9 dBi at the main lobe is preferable. An array can be used to achieve a higher gain instead of a single antenna;

- Reduced size of the antennas can improve the portability of the system;

- The use of low cost materials.

Author Contributions: All the authors have contributed to this paper. C.G. and C.L. did the full research, its discussion, and wrote the manuscript. P.P. and J.V. did the technical revision of this manuscript and supervised all the research work.

Abbreviations

The following abbreviations are used in this manuscript:

AR	Axial Ratio
CMOS	Complementary Metal-Oxide-Semiconductor
CP	Circular Polarization
CW	Continuous Wave
EM	Electromagnetic
FMCW	Frequency-Modulated Continuous Wave
GSA	Ground Surround Antenna
HPBW	Half-Power Beamwidth
IQ	In-phase and Quadrature
LHCP	Left-Hand Circular Polarization
LNA	Low Noise Amplifier
LP	Linear Polarization

LTCC	Low-Temperature Co-Fired Ceramic
NCVS	Non-contact Vital Signs
PCB	Printed Board Circuit
QHC	Quadrature Hybrid Coupler
RCS	Radar-Cross Section
RF	Radio-Frequency
RHCP	Right-Hand Circular Polarization
RX	Reception
SDR	Software Defined Radio
SIW	Substrate Integrated Waveguide
SNR	Signal-to-Noise Ratio
SNIR	Signal-to-Noise-plus-Interference Ratio
TX	Transmission
UWB	Ultra-Wide Band

References

1.	Mpanda, R.S.; Liang, Q.; Xu, L.; Lin, Q.; Shi, J. Investigation on Various antenna design techniques for Vital Signs Monitoring. In Proceedings of the Cross Strait Quad-Regional Radio Science and Wireless Technology Conference (CSQRWC), Xuzhou, China, 21–24 July 2018. [CrossRef]
2.	Gouveia, C.; Malafaia, D.; Vieira, J.N.; Pinho, P. Bio-radar performance evaluation for different antenna designs. *URSI Radio Sci. Bull.* **2018**, *364*, 30–38.
3.	Boric-Lubecke, O.; Lubecke, V.M.; Droitcour, A.D.; Park, B.-K.; Singh, A. *Doppler Radar Physiological Sensing*; John Wiley & Sons: Cambridge, UK, 2015.
4.	Skolnik, M.I. *Introduction to Radar Systems*; Mc Grow-Hill: New York, NY, USA, 2001.
5.	Droitcour, A.D. Non-Contact Measurement of Heart and Respiration Rates with a Single-Chip Microwave Doppler Radar. Ph.D. Thesis, Stanford University, Stanford, CA, USA, 2006.
6.	Taylor, J.D. *Ultra-Wideband Radar Technology*; CRC Press: Boca Raton, FL, USA, 2000.
7.	Das, V.; Boothby, A.; Hwang, R.; Nguyen, T.; Lopez, J.; Lie, D.Y.C. Antenna evaluation of a non-contact vital signs sensor for continuous heart and respiration rate monitoring. In Proceedings of the IEEE Topical Conference on Biomedical Wireless Technologies, Networks, and Sensing Systems (BioWireleSS), Santa Clara, CA, USA, 15–18 January 2012; pp. 13–16.
8.	Fletcher, R.; Han, J. Low-cost differential front-end for Doppler radar vital sign monitoring. In Proceedings of the IEEE MTT-S International Microwave Symposium Digest, Boston, MA, USA, 7–12 June 2009; pp. 1325–1328.
9.	Boothby, A.; Hwang, R.; Das, V.; Lopez, J.; Lie, D.Y.C. Design of Axial-mode Helical Antennas for Doppler-based continuous non-contact vital signs monitoring sensors. In Proceedings of the IEEE Radio and Wireless Symposium, Santa Clara, CA, USA, 15–18 January 2012; pp. 87–90.
10.	Schäfer, S.; Diewald, A.R.; Schmiech, D.; Müller, S. One-dimensional Patch Array for Microwave-based Vital Sign Monitoring of Elderly People. In Proceedings of the 19th International Radar Symposium (IRS), Bonn, Germany, 20–22 June 2018; pp. 1–10.
11.	Shen, T.; Kao, T.J.; Huang, T.; Tu, J.; Lin, J.; Wu, R. Antenna Design of 60-GHz Micro-Radar System-In-Package for Noncontact Vital Sign Detection. *IEEE Antennas Wirel. Propag. Lett.* **2012**, *11*, 1702–1705. [CrossRef]
12.	Lan, S.; Xu, Y.; Chu, H.; Qiu, J.; He, Z.; Denisov, A. A Novel 24 GHz Microstrip Array Module Design for Bioradars. In Proceedings of the International Symposium on Antennas and Propagation (ISAP), Hobart, Australia, 9–12 November 2015; pp. 1–3.
13.	Lan, S.; Duan, L.; He, Z.; Yang, C.; Denisov, A.; Ivashov, S.; Anishchenko, L. A 77 GHz Bioradar Antenna Module Design Using Microstrip Arrays. In Proceedings of the IEEE International Symposium on Antennas and Propagation (APSURSI), Fajardo, Puerto Rico, 26 June–1 July 2016; pp. 1177–1178.
14.	Chioukh, L.; Boutayeb, H.; Deslandes, D.; Wu, K. Noise and sensitivity analysis of harmonic radar system for vital sign detection. In Proceedings of the IEEE MTT-S International Microwave Workshop Series on

RF and Wireless Technologies for Biomedical and Healthcare Applications (IMWS-BIO), Singapore, 9–11 December 2013; pp. 1–3.

15. Park, J.-H.; Jeong, Y.-J.; Lee, G.-E.; Oh, J.-T.; Yang, J.-R. 915-MHz Continuous-Wave Doppler Radar Sensor for Detection of Vital Signs. *Electronics* **2019**, *8*, 855. [CrossRef]

16. Gu, C.; He, Y.; Zhu, J. Noncontact Vital Sensing With a Miniaturized 2.4 GHz Circularly Polarized Doppler Radar. *IEEE Sens. Lett.* **2019**, *3*, 1–4. [CrossRef]

17. Kim, J.-G.; Sim, S.-H.; Cheon, S.; Hong, S. 24 GHz circularly polarized Doppler radar with a single antenna. In Proceedings of the European Microwave Conference, Paris, France, 25–27 October 2005; pp. 1382–1386.

18. Bo, H.; Fu, Q.; Xu, L.; Lure, F.; Dou, Y. Design and implementation of a 2.45 GHz RF sensor for non-contacting monitoring vital signs. In Proceedings of the Computing in Cardiology Conference (CinC), Vancouver, BC, Canada, 11–14 September 2016; pp. 1113–1116.

19. Chuang, H.; Kuo, H.; Lin, F.; Huang, T.; Kuo, C.; Ou, Y. 60-GHz Millimeter-Wave Life Detection System (MLDS) for Noncontact Human Vital-Signal Monitoring. *IEEE Sens. J.* **2012**, *12*, 602–609. [CrossRef]

20. Myoung, S.-S.; Park, J.-H.; Yook, J.-G.; Jang, B.-J. 2.4 GHz Bio-radar System with Improved Performance by Using Phase-Locked Loop. *Microw. Opt. Technol. Lett.* **2010**, *52*, 2074–2076. [CrossRef]

21. An, Y.-J.; Hong, Y.-P.; Jang, B.-J.; Yook, J.-G. Comparative Study of 2.4 GHz and 10 GHz Vital Signal Sensing Doppler Radars. In Proceedings of the 40th European Microwave Conference, Paris, France, 28–30 September 2010; pp. 501–504.

22. Myoung, S.-S.; An, Y.-H.; Yook, J.-G.; Jang, B.-J.; Moon, J.-H. A Novel 10 GHz Super-Heterodyne Bio-Radar System Based on a Frequency Multiplier and Phase-Locked Loop. *Prog. Electromagn. Res. C* **2010**, *19*, 149–162. [CrossRef]

23. Rabbani, M.S.; Ghafouri-Shiraz, H. Ultra-Wide Patch Antenna Array Design at 60 GHz Band for Remote Vital Sign Monitoring with Doppler Radar Principle. *J. Infrared Millim. Terahertz Waves* **2017**, *38*, 548–566. [CrossRef]

24. Van, N.T.P.; Tang, L.; Minh, N.D.; Hasan, F.; Mukhopadhyay, S. Extra Wide Band 3D Patch Antennae System Design for Remote Vital Sign Doppler Radar Sensor Detection. In Proceedings of the Eleventh International conference on Sensing Technology (ICST), Sydney, Australia, 4–6 December 2017; pp. 1–5.

25. Hsu, T.W.; Tseng, A.C.H. Compact 24 GHz Doppler Radar Module for Non-Contact Human Vital Sign Detection. In Proceedings of the International Symposium on Antennas and Propagation, Okinawa, Japan, 24–28 October 2016; pp. 994–995.

26. Xiao, Y.; Li, C.; Lin, J. A Portable Noncontact Heartbeat and Respiration Monitoring System Using 5-GHz Radar. *IEEE Sens. J.* **2007**, *7*, 1042–1043. [CrossRef]

27. Huang, J.; Tseng, C. A 5.8-GHz radar sensor chip in 0.18-μm CMOS for non-contact vital sign detection. In Proceedings of the IEEE International Symposium on Radio-Frequency Integration Technology (RFIT), Taipei, Taiwan, 24–26 August 2016; pp. 1–3.

28. Ma, X.; Wang, Y.; Song, W.; You, W.; Lin, J.; Li, L. A 100-GHz Double-Sideband Low-IF CW Doppler Radar in 65-CMOS for Mechanical Vibration and Biological Vital Sign Detections. In Proceedings of the IEEE MTT-S International Microwave Symposium, Boston, MA, USA, 2–7 June 2019; pp. 136–139.

29. Liang, Q.; Mpanda, R.S.; Wang, X.; Shi, J.; Xu, L. A Printed Dipole Array Antenna for Non-contact Monitoring System. In Proceedings of the Cross Strait Quad-Regional Radio Science and Wireless Technology Conference (CSQRWC), Xuzhou, China, 21–24 July 2018. [CrossRef]

30. Balanis, C.A. *Antenna Theory: Analysis and Design*, 4th ed.; John Wiley & Sons: Hoboken, NJ, USA, 2016; pp. 474–953.

31. Chan, K.K.; Tan, A.E.; Rambabu, K. Circularly Polarized Ultra-Wideband Radar System for Vital Signs Monitoring. *IEEE Trans. Microw. Theory Tech.* **2013**, *61*, 2069–2075. [CrossRef]

32. Nosrati, M.; Tavassolian, N. Experimental Study of Antenna Characteristic Effects on Doppler Radar Performance. In Proceedings of the IEEE International Symposium on Antennas and Propagation & USNC/URSI National Radio Science Meeting, San Diego, CA, USA, 9–14 July 2017; pp. 209–211.

33. Nosrati, M.; Tavassolian, N. Effects of Antenna Characteristics on the Performance of Heart Rate Monitoring Radar Systems. *IEEE Trans. Antennas Propag.* **2017**, *65*, 3296–3301. [CrossRef]

34. Li, C.; Lin, J. Complex signal demodulation and random body movement cancellation techniques for non-contact vital sign detection. In Proceedings of the IEEE MTT-S International Microwave Symposium Digest, Atlanta, GA, USA, 15–20 June 2008; pp. 567–570.

35.   Anishchenko, A.; Razevig, V.; Chizh, M. Blind Separation of Several Biological Objects Respiration Patterns by Means of a Step-Frequency Continuous-Wave Bioradar. In Proceedings of the IEEE International Conference on Microwaves, Antennas, Communications and Electronic Systems (COMCAS), Tel-Aviv, Israel, 13–15 November 2017; pp. 1–4.

36.   Remote Sensing Laboratory; Bauman Moscow State Technical University. BioRASCAN Radar for Detection and Diagnostic Monitoring of Humans. Available online: http://www.rslab.ru/english/product/biorascan/ (accessed on 30 September 2019).

37.   RFBeam. K-LC5 High Sensitivity Dual-Channel Transceiver. Available online: https://www.rfbeam.ch/ product?id=9 (accessed on 30 September 2019).

38.   Anishchenko, L. Challenges and Potential Solutions of Psychophysiological State Monitoring with Bioradar Technology. *Diagnostics* **2018**, *8*, 73. [CrossRef] [PubMed]

39.   RFBeam. K-LC2 Dual Channel Radar Transceiver. Available online: https://www.rfbeam.ch/product?id=5 (accessed on 28 October 2019).

40.   Li, P.; Hou, N. A Portable 24 GHz Doppler Radar System for Distant Human Vital Sign Monitoring. In Proceedings of the 5th International Conference on Information Science and Control Engineering, Zhengzhou, China, 20–22 July 2018; pp. 1050–1052.

41.   Malafaia, D.; Oliveira, B.; Ferreira, P.; Varum, T.; Vieira, J.; Tomé, A. Cognitive bio-radar: The natural evolution of bio-signals measurement. *J. Med. Syst.* **2016**, *40*, 219. [CrossRef] [PubMed]

42.   Zhu, F.; Wang, K.; Wu, K. A Fundamental-and-Harmonic Dual-Frequency Doppler Radar System for Vital Signs Detection Enabling Radar Movement Self-Cancellation. *IEEE Trans. Microw. Theory Tech.* **2018**, *66*, 5106–5118. [CrossRef]

43.   Gu, C.; Li, C.; Lin, J.; Long, J.; Huangfu, J.; Ran, L. Instrument-Based Noncontact Doppler Radar Vital Sign Detection System Using Heterodyne Digital Quadrature Demodulation Architecture. *IEEE Trans. Instrum. Meas.* **2010**, *59*, 1580–1588.

44.   Nieh, C.; Lin, J. Adaptive beam-steering antenna for improved coverage of non-contact vital sign radar detection. In Proceedings of the IEEE MTT-S International Microwave Symposium (IMS2014), Tampa, FL, USA, 1–6 June 2014; pp. 1–3.

45.   Nosrati, M.; Shahsavari, S.; Lee, S.; Wang, H.; Tavassolian, N. A Concurrent Dual-Beam Phased-Array Doppler Radar Using MIMO Beamforming Techniques for Short-Range Vital-Signs Monitoring. *IEEE Trans. Antennas Propag.* **2019**, *67*, 2390–2404. [CrossRef]

46.   Tseng, C.-H.; Chao, C.-H. Noncontact vital-sign radar sensor using metamaterial-based scanning leaky-wave antenna. In Proceedings of the IEEE MTT-S International Microwave Symposium (IMS), San Francisco, CA, USA, 22–27 May 2016; pp. 1–3.

47.   Alemaryeen, A.; Noghanian, S.; Fazel-Rezai, R. Antenna Effects on Respiratory Rate Measurement Using a UWB Radar System. *IEEE J. Electromagn. RF Microw. Med. Biol.* **2018**, *2*, 87–93. [CrossRef]

48.   Schantz, H.G. Introduction to Ultra-wideband Antennas. In Proceedings of the IEEE Conference on Ultra Wideband Systems and Technologies, Reston, VA, USA, 16–19 November 2003; pp. 1–9.

49.   Park, Z.; Li, C.; Lin, J. A Broadband Microstrip Antenna With Improved Gain for Noncontact Vital Sign Radar Detection. *IEEE Antennas Wirel. Propag. Lett.* **2009**, *8*, 939–942. [CrossRef]

50.   Tang, T.; Chuang, Y.; Lin, K. A Narrow Beamwidth Array Antenna Design for Indoor Non-contact Vital Sign Sensor. In Proceedings of the IEEE International Symposium on Antennas and Propagation, Chicago, IL, USA, 8–14 July 2012; pp. 1–2.

51.   Lv, H.; Liu, M.; Jiao, T.; Zhang, Y.; Yu, X.; Li, S.; Jing, X.; Wang, J. Multi-target Human Sensing via UWB Bio-radar Based on Multiple Antennas. In Proceedings of the IEEE International Conference of IEEE Region 10 (TENCON 2013), Xian, China, 22–25 October 2013; pp. 1–4.

52.   Mattia, V.D.; Petrini, V.; Pallotta, E.; Leo, A.D.; Pieralisi, M.; Manfredi, G.; Russo, P.; Primiani, V.M.; Cerri, G.; Scalise, L. Design and Realization of a Wideband Antenna for Non-contact Respiration Monitoring in AAL Application. In Proceedings of the IEEE/ASME 10th International Conference on Mechatronic and Embedded Systems and Applications (MESA), Senigallia, Italy, 10–12 September 2014; pp. 1–4.

53.   Li, C.; Xiao, Y.; Lin, J. Design Guidelines for Radio Frequency Non-contact Vital Sign Detection. In Proceedings of the 29th Annual International Conference of the IEEE Engineering in Medicine and Biology Society, Lyon, France, 23–26 August 2007; pp. 1651–1654.

14

Design of a Wideband L-Shape Fed Microstrip Patch Antenna Backed by Conductor Plane for Medical Body Area Network

Chai-Eu Guan * and Takafumi Fujimoto

Graduate School of Engineering, Nagasaki University, Nagasaki 852-8521, Japan
* Correspondence: guan@nagasaki-u.ac.jp

Abstract: This paper describes a compact patch antenna intended for medical body area network. The antenna is fed using a proximity coupling scheme to support the antenna that radiates in the free space and on the human body at the 2.45 GHz ISM band. The conductor plane is placed 2 mm or $0.0163\lambda_0$ (λ_0 is free space wavelength at 2.45 GHz) below the antenna to reduce backward radiation to the human body. Separation distance must be kept above 2 mm, otherwise, gain of the proposed antenna decreases when antenna is situated on the human body. The L-shape feed line is introduced to overcome impedance mismatch caused by the compact structure. The coupling gap between the proposed antenna and the length of the L-shape feed line are optimized to generate dual resonances mode for wide impedance bandwidth. Simulation results show that specific absorption rate (SAR) of the proposed antenna with L-shape feed line is lower than conventional patch antenna with direct microstrip feed line. The proposed antenna achieves impedance bandwidth of 120 MHz (4.89%) at the center frequency of 2.45 GHz. The maximum gain in the broadside direction is 6.2 dBi in simulation and 5.09 dBi in measurement for antenna in the free space. Wide impedance bandwidth and radiation patterns insensitive to the presence of human body are achieved, which meets the requirement of IoT-based wearable sensor.

Keywords: low profile antenna; impedance matching; medical body area network; dual resonances

1. Introduction

As the awareness for preventive healthcare grows, developments of antennas for wireless body area networks (WBANs) have been increased gradually. Remote monitoring in medical applications includes wearable devices available in various forms: implanted, body-centric, and textile-based sensors. Several frequency bands have been allocated for WBAN communication systems, which include industrial, scientific, and medical (ISM) band (2.4–2.48 GHz), and ultra wideband (UWB, 3.1–10.6 GHz) [1]. Transient characterization of body-centric wireless communications was conducted on UWB body-worn antennas to detect pulses at various postures of test subject. Preservation of the shape of the received pulse was demonstrated in [2]. An on-body propagation channel for hearing aids and its link loss model at 2.45 GHz ISM band was proposed [3]. In recent years, 2.45 GHz band has been allocated by FCC and ETSI (European Standards Organization) for the medical body area network (MBAN). The MBAN system is intended for vital physiology parameters monitoring such as blood pressure, electrocardiogram (ECG), and glucose level [4].

In smart MBAN application, antennas used for off-body communication include interaction between sensors attached to or implanted in human body to external terminal. A study by Hall et al. [5] shows that antenna in the vicinity of human body suffers from unstable surface currents due to the near field coupling with the human body. The body area communication network requires antenna to have stable reflection coefficient and radiation pattern, irrespective of its working environment.

Microstrip patch antennas reported in the literatures have a narrowband with high Q-factor, thus the input impedance of the antennas was susceptible to proximity coupling with the human body, resulting in poor radiation efficiency.

Various antenna designs have been proposed for MBAN applications such as planar inverted-F antennas (PIFAs) [6,7], electromagnetic bandgap-backed (EBG) monopole antenna [8], dual-mode switchable antennas [9,10] and stacked antennas [11–13]. Most of the reported antennas were driven by coaxial probes affixed orthogonally to the body surface. The protruded SMARF (subminiature version a radio frequency) connector is not suitable for wearable devices. Radiation pattern diversity in the dual mode switchable antenna raises a unique power requirement for the antenna to operate in off-body communication links [9,10]. Low profile printed antennas backed by EBG-reflector consist of a periodic structure that reduces electromagnetic leakage to the human body, but its operation bandwidth is limited by the reflection phase characteristic of the periodic EBG-structure [8]. Apart from the EBG-based reflector, a conductor plate can be used to reduce human body loading effect, but the separation gap between microstrip patch antenna (MSA) and conductor plane must be larger than a quarter wavelength, $\lambda_0/4$ (λ_0 is a wavelength in free space). This is due to out-of-phase reflected signal from conductor plane that cancels out a normally incident plane, which causes impedance mismatch at the input port of the antenna [14]. So, how could low profile structure (air gap less than $\lambda_0/4$) be achieved in MSA backed by a conductor plate, without compromising on antenna height and performance?

In this work, a novel L-shape proximity feeding scheme is proposed to obtain a low-profile antenna backed by a conductor plate. Wideband frequency response is achieved through dual resonances mode from L-shape feed line and the antenna. The length and coupling gap of the L-shape feed line are optimized to overcome the impedance mismatch caused by the small separation gap between the antenna and conductor plate. Antenna performances are evaluated in the aspect of reflection coefficient, radiation pattern, specific absorption rate in the free space, and on the human body.

2. Antenna Design

Figure 1a–c shows the direct-fed microstrip patch antenna (D-fed Rect-MSA), the proximity-fed rectangular MSA (P-fed Rect-MSA), and the proximity-fed I-shape MSA (P-fed I-shape MSA). First of all, the same patch size is used for D-fed Rect-MSA and P-fed Rect-MSA as shown in Figure 1a,b. Next, the length l_a of P-fed Rect-MSA is reduced to I-shape (Figure 1c) to decrease patch size. Last, impedance matching of the input port of P-fed Rect-MSA is achieved by using L-shape feed line. In the experimental results, the proposed P-fed I-shape MSA in Figure 1c is compared with the D-fed Rect-MSA in Figure 1a for antenna performance in the free space and on human body. Design and analysis of D-fed Rect-MSA, P-fed Rect-MSA and P-fed I-shape MSA are performed using FDTD-based full wave electromagnetic simulation software Remcom XFdtd.

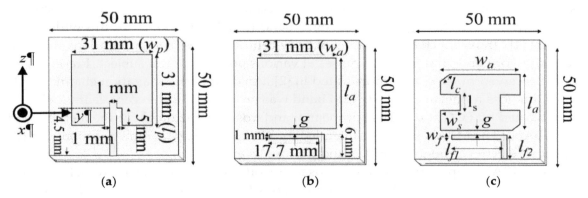

Figure 1. Geometry of the direct-fed (D-fed) antenna and the proximity-fed (P-fed) antennas. (**a**) D-fed Rect-MSA; (**b**) P-fed Rect-MSA; (**c**) proposed P-fed I-shape MSA.

2.1. Antenna Design and L-Shape Feeding Scheme

The patch antenna of D-fed Rect-MSA and P-fed Rect-MSA is set to $0.253\lambda_0 \times 0.253\lambda_0$ at 2.45 GHz ($w_p = w_a$ and $l_p = l_a$, refer to Figure 1a,b) for comparison purpose. Next, the patch antenna of the proposed P-fed Rect-MSA, which exhibits broad impedance bandwidth, is reduced to become I-shape as shown in Figure 1c. After the size of the antenna is reduced, a drastic increase of the resistance and reactance of the input port of the antenna at 2.45 GHz are observed. Therefore, L-shape feeding scheme is proposed and used to match impedance between P-fed I-shape MSA and feed line. The D-fed Rect-MSA and P-fed I-shape MSA (proposed) are fabricated on 3.2 mm substrate with dielectric constant $\varepsilon_r = 3.3$, and $\tan \delta = 0.003$ for comparison purpose. The resonance frequency from L-shape feed line is estimated, as follows:

$$f_{L\text{-}shape} = \frac{1}{2\pi \sqrt{(L_f C_g)}} \tag{1}$$

L_f is the total inductance from l_{f1} and l_{f2} of the L-shape feed line; C_g is the capacitance associated with the coupling gap g shown in Figure 1b,c. The effect of the coupling gap on resonance frequency in Equation (1) will be discussed in simulation results.

The design procedure starts by setting antenna resonance frequency to 2.45 GHz. In Figure 1b,c, the L-shape feed line is proposed and placed in the vicinity of P-fed Rect-MSA to generate two resonance frequencies in order to obtain wide impedance bandwidth. Next, the patch size of P-fed Rect-MSA is reduced to become P-fed I-shape MSA, which is suitable for wearable sensors. To reduce the human body loading effect on the antenna performance, a conductor plate is placed on the bottom of the antenna. The separation gap between the conductor plate and antenna is kept small to realize low profile structure. However, the low profile structure leads to a poor reflection coefficient in the proposed P-fed I-shape MSA backed by the conductor plate. To improve the reflection coefficient, key parameters such as coupling length l_{f1} and coupling gap g of L-shape proximity feeding scheme are studied to reduce high impedance at the input port.

2.2. Parameter Analysis to Achieve Wideband Response

At the beginning of the antenna design, an identical patch antenna that resonates at 2.45 GHz is chosen for D-fed Rect-MSA and P-fed Rect-MSA. The antennas are then fed by difference feeding techniques for comparison purpose. Figure 2 shows simulated reflection coefficients for D-fed Rect-MSA and P-fed Rect-MSA in the free space. By using the proximity feeding scheme presented in this work, the simulated impedance bandwidth of P-fed Rect-MSA is 1.8 times wider than the one in D-fed Rect-MSA, owning to the dual resonances mode generated by the L-shape feed line and the patch antenna. To obtain a wide impedance bandwidth, the physical length l_{f1} of the L-shape feed line of P-fed Rect-MSA is optimized to generate lower resonant mode from L-shape feed line at 2.4 GHz.

After the length l_a of P-fed Rect-MSA in Figure 1b is reduced from 31 mm ($0.253\lambda_0$) to 27 mm ($0.221\lambda_0$), the upper resonant frequency is shifted from 2.45 GHz to 2.72 GHz as illustrated in Figure 3a. Following the length reduction (l_a) of P-fed Rect-MSA, the upper and lower resonant frequencies fall apart (dashed black line in Figure 3a). Figure 3b also demonstrates shift in upper resonance frequency and drastic increase of Ohmic resistance at 2.45 GHz ISM band after the length l_a of rectangular P-fed Rect-MSA is reduced (from 31 mm to 27 mm). To overcome the impedance mismatch between the antenna and the L-shape feed line at 2.45 GHz, the center body of P-fed Rect-MSA is truncated to become P-fed I-shape MSA. As a result, the upper resonant frequency is shifted back to 2.45 GHz ISM band and the two resonant frequencies become close to each other again (solid red line showed in Figure 3a). By increasing the length l_s of truncated section in P-fed I-shape MSA, Ohmic resistance at the input port is matched to 50 Ω at 2.45 GHz ISM band, as depicted in blue and red lines of Figure 3b. The dual resonances mode with good reflection coefficient is achieved when l_s is increased to 5.6 mm. Then, the center frequency of P-fed I-shape MSA is adjusted to 2.45 GHz by truncating upper-left and lower-right corners ($l_c = 2.4$ mm) of the P-fed I-shape MSA.

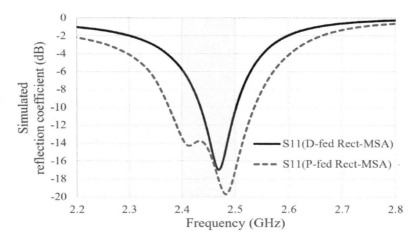

Figure 2. Simulated reflection coefficient of D-fed (direct-fed) Rect-MSA and P-fed (proximity-fed) Rect-MSA ($g = 0.5$ mm, $l_a = 27$ mm).

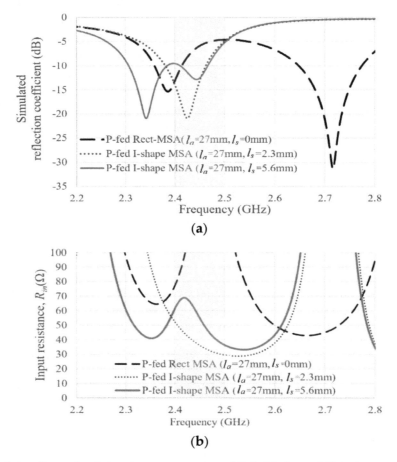

Figure 3. Length l_s is adjusted to bring input resistance of P-fed I-shape MSA close to 50 Ω at 2.45 GHz ISM band, after P-fed Rect-MSA in dashed black line ($g = 0.5$ mm, $l_a = 31$ mm, $l_s = 0$ mm) is miniaturized to become P-fed I-shape MSA ($g = 0.5$ mm, $l_a = 27$ mm, l_s varies from 2.3 mm to 5.6 mm). (**a**) Simulated reflection coefficient; (**b**) Input resistance R_{in}.

Figure 4 shows optimization of the coupling gap g between the I-shape patch and the L-shape feed line of P-fed I-shape MSA in order to improve reflection coefficient. A wide impedance bandwidth is realized by adjusting the coupling gap g of the L-shape feed line to reduce inductive reactance at the input port of the proposed antenna. After the coupling gap g is reduced from 0.8 mm to 0.2 mm, inductive reactance is reduced to half of its initial value at 2.45 GHz ISM band as depicted in Figure 4b. As a result, wide impedance bandwidth with reflection coefficient less than 10 dB criterion from 2.36

to 2.48 GHz is obtained in the free space (dashed black line in Figure 4a). From parametric analysis in Figures 2–4, the process of obtaining wideband characteristic through L-shape proximity feeding scheme can be summarized, as follows: (1) length l_s of truncated section of P-fed I-shape MSA is increased to bring upper and lower resonance frequencies to be close to each other. (2) Coupling gap g is decreased to reduce inductance of the input impedance, so that wide impedance bandwidth with reflection coefficient less than 10 dB criterion could be achieved.

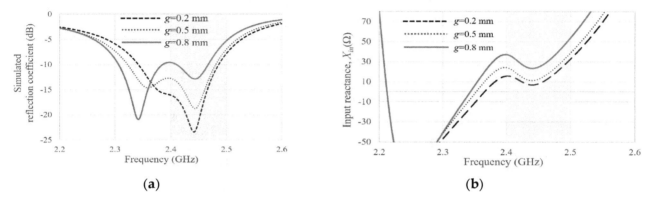

Figure 4. Input impedance matching in P-fed I-shape MSA through coupling gap g adjustment (refer to Table 2 for dimensions of the antenna). **(a)** Simulated reflection coefficient; **(b)** input reactance, X_{in}.

The Smith chart in Figure 5 describes impedance matching of the proposed P-fed I-shape MSA using the L-shape feed line. After the proposed antenna size is reduced, the input impedance of the antenna increases drastically, especially resistance at the input port of the antenna at 2.45 GHz (Figure 3). Next, the coupling gap between the L-shape feed line and the antenna is reduced to 0.5 mm to make the proposed antenna to become less inductive (Figure 4). Finally, two corners of the proposed antenna are truncated to move input impedance at 2.45 GHz closer to circle centered at the coordinate $(0, 0)$ where $Z = 50\ \Omega$, as shown by yellow node in the smith chart.

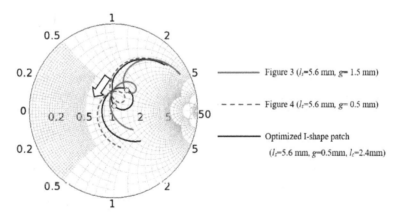

Figure 5. Input impedance optimization after the size of the antenna is changed from rectangular shape (Figure 1b) to I-shape MSA (Figure 1c).

2.3. Conductor Plate in Low Profile Antenna Affixed to Human Body

Figure 6 shows the geometry of P-fed I-shape MSA and simulation setup for antennas affixed to the human body. The proposed antenna is placed on above $200 \times 200\ mm^2$ tissue model which consists of skin, fat, and muscle, mimicking human tissue. Table 1 shows dielectric properties obtained from IT'IS Foundation for human tissue model. To reduce backward radiation to the human body, the conductor plate is proposed and placed below the antenna. D-fed Rect-MSA and P-fed I-shape

MSA, which are placed in the free space and on human body, are simulated in the following conditions: without the conductor plate, and with the conductor plate. When the 50×50 mm^2 conductor plate is placed 2 mm below the P-fed I-shape MSA, the antenna has a poor reflection coefficient. The impedance matching technique through the L-shape feed line discussed in Section 2.2 is repeated until the input impedance of the proposed antenna is approximately 50 Ω at 2.45 GHz. Compared with the D-fed Rect-MSA, the proposed P-fed I-shape MSA has more key parameters (coupling gap g, and truncated section of I-shape antenna, l_s) to tune for impedance matching.

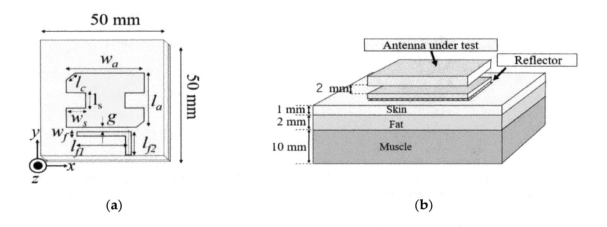

(a)   (b)

Figure 6. Geometry of the proposed P-fed I-shape MSA. (**a**) Proposed antenna after downsizing; (**b**) simulation setup on the multi-layer tissue model.

Table 1. Multilayer human tissue model [15].

Tissue	Relative Permittivity ε_r	Electric Conductivity (S/m)	Tissue Thickness (mm)
Skin	38.1	1.43	1 mm
Fat	10.8	0.26	2 mm
Muscle	52.8	1.74	10 mm

Figure 7 shows the effect of separation distance between the proposed P-fed I-shape MSA and conductor plate on the gain performance of the antenna. The separation distance is varied from 1 mm to 4 mm, then the gain of antenna in the free space and on human body are observed. Figure 7a shows the simulated gain of the antenna on the multi-layer tissue model (when the separation distance is 4 mm) is above 6 dBi across 2.45 GHz ISM band, which is higher than the gain of the antenna in the free space. Low gain in the free space is due to limited size (50×50 mm^2) of the ground plane and reflector used. However, the increase of contact area between the human body and reflector enlarges the conductor plate electrically. A separation distance of 2 mm (Figure 7b) is suitable for this work, since the gain difference between the antenna in the free space and on the human body is less than 1 dB at 2.45 GHz ISM band. When the separation distance is reduced further to 1 mm, the human body loading effect reduces the gain of the antenna on the human body (Figure 7c). The gain difference between the antenna on the human body and antenna in the free space is more than 1 dB. For MBAN application, it is recommended to keep the gain of the antenna consistent in the free space and on human tissue. Otherwise, the gain difference corresponds to the working environment, which may raise the demand of the RF power amplifier with auto gain control (AGC) feature for body area network communication. Therefore, the separation distance of 2 mm is chosen in this work to obtain stable gain for the antenna in the free space and on the human body.

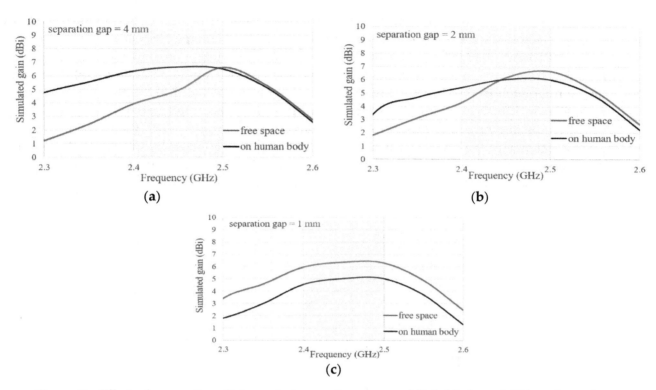

Figure 7. Effect of separation distance between the proposed P-fed I-shape MSA and conductor plate, simulated in the free space and on the human body; size of the antenna and conductor plate is 50 mm × 50 mm. **(a)** Separation distance = 4 mm; **(b)** separation distance = 2 mm; **(c)** separation distance = 1 mm.

Figure 8 shows the simulated reflection coefficient for the D-fed Rect-MSA and P-fed I-shape MSA in the free space and on the multi-layer tissue model. The operation bandwidth of the proposed P-fed I-shape MSA at 2.45 GHz ISM band is wider than the D-fed Rect-MSA, irrespective of the working environment. The D-fed Rect-MSA suffered a 30 MHz frequency shift after the antenna is moved from the free space to the human body. It is worth mentioning that the proposed P-fed I-shape MSA exhibits reflection coefficient less sensitive to human body loading effect. The resonance frequency of the proposed P-fed I-shape MSA experiences only 10 MHz frequency shift at upper frequency. It is worth noting that the reactance component of the proposed L-shape feed line is more robust to the change in the working environment from the free space to the human body, which satisfies the requirements of MBAN applications.

Figure 9a shows the influence of the conductor plate's size on the gain response of the proposed antenna in the free space. As can be seen, the conductor plate acts as a reflector for P-fed I-shape at certain frequencies, start from 2.45 GHz to 2.5 GHz. The size of the conductor plate determines front-to-back (F/B) lobe ratio, where the antenna backed by larger conductor plate has higher gain than the one backed by 50 × 50 mm^2 conductor plate (Figure 9a). Since this work focuses on the compact antenna design, the 50 × 50 mm^2 ($0.4\lambda_0 \times 0.4\lambda_0$) conductor plate is selected because it has stable gain at center frequency (2.45 GHz) for the antenna in the free space and on human body. Figure 9b shows the contribution of the conductor plate on the gain of P-fed I-shape MSA when it is placed on human body and simulated as follows: (1) P-fed I-shape MSA with conductor plate; (2) P-fed I-shape MSA without conductor plate. For the P-fed I-shape MSA without the conductor plate, gain degradation exceeds 1.5 dB due to human body loading effect, while the gain of the P-fed I-shape MSA backed by the conductor plate is slightly affected. Additionally, gain variation in P-fed I-shape MSA backed by the conductor plate when the working environment is changed (from the free space to the human body) is less than 1 dB from 2.42 GHz to 2.5 GHz.

Figure 10a shows the radiation efficient of the proposed P-fed I-shape MSA and D-fed Rect-MSA in the free space. Across the 2.45 GHz ISM band, the proposed P-fed I-shape MSA and D-fed Rect-MSA have radiation efficiency over 90%. As for the D-fed Rect-MSA, radiation efficiency starts to decrease when operating frequency is located below or above the center frequency (2.46 GHz). Loss due to impedance mismatch in the direct microstrip feed line of D-fed Rect-MSA limits beamwidth of the antenna. Figure 10b shows radiation efficiency of the proposed P-fed I-shape MSA and D-fed Rect-MSA when the antennas are affixed to the human body. Compared with the D-fed Rect-MSA, the radiation efficiency of the proposed P-fed I-shape MSA is less affected by human body loading. Although decrease in radiation efficiency of the P-fed I-shape MSA affixed to human body is observed, the antenna's radiation efficiency exceeds 80% from 2.44 GHz to 2.54 GHz.

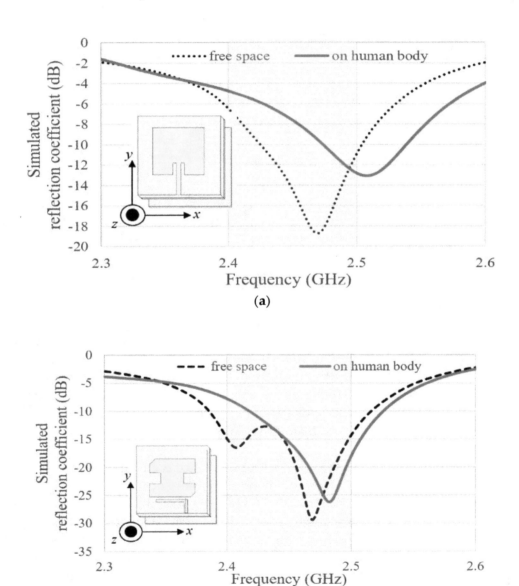

Figure 8. Shift in resonance frequency due to human body loading effect. **(a)** D-fed Rect-MSA; **(b)** proposed P-fed I-shape MSA (refer to Table 2 for the dimensions of the related antennas).

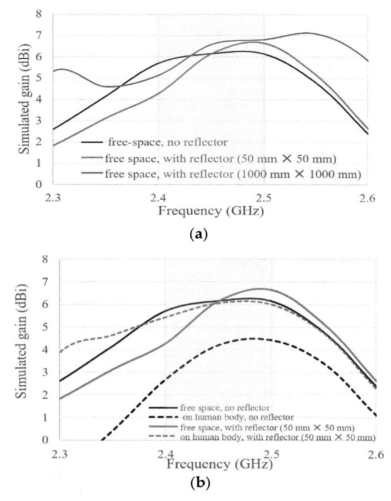

Figure 9. Influence of the conductor plate size on the gain of the proposed antenna (refer to Table 2 for dimensions of the proposed antenna). (**a**) P-fed I-shape MSA in the free space only; (**b**) P-fed I-shape MSA in the free space and on the human body.

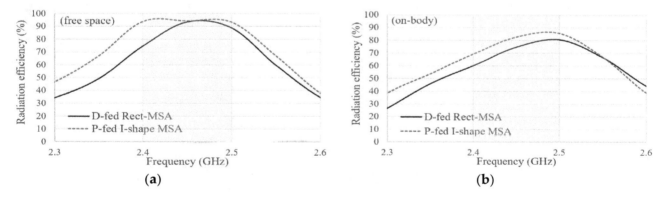

Figure 10. Radiation efficient of P-fed I-shape MSA and D-fed Rect-MSA backed by conductor plate. (**a**) Antennas in the free space; (**b**) antennas on the human body.

Figure 11 shows simulated 3D radiation patterns for the P-fed I-shape MSA backed by conductor plate in the free space and on the human body at 2.45 GHz, in Figure 11a,b respectively. The back lobe level decreases when the operating environment is changed from the free space to human body. For the antenna affixed onto the human body, the area of contact between the conductor plate of the P-fed I-shape MSA increases, thus the conductor plate becomes electrically large to suppress backward radiation. The gain scale in Figure 11 shows gain of the antenna on the human body remains at 6.2 dBi regardless of the operating environment, however, the back lobe level is suppressed when the antenna is moved from the free space (Figure 11a) to the human body (Figure 11b). Next, the specific absorption rate is used to calculate average radiated power absorbed by human tissue due to electromagnetic (EM) leakage from the proposed antenna.

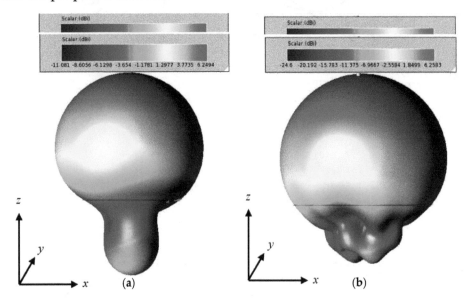

Figure 11. 3D radiation patterns of the P-fed I-shape MSA backed by conductor plate at 2.45 GHz in xz-plane. (**a**) Antenna in the free space; (**b**) antennas on the human body.

2.4. Specific Absorption Rate (SAR)

Amount of EM energy absorbed by the human body could be measured through specific absorption rate (SAR). According to FCC regulations [16] and IEEE C95.1 [17], EM energy absorbed by the human body should not exceed 1.6 W/kg averaged over 1 g of tissue. Figure 12 shows the SAR simulation setup for the proposed antenna (backed by conductor plate) placed on multi-layer tissue. For SAR evaluation, an identical simulation environment is used to evaluate SAR for D-fed Rect-MSA and the proposed P-fed I-shape MSA, i.e., 50×50 mm^2 conductor plate size is used. It is worthy to mention that the maximum SAR in the P-fed I-shape MSA is 0.6414 W/kg, which is lower than SAR = 1.524 W/kg in D-fed Rect-MSA when the incident power density at the antenna surface is 100 mW/cm^2 (based on FCC-regulated maximum power density). Low SAR in the proposed P-fed I-shape MSA is in agreement with simulation results in Figure 11b, which shows that the back lobe is suppressed when the P-fed I-shape MSA backed by the conductor plate is moved from the free space to the human body. Next, the P-fed I-shape MSA and D-fed Rect-MSA are fabricated, measured, and compared to verify the robustness of the antennas against human body loading effect.

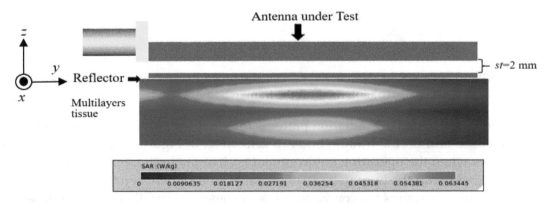

Figure 12. Specific absorption rate (SAR) simulation setup to calculate electromagnetic energy absorbed by human tissue.

3. Fabrications

The proposed P-fed I-shape MSA and D-fed Rect-MSA were fabricated for comparison purpose. The antennas were fabricated on polyphenylene ether (PPE) substrate (inset of Figure 13). Styrofoam is then inserted as a spacer between antennas (substrate thickness $t_a = 3.2$ mm) and conductor plates (substrate thickness $t_r = 0.8$mm). The total antenna height, including the conductor plate and styrofoam ($st = 2$ mm), is 6 mm. The final dimensions of the P-fed I-shape MSA are listed in Table 2. SMA RF connectors were mounted on the rear side of the antennas to ensure that the conductor plate on the bottom of the P-fed I-shape MSA and D-fed Rect-MSA is in contact with the human body.

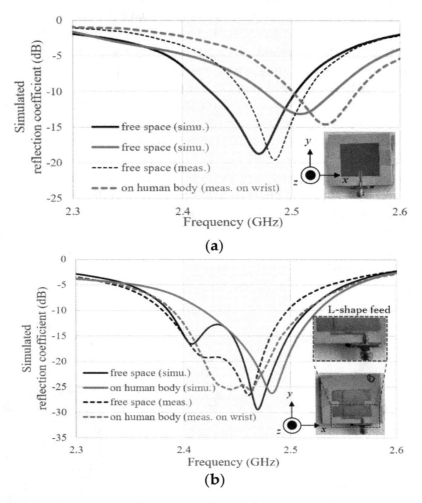

(a)

(b)

Figure 13. Simulated and measured reflection coefficient of the antenna in the free space and on human body. **(a)** D-fed Rect-MSA; **(b)** P-fed I-shape MSA.

Table 2. Dimension of the proposed P-fed I-shape MSA of Figure 6 (units in mm).

w_a	l_a	w_s	l_s	w_f	l_{f1}	l_{f2}	g	l_c	st
31.4	27.0	6.0	5.6	1.0	17.7	6.0	0.5	2.4	2.0

4. Experimental Results

The proposed antenna (P-fed I-shape MSA) is compared with D-fed Rect-MSA for the antenna in the free space and on the human body. Keysight 8719D vector network analyzer (50 MHz to 13.5 GHz) was used to measure reflection coefficients of the antennas in the free space and on human body. To investigate the human body loading effect, the proposed antenna was affixed on different parts of the human body by using adhesive tape. Radiation patterns and antenna gain were measured and validated in the anechoic chamber. Horn antenna (Microwave Factory MDH0118) was used as pre-calibrated standard gain antenna to measure the radiation patterns of the proposed antenna. The gain of the proposed antenna is measured and calculated by using the gain comparison method.

4.1. Human Body Loading Effect on the Antenna Reflection Coefficient

Figure 13 shows the reflection coefficients of the proposed P-fed I-shape MSA and D-fed Rect-MSA in the free space and on the human body. As can be seen in the figure, measured reflection coefficients of the antennas in the free space and on the human body are agreed with simulation ones. Comparison of Figure 13a with Figure 13b demonstrates that, in the presence of the human body, the L-shape proximity feeding scheme introduced in this work is more robust to the change of working environment, compared with direct feeding scheme in D-fed Rect-MSA. In simulation results, lower resonance frequency of P-fed I-shape MSA disappears when the antenna is affixed onto human tissue model. Contrarily, the measurement result shows lower resonance frequency and is not affected when the antenna is affixed to the wrist. The disagreement in the lower resonance frequency between simulation and measurement happens because the area of the wrist exposed to the electromagnetic wave is smaller than than in the human tissue model (200 mm × 200 mm) used in the simulation. To further verify the measured result, extra measurements on the reflection coefficients of the antenna were made on different parts of human body.

Figure 14 shows the measured reflection coefficient of the P-fed I-shape MSA affixed to various positions on the human body. It can be observed that measured reflection coefficient is insensitive to being situated in close proximity to the human body. Reflection coefficient is only slightly affected when the antenna is moved from wrist to the upper arm and chest. The proposed antenna has impedance bandwidth (satisfies reflection coefficient less than 10-dB criterion) of 120 MHz at 2.45 GHz ISM band, which is wider than D-fed Rect-MSA.

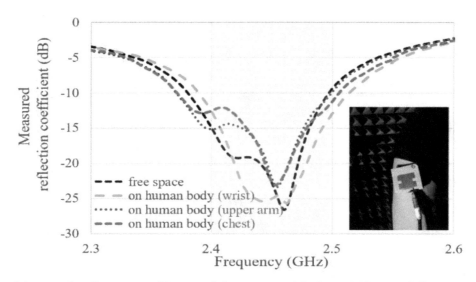

Figure 14. Measured reflection coefficient of the proposed I-shape MSA on difference parts of the human body.

4.2. Radiation Patterns and Measured Gain

Figures 15 and 16 shows normalized radiation patterns for the proposed P-fed I-shape MSA at xz plane ($\varphi = 0°$) and yz plane ($\varphi = 90°$) at 2.45 GHz. In Figures 15 and 16, all dataset in xz plane (co-polarized E_φ) and yz plane (co-polarized E_θ) are divided by the peak gain at their respective planes. Next, the normalized gain is plotted starting from the maximum value of 0 dB. Figure 15a,b shows, for the case of antenna in the free space, the measured radiation patterns coincidence with the simulated ones at broadside scenario along the positive z-axis. Simulated gain is 6.2 dBi and the measured gain is 5.09 dBi at 2.45 GHz along z-axis. Measured radiation patterns in both planes (xz-plane and yz-plane) are agreed with the simulations results for the antenna placed in the free space. By considering the reliability of the simulation results, broadside radiation patterns coincident to the simulation results are expected for the antenna affixed to the human body.

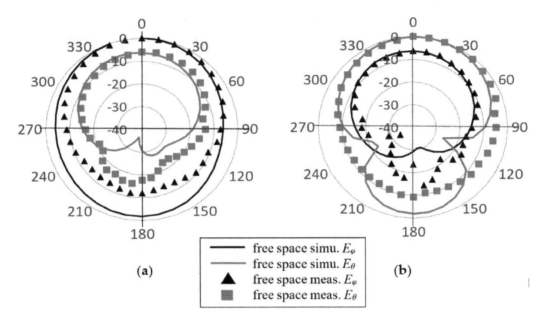

(a)   (b)

	free space simu. E_φ
	free space simu. E_θ
▲	free space meas. E_φ
■	free space meas. E_θ

Figure 15. Simulated and measured radiation patterns of P-fed I-shape MSA at 2.45 GHz for antenna used in off-body communication. (**a**) xz-plane; (**b**) yz-plane.

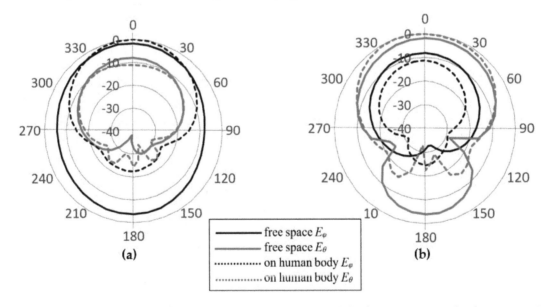

(a)   (b)

	free space E_φ
	free space E_θ
·············	on human body E_φ
·············	on human body E_θ

Figure 16. Radiation patterns of P-fed I-shape MSA at 2.45 GHz for antenna in the free space (solid) and on human body (dashed). (**a**) xz plane; (**b**) yz plane.

Figure 16 shows the dominant electric fields in the observation planes (E_φ in xz plane, E_θ in yz plane) for the antenna in the free space and on the human body. The proposed P-fed I-shape MSA has stable antenna gain at broadside, even when the antenna is being situated on the human body. The radiation pattern of the antenna on the human body is unidirectional, which is similar to radiation pattern of the antenna in the free space. The E_φ and E_θ components of the P-fed I-shape MSA in free space are slightly lower than the one affixed to the human body, due to the high back lobe of the antenna placed in the free space, which has a limited area of conductor plate (50×50 mm^2). When the proposed antenna is affixed to the human body, the contact area between the conductor plate and the human body becomes electrically large, therefore the back lobe level is suppressed.

Table 3 shows comparisons of the proposed antenna with other state-of-art antennas for the body area network. There are distinct differences between unidirectional antennas and omnidirectional UWB antennas in terms of gain characteristic and impedance bandwidth, therefore antennas with unidirectional characteristics are selected for comparison. As can be seen in the table, the proposed antenna has an operation bandwidth almost the same as EBG-backed antenna. Wide impedance bandwidth is achieved in the work, owing to extra resonance from L-fed feed line. For low profile antennas discussed, the size of conductor plate in the proposed work is slightly smaller than a metasurface that consists of periodic structure [7,8]. Gain difference between the proposed antenna in the free space and on the human body is less than 1 dB across 2.45 GHz ISM band. Air gap of 2 mm $(0.016\lambda_0)$ in the proposed antenna is smaller than air gap in EBG-backed antennas in [7,8]. Regardless of the operating environment, gain of the proposed antenna has a stable radiation pattern which makes its suitable to be deployed on the human body and remote terminals.

Table 3. Performance comparison with other state-of-art wearable antennas.

Ref.	Size (mm^3) $w \times l \times h$ *1	Size Comparison	f_0 (GHz)	Radiation Pattern	BW (MHz)	Max. Gain *2 (dBi)
[7]	$83 \times 83 \times 7$	321.5%	2.45	Unidirectional	130	4.5
[8]	$68 \times 38 \times 6.6$	113.7%	2.45	Unidirectional	119.6	6.88
[9]	$\pi(24)^2 \times 3.2$	38.6%	2.45	Unidirectional	25	3.83
This work	$50 \times 50 \times 6$	100%	2.45	Unidirectional	120	5.09

*1 Total height of antenna includes antenna thickness, air gap and EBG (electromagnetic bandgap) plate or conductor plate. *2 Measured peak gain for antenna in the free space.

5. Conclusions

A compact microstrip patch antenna backed by a conductor plane has been presented. The conductor plate functions as reflector, and is placed below the antenna to suppress EM radiation to the human body. The proposed antenna with the L-shape proximity feeding scheme generates a dual-resonances mode for a wide impedance bandwidth. Additionally, the proposed feeding scheme offers more parameters for impedance matching in order to realize low profile antenna. A prototype was designed, fabricated, and compared with the conventional direct-fed patch antenna. Measured reflection coefficient and radiation characteristics of the proposed antenna agree well with the simulation results, which shows wide impedance bandwidth at 2.45 GHz ISM band for antenna working in the free space and on the human body. The calculated SAR value below FCC limitation demonstrates that the proposed antenna is suitable for smart MBAN applications. For future works on the body-centric antenna, study on the reflection phase corresponds to the separation distance between the antenna and conductor plate is required to find achievable maximum gain. Additionally, fabrication of a phantom mimicking the characteristic of human tissue is required for radiation pattern measurement of the body-centric antenna affixed to the human body.

Author Contributions: This article is contributed by 2 authors; their individual contributions are listed out, as follows: Conceptualization, C.-E.G. and T.F.; methodology, C.-E.G. and T.F.; software, C.-E.G.; validation, C.-E.G.; formal analysis, C.-E.G.; investigation, C.-E.G. and T.F.; resources, C.-E.G.; data curation, C.-E.G. and T.F.; writing—original draft preparation, C.-E.G.; writing—review and editing, C.-E.G. and T.F.; visualization, C.-E.G.; supervision, T.F.; project administration, C.-E.G. and T.F.; funding acquisition, C.-E.G. All authors have read and agreed to the published version of the manuscript.

References

1. Yan, S.; Soh, P.-J.; Vandenbosch, G.A.E. Wearable Ultrawideband Technology—A Review of Ultrawideband Antennas, Propagation Channels, and Applications in Wireless Body Area Networks. *IEEE Access* **2018**, *6*, 42177–42185. [CrossRef]
2. Alomainy, A.; Sani, A.; Rahman, A.; Santas, J.G.; Hao, Y. Transient Characteristics of Wearable Antennas and Radio Propagation Channels for Ultrawideband Body-Centric Wireless Communications. *IEEE Trans. Antennas Propag.* **2009**, *57*, 875–884. [CrossRef]
3. Chandra, R.; Johansson, A.J. A Link Loss Model for the On-Body Propagation Channel for Binaural Hearing Aids. *IEEE Trans. Antennas Propag.* **2013**, *61*, 6180–6190. [CrossRef]
4. Wang, D.; Evans, D.; Krasinski, R. IEEE 802.15.4J: Extend IEEE 802.15.4 radio into the MBAN spectrum [Industry Perspectives]. *IEEE Wirel. Commun.* **2012**, *19*, 4–5. [CrossRef]
5. Hall, P.S.; Hao, Y.; Nechayev, Y.I.; Alomainy, A.; Constantinou, C.; Parini, C.; Kamarudin, M.R.; Salim, T.Z.; Hee, D.T.M.; Dubrovka, R.; et al. Antennas and propagation for on-body communication systems. *IEEE Antennas Propag. Mag.* **2007**, *49*, 41–58. [CrossRef]
6. Lin, C.; Saito, K.; Takahashi, M.; Ito, K. A Compact Planar Inverted-F Antenna for 2.45 GHz On-Body Communications. *IEEE Trans. Antennas Propag.* **2012**, *60*, 4422–4426. [CrossRef]
7. Gao, G.; Hu, N.; Wang, S.; Yang, C. Wearable planar inverted-F antenna with stable characteristic and low specific absorption rate. *Microw. Opt. Technol. Lett.* **2018**, *60*, 876–882. [CrossRef]
8. Abbasi, M.A.B.; Nikolaou, S.S.; Antoniades, M.A.; Stevanović, M.N.; Vryonides, P. Compact EBG-Backed Planar Monopole for BAN Wearable Applications. *IEEE Trans. Antennas Propag.* **2017**, *65*, 453–463. [CrossRef]
9. Tong, X.; Liu, C.; Liu, X.; Guo, H.; Yang, X. Switchable ON-/OFF-Body Antenna for 2.45 GHz WBAN applications. *IEEE Trans. Antennas Propag.* **2018**, *66*, 967–971. [CrossRef]
10. Mendes, C.; Peixeiro, C. A Dual-Mode Single-Band Wearable Microstrip Antenna for Body Area Networks. *IEEE Antennas Wirel. Propag. Lett.* **2017**, *16*, 3055–3058. [CrossRef]
11. Yang, D.; Hu, J.; Liu, S. A Low-Profile UWB Antenna for WBAN Applications. *IEEE Access* **2018**, *6*, 25214–25219. [CrossRef]
12. Zhu, X.Q.; Guo, Y.X.; Wu, W. Miniaturized Dual-Band and Dual-Polarized Antenna for MBAN Applications. *IEEE Trans. Antennas Propag.* **2016**, *64*, 2805–2814. [CrossRef]
13. Chandran, A.R.; Timmons, N.; Morrison, J. Stacked circular patch antenna for medical BAN. In Proceedings of the 2014 Loughborough Antennas and Propagation Conference (LAPC), Loughborough, UK, 10–11 November 2014; pp. 22–25.
14. Fan, Y.; Yahya, R.S. Reflection Phase Characterizations of the EBG Ground Plane for Low Profile Wire Antenna Applications. *IEEE Trans. Antennas Propag.* **2003**, *51*, 2691–2703.
15. Dielectric Properties of Tissues. Available online: https://itis.swiss/virtual-population/tissue-properties/database/dielectric-properties/ (accessed on 24 September 2019).
16. Means, D.L.; Chan, K.W. *Evaluating Compliance with FCC Guidelines for Human Exposure to Radiofrequency Electromagnetic Fields*; Federal Communications Commission Office of Engineering, and Technology: Washington, DC, USA, 2001.
17. *IEEE Standard for Safety Levels with Respect to Human Exposure to Radio Frequency Electromagnetic Fields 3 kHz to 300 GHz*; IEEE Standard C95.1-2005; IEEE: Piscataway, NJ, USA, 2005.

15

Spectrum Occupancy Measurements and Analysis in 2.4 GHz WLAN

Adnan Ahmad Cheema [1] and Sana Salous [2,*]

[1] School of Engineering, Ulster University, Jordanstown BT37 0QB, UK

[2] Department of Engineering, Durham University, Durham DH1 3LE, UK

* Correspondence: sana.salous@durham.ac.uk

Abstract: High time resolution spectrum occupancy measurements and analysis are presented for 2.4 GHz WLAN signals. A custom-designed wideband sensing engine records the received power of signals, and its performance is presented to select the decision threshold required to define the channel state (busy/idle). Two sets of measurements are presented where data were collected using an omni-directional and directional antenna in an indoor environment. Statistics of the idle time windows in the 2.4 GHz WLAN are analyzed using a wider set of distributions, which require fewer parameters to compute and are more practical for implementation compared to the widely-used phase type or Gaussian mixture distributions. For the omni-directional antenna, it was found that the lognormal and gamma distributions can be used to model the behavior of the idle time windows under different network traffic loads. In addition, the measurements show that the low time resolution and angle of arrival affect the statistics of the idle time windows.

Keywords: cognitive radio; spectrum occupancy; dynamic spectrum access; time resolution; directional; sensing engine

---

1. Introduction

The vision of 5G network (5GN) encapsulates many application areas, e.g. mobile broadband, connected health, intelligent transportation, and industry automation [1]. To entertain such a wide variety of applications, telecom manufacturers and standardization bodies require the 5GN to support a few Gbps data rates and low latency to a fraction of a millisecond. However, the bottleneck to achieve such requirements will depend on a better understanding of the radio propagation channel in the millimeter wave band to meet such critical constraints [2].

Another aspect of the 5GN is to provide coexistence and improve the spectrum utilization below the 6 GHz band by using concepts of the cognitive radio (CR) network [3]. In this paper, a CR module, which can perform sensing and detection of the spectrum holes, is referred to as "Sensing Engine" (SE), and the time to sense a snapshot of the bandwidth is defined as time resolution. In a CR network, unlicensed users can access the spectrum holes in time, frequency and space or any of their combinations [4–6], provided they cause no interference [4,7].

Opportunistic spectrum access [8] broadly defines the approaches which can enable unlicensed users to find spectrum holes when licensed users are not active. These approaches will improve spectrum utilization and overcome emerging spectrum demands. In future wireless networks, these approaches will be very important to provide spectrum access in networks like Internet of Things [9,10], 5G [11], device-to-device [12], and drones assisted [13]. However, one of the fundamental challenges is to reliably detect spectrum holes and develop models to predict their occurrence along with the idle time windows (ITWs), a continuous fraction of time when the licensed users of the network are not active. These models will be helpful to decide the optimum spectrum allocation based on unlicensed users' requirements (e.g. data rates) and/or to improve spectrum utilization.

The 2.4 GHz industrial, scientific and medical (ISM) band is widely used to provide wireless communication using technologies like WLAN, Bluetooth and Zigbee. Several studies [14–18] in this band have demonstrated that spectrum utilization is very low and this can be helpful to meet spectrum demands of future wireless networks by using concepts of opportunistic spectrum access. However, considering the 2.4 GHz ISM band specifications where the signal duration can be on the order of 100–200 μsec with the narrowest frequency resolution bandwidth requirement as low as 1 MHz [19], an accurate characterization of ITWs is a fundamental challenge. In this paper, our focus is to model the ITWs in the 2.4 GHz WLAN, which is widely-used technology in public and private networks to connect licensed users, by conducting high-resolution spectrum occupancy measurements according to band specifications. It is expected that these measurement-based models will be useful for users in future wireless networks to access 2.4 GHz WLAN spectrum without causing any interference.

For opportunistic spectrum access in the 2.4 GHz WLAN band, a two-state (idle/busy) continuous-time semi-Markov model was proposed in Refs. [6,20,21] where empirical distributions of the ITWs were fitted with the exponential (EX), generalized Pareto (GP) and phase-type distributions (e.g., Hyper Erlang). Although phase type distributions, which are complex to compute due to the high number of parameters, provide an excellent fit, comparatively less complex distributions like the GP can also provide a good fit. In Refs. [6,20], a narrowband vector signal analyzer (VSA) was used to perform high time resolution measurements in a 2.4 GHz WLAN channel, where data packets were generated artificially to stimulate network traffic in an interference controlled environment. This approach facilitates the identification of ITW from a known pattern of transmitted data packets. In Ref. [21], four narrowband sensors were used to monitor the 2.4 GHz WLAN traffic. To sense all 2.4 GHz WLAN channels, the span was divided into 16 channels and each channel was traversed sequentially after 8.192 seconds. Due to this long traverse time, concurrent and continuous network traffic is not possible to monitor in all channels. In Refs. [22,23], a spectrum analyzer (SA) was used to conduct low time resolution (a second or more) measurements in the 2.4 GHz WLAN band where Geometric and GP distributions, respectively, provide an excellent fit for the ITWs.

It is important to highlight that due to such low time resolution, 2.4 GHz WLAN signals smaller than the time resolution will not be detected, which will lead to unrealistic longer ITWs and relating models will generate interference to licensed users. In Ref. [24], high time resolution measurements were performed in the 2.4 GHz WLAN band, and Gaussian mixture distribution (based on 4 components) was found to provide an excellent fit for ITWs. However, this distribution required a higher number of parameters for computation.

Table 1 provides a summary of the measurements in the 2.4 GHz WLAN band and distributions used to model the ITWs for a two-state continuous time semi-Markov model. The order of the distribution is provided from excellent fit (marked in bold) to worst. In this paper, we model the statistics of ITW for 2.4 GHz WLAN signals, recorded in real network traffic at high time resolution using a custom-designed SE [18]. In addition, statistics of the ITWs are modelled using numerically simple distributions (e.g. Weibull (WB), Gamma (GM) and lognormal (LN)) which require fewer parameters and are more practical for implementation compared to phase type or Gaussian mixture distributions.

For comparison with existing work, the GP distribution was used, which also required few parameters and was found to provide good fit in most cases. Moreover, the effect of using different time resolutions on the statistics of ITW are investigated.

Table 1. Summary of the existing measurements to model ITWs in the 2.4 GHz WLAN band.

Ref.	Time Resolution	SE and Bandwidth	Network Traffic	Spectrum Sensing	Distribution(s) for ITW
[6]	High	VSA and N	Artificial	P and F	**GP**
[20]	High	VSA and N	Artificial	P and F	**HE**, GP, EX
[21]	Low	TMote Sky and N	Real	P	**HX**, HE, GP, EX
[23]	Low	SA and W	Real	P	**Geometric**, Bernoulli, LN
[22]	Low	SA and W	Real	P	**GP**, EX, GM, LN, WB
[24]	High	SDR and N	Artificial and Real	P	**G**, GP, HE and EX

Ref.: reference; N: Narrowband; W: Wideband; P: power; F: Feature-based; HE: Hyper-Erlang; HX: Hyper-Exponential; G: Gaussian mixture; SDR: software defined radio.

Directional antennae were used in Refs. [25–27] to find the effect of the angular dimension on the spectrum occupancy. These measurements were conducted using considerably low time resolution per antenna or angle, ranging from 6 seconds to over a minute, which makes it difficult to capture short duration signals and can lead to unrealistic ITW. To the best of our knowledge, these are the only measurement-based papers, which investigate the effect of angular dimension on the spectrum occupancy. However, no further analysis of the statistics of ITWs was provided. In this paper, we also investigate the effect of the angular dimension on the statistics of the ITW.

Although previously different SEs and distributions were used to model the statistics of ITW, they have the following shortcomings:

- The empirical distribution of ITW based on the high time resolution measurements was previously modelled using phase type or Gaussian mixture distributions which require higher parameters for computation and are not feasible for implementation. In addition, low time resolution-based measurements tend not to detect 2.4 GHz WLAN signals, which leads to unrealistic statistics of ITWs. Moreover, the statistics of the ITW were not presented for individual WLAN channels which could have different distributions for ITW. In this paper, in addition to the GP distribution, which was found to provide good fit compared to phase type or Gaussian mixture distributions and required few parameters for computation, similar numerically simple distributions like WB, GM, EX, and LN are also tested and shown to provide appropriate fit under certain traffic load conditions. The analysis is also presented for each of the concurrently measured WLAN channels.
- This work also investigates the effect of time resolution on the statistics of the ITW, which is vital to understand the accuracy of the width of the ITW in relation to set time resolution of the SE.
- This work first investigates the effect of the angular dimension and time resolution per angle on the presence of 2.4 GHz WLAN signals. Then, using the OR hard combining technique the statistics of the ITW per WLAN channel are also analyzed.

In this paper, Section 2 provides a short introduction of the custom-designed SE and related performance. The measurements setup is provided in Section 3 followed by the data analysis methodology in Section 4. The results of the measurements using an omni-directional antenna are discussed in Section 5. In Section 6, the effect of the angular dimension on the ITW is presented with conclusions in Section 7.

2. SE and Performance in 2.4–2.5 GHz

The SE was developed at Durham University and can operate in two frequency bands: (a) 0.25–1 GHz and (b) 2.2–2.95 GHz with a time resolution as high as 204.8 μsec. The logged in data from the SE can be monitored online or stored for offline processing where further filtering is applied followed by a double Fast Fourier transform (FFT) to compute the received power. The architecture and implementation details of the SE can be found in Ref. [18].

Before performing measurements in the desired band, the SE was calibrated for all the gains and losses, which is essential to get the correct value of the received power. To quantify the sensitivity and instantaneous dynamic range (IDR) of the SE, a continuous wave (CW) signal was fed into the SE via an attenuator which was gradually increased until the received signal could not be distinguished from the noise floor. The data were logged in at 80 MHz with a time resolution of 204.8 μsec for 100 MHz sensing bandwidth centered at 2.45 GHz. The raw data were filtered using a high-order Gaussian window to get 400 kHz frequency resolution bandwidth. Figure 1 shows the received signal power against the detected signal to noise ratio (SNR) in the sensed bandwidth. Table 2 summarizes the measured performance parameters of the SE where the noise figure (NF), the difference between the theoretical and measured noise floor, was found to be about ~11 dB. The measured sensitivity was found to be −95 dBm (with at least 12 dB SNR) and an IDR value of 33 dB was achieved. These performance parameters are helpful to find the noise-free region for spectrum holes detection. More details are provided in Sections 5 and 6.

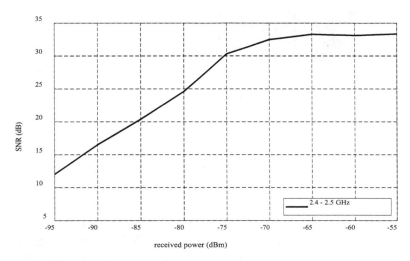

Figure 1. Received Power versus SNR in 2.4–2.5 GHz.

Table 2. Performance parameters of SE in 2.4–2.5 GHz.

Centre Frequency MHz	Bandwidth MHz	Noise Floor dBm	NF dB	Sensitivity		IDR dB
				dBm	SNR (dB)	
2450	100	−107.03	10.95	−95	12.03	33.32

3. Measurement Setup

To analyze the occupancy of the 2.4 GHz WLAN signal, the bandwidth of the SE was configured to 100 MHz centered at 2.45 GHz with a time resolution of 204.8 μsec. A custom-designed wideband omni-directional discone antenna, placed at 1.5 m above ground, as shown in Figure 2a, was used in the measurements. The measurements were taken during working hours from 01:15 pm to 01:35 pm in an indoor environment. For the directional measurements, three commercial log periodic vertically-polarized antennae (see Figure 2b) were used with beam widths of 55 degrees. The antennae were placed at 1.5 m above ground with an angular separation of 90 degrees. Due to having three antennae, switching at 204.8 μsec, the time resolution per antenna or angle was 819.2 μsec, which corresponds to the switching time between three antennae, and an additional reference sweep was taken to get the correct antenna switching sequence in each data file. The measurements were performed in the same environment from 05:20 pm to 05:40 pm.

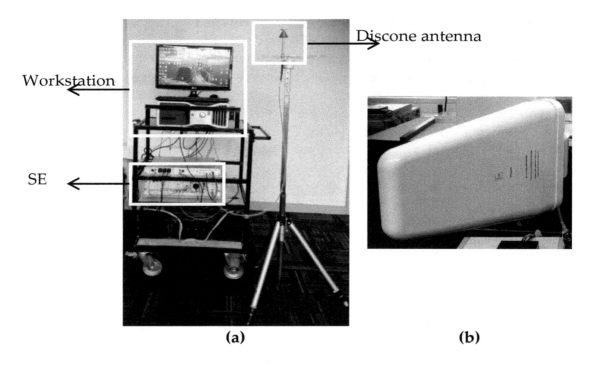

Figure 2. Measurement setup: (**a**) discone antenna, (**b**) log periodic antenna.

At the beginning of each measurement, the radio frequency (RF) attenuator and signal conditioning (SC) gains were calculated based on the sampled data and recorded to calibrate the received power. In both sets of measurements, the raw data were acquired with an 80 MHz sampling rate in multiple files where each file contains 2 seconds of data. Over 976,000 snapshots were collected for both omni-directional and directional setups and processed with 400 kHz frequency resolution by applying a high-order Gaussian window which is sufficient to detect 2.4 GHz WLAN signals or any other available short duration signals in the 2.4 GHz ISM band.

4. Data Analysis Methodology

To estimate the detection of signals in the sensed bandwidth, the energy detection [28] approach was used where the received power was compared with a predefined threshold to define the state of the channel. A channel was considered in a busy state if the received power was above the predefined threshold and otherwise considered in the idle state. Based on this, a binary time series was created in which '1' represents the busy state (presence of signal) and '0' represents the idle state (absence of signal). To find the spectrum utilization, the duty cycle (DC) was calculated based on the binary time series using Equation (1):

$$DC = \text{Total time occupied by busy states/Total time occupied by both states} \qquad (1)$$

The DC is an important parameter and its lower values indicate the availability of the ITWs.

The duration of consecutive 0's i.e., ITW, in the binary time series was computed along with its empirical cumulative distribution function (CDF). The empirical distribution was fitted with GP, WB, EX, GM and LN distributions. The associated parameters (shape: k, scale: δ and location: μ) were found based on the maximum likelihood techniques and used to compute the mean (M) for the fitted distribution. Table 3 summarizes the distribution functions along with the mean formula for the respective distribution.

Table 3. Distribution functions.

Distribution	Function and Mean	Distribution	Function and Mean
GP	$F(t_w) = 1 - \left(1 + k\left(\frac{t_w}{\delta}\right)\right)^{-\frac{1}{k}}$ $M = \frac{\delta}{1-k}$ where $k < 1$	EX	$F(t_w) = 1 - \exp\left(-\frac{t_w}{\delta}\right)$ $M = \delta$
WB	$F(t_w) = 1 - \exp\left(-\frac{t_w}{\delta}\right)^k$ $M = \delta\Gamma\left(1 + \frac{1}{k}\right)$	GM	$F(t_w) = \frac{1}{\Gamma(k)}\gamma\left(k, \frac{t_w}{\delta}\right)$ $M = k\delta$
LN	$F(t_w) = \frac{1}{2}\left(1 + \mathrm{erf}\left(\frac{\ln t_w - \theta}{\sqrt{2}\delta}\right)\right)$ $M = \exp\left(\theta + \frac{\delta^2}{2}\right)$	-	-

t_w denotes idle time windows ($t_w \geq 0$), exp( ) is exponential function, Γ( ) is Gamma function, γ( ) is lower incomplete gamma function and erf( ) is the error function.

To measure the goodness of fit between the empirical and used distributions, Kolmogorov Smirnov (KS) test was used, and the distribution with minimum KS distance was chosen as the best representative of the empirical distribution.

5. Omni-Directional Antenna Measurements

This section provides analysis of the measurements taken with the omni-directional antenna. Figure 3 displays the time-frequency map of the WLAN traffic over the duration of 1 second, where −97 dBm is chosen as the decision threshold which is 10 dB above the measured noise floor and found to be the noise-free region.

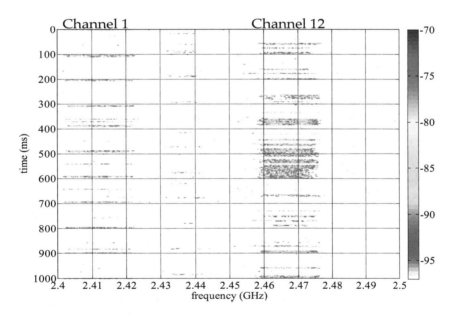

Figure 3. Time-frequency map from 2.4 GHz to 2.5 GHz.

Figure 3 shows that most of the users' activity was present in the WLAN channels '1' and '12'. There are different WLAN packets with different durations. For example, channel '1' did not remain in the busy state all the time. It remains in the idle state for various durations as represented by the 'white' color in the time-frequency map. Thus, by exploiting the idle state of the channel it can be accessed in the time domain by CR users. To find the statistics of the ITW, both channels were further processed to get the binary time series as shown in Figures 4 and 5. Based on the binary time series, the DC values were found to be 7.4013% in channel '1', 11.5071% in channel '12' and 4.6449% over full-sensed bandwidth. These lower DC values indicate that both channels are highly underutilized and can be used for opportunistic spectrum access.

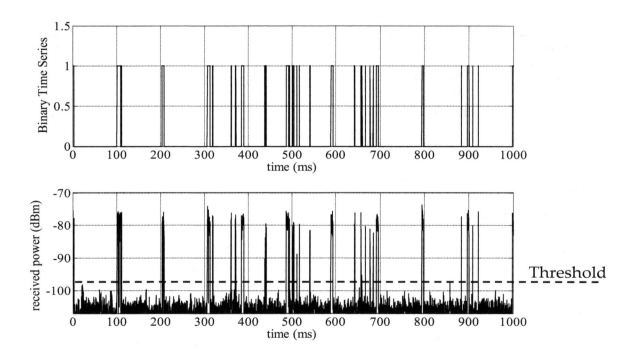

Figure 4. Mapping from received power to Binary Time Series in channel '1'.

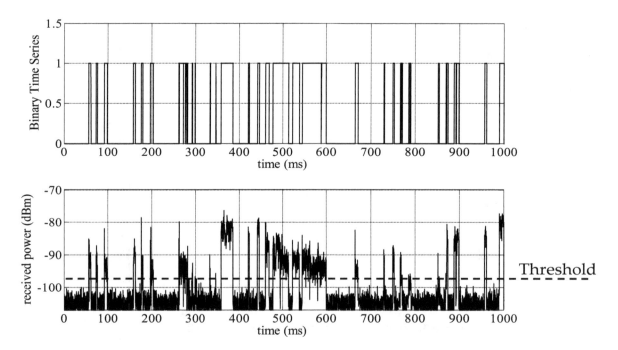

Figure 5. Mapping from received power to Binary Time Series in channel '12'.

Figure 6 shows the empirical CDF of the ITW found in channel '1'. The gamma distribution provides the best fit with minimum KS distance. The estimated parameters are $k = 0.4898$, $\delta = 73.8318$ and the mean $M = 36.1628$ ms. Apart from this, the log normal distribution also provides the second best fit. It is also important to observe that the empirical CDF tends to increase rapidly in the interval (95 ms to 100 ms). The reason for such long windows is because the channel was in the idle state most of the time and the occupancy state was changing due to the arrival of the beacon packet which was observed every 100 ms. Figure 4 also shows that the beacon packets were detected at t = 0, 100, 200 and 800 ms.

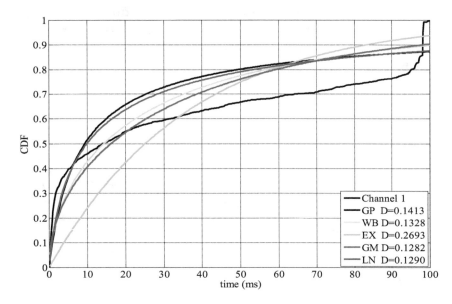

Figure 6. Empirical CDF versus fitted distribution of channel '1'.

Figure 7 shows the empirical CDF of the ITW in channel '12' where the GP distribution provides the best fit with estimated parameters k = 1.1620 and δ = 3.285. Although it provides the best fit, since k > 1, the mean of the distribution is not finite. To calculate the mean value, the WB distribution parameters are used, which provide the second best fit to the empirical CDF. The estimated parameters for this distribution are k = 0.6255 and δ = 9.0156 and the mean M = 12.8766 ms. By comparing both empirical CDFs, it is observed that channel '1' has less traffic load which provides longer ITW, so it is most suitable for CR users. Another key observation is that, it is not possible to have ITW longer than 100 ms since the WLAN access point was periodically transmitting a beacon packet approximately every 100 ms.

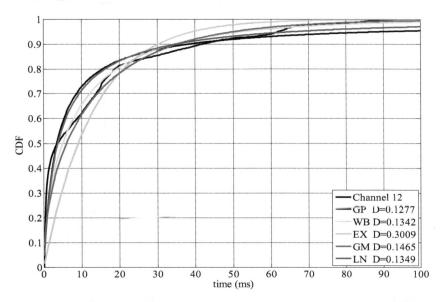

Figure 7. Empirical CDF versus fitted distribution of channel '12'.

Since the SE uses a frequency sweep, each frequency point was traversed after the configured time resolution. To investigate the effect of low time resolution on the statistics of ITW, the time series of both channels were resampled individually in the time domain with 1.6384 ms and 3.2768 ms time resolutions. This illustrates the effect on the ITW when the network traffic is sensed using a low time resolution SE. Figures 8 and 9 show the effect of different time resolutions on channels '1' and '12', where reducing the time resolution tends to increase the duration of the ITW. This apparent increase is due to the SE inability to detect packets with a duration less than the time resolution. Thus, measurements performed using low time resolution SE do not provide reliable data for modelling time-based opportunistic spectrum access.

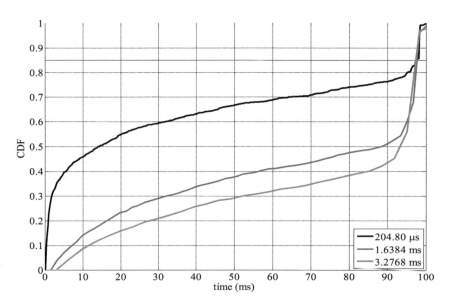

Figure 8. Effect of time resolution on empirical CDF of channel '1'.

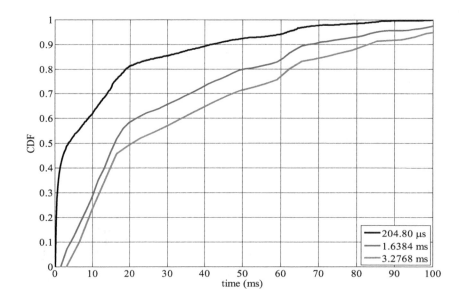

Figure 9. Effect of time resolution on empirical CDF of channel '12'.

The empirical distributions can be further utilized to study how the variability of the load in a channel affects the fitted distributions and their parameters. Table 4 summarizes the KS distance between the empirical and fitted distributions (best distributions are marked bold). The GP distribution is found to have the best fit for channel '1' while the LN distribution provides the best fit for channel '12' for time resolutions of 1.6384 ms and 3.2768 ms, respectively.

Table 4. KS distance for fitted distribution for different time resolutions.

CDF.	Channel 1			Channel 12		
	204.8 µs	1.6384 ms	3.2768 ms	204.8 µs	1.6384 ms	3.2768 ms
GP	0.1413	**0.2412**	**0.2741**	**0.1277**	0.1054	0.1507
WB	0.1328	0.2578	0.2879	0.1342	0.1321	0.1560
EX	0.2693	0.2492	0.2969	0.3009	0.1067	0.1522
GM	**0.1282**	0.2537	0.2900	0.1465	0.1408	0.1724
LN	0.1290	0.2449	0.2903	0.1349	**0.0796**	**0.1438**

For both channels, the mean ITW were calculated for comparison. Table 5 summarizes the parameters and the mean value for the distributions which provided the best fit. The mean value tends to increase with the drop in traffic load as the channels remain in the idle state for longer time windows. These results show that in previous work [20,24], where the GP distribution was shown to provide the second best fit, the LN and GM distributions can also be used to model the behavior of the ITW for different network traffic loads. Moreover, by having wideband detection capability, a CR user can monitor the traffic load on particular or multiple channels and can concurrently exploit the idle states in multiple channels or radio technologies (e.g., 2.4 GHz WLAN and Bluetooth).

Table 5. Parameters for best fitted distribution.

Parameters	Channel 1			Channel 12		
	204.8 µs	1.6384 ms	3.2768 ms	204.8 µs	1.6384 ms	3.2768 ms
k	0.4898	−0.4276	−0.4968	0.6255	-	-
δ	73.8318	87.5879	102.667	9.0156	0.9796	0.9142
µ	-	-	-	-	2.9397	3.2385
M	36.1628	61.3533	68.5910	12.8766	30.5544	38.7210

6. Directional Measurements

The influence of the angle of arrival on the statistics of the ITW is studied from directional measurements. Three antennae (A1, A2 and A3) were used for this indoor measurement. Figure 10 shows the effect of different directions on the received power in channel '1' where it can be observed that antenna A3 has a higher received power compared to the other two antennae. Similarly, antenna A1 has a higher received power for channel '12' as shown in Figure 11. While it is expected that the signal power varies with the angle of arrival, if at a certain direction the received power is lower than the decision threshold due to propagation losses, the channel may be considered in the idle state. Further to this, the time resolution per antenna also affects the state of the channel. For example, consider channel '12' on antenna A3, where the received power dropped by more than 10 dB and it could no longer detect short duration signals.

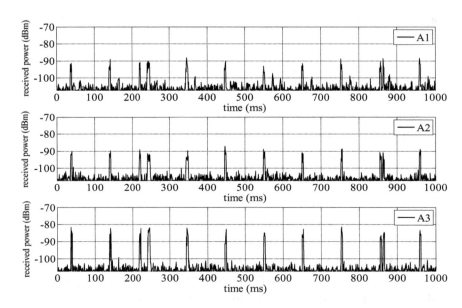

Figure 10. Effect of directional antenna on received power and short duration packets in channel '1'.

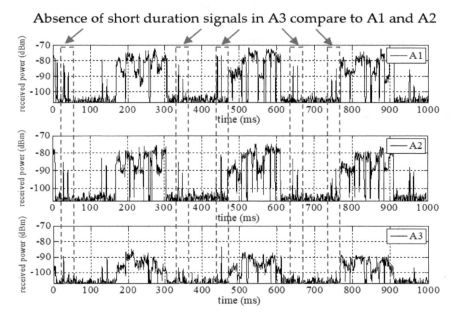

Figure 11. Effect of directional antenna on received power and short duration packets in channel '12'.

A decision threshold of −97 dBm, 10 dB above the measured noise floor, was used per direction to find the respective binary time series. Table 6 compares the DC values in relation to different directional antennas and gives a summary of the DC values based on the omni-directional antenna. To remove the uncertainty in the occupancy state due to the variation in the received power or the effect of time resolution per antenna, the binary time series were combined using the OR hard combining data fusion technique, where a channel is considered in the busy state if it is detected in any direction. This series is used to compute the empirical CDF of the ITW. Figure 12 shows the empirical CDF of both channels and Table 7 summarizes the KS distance for fitted distributions (best distributions are marked in bold). The Weibull distribution provides the best fit for channel '1' with parameters k = 1.1070, δ = 51.7498 and M = 49.8307 ms. Channel '12' empirical CDF is fitted best with the lognormal distribution with parameters δ = 1.4980, μ = 2.6644 and M = 44.0975 ms. Thus, the analysis shows that the angular dimension can influence the statistics of the idle window due to propagation losses and the time resolution per antenna used to acquire the data. To compensate for these effects, decisions can be made using the OR-combining technique. However, the time domain-based directional opportunistic spectrum access can be vital in cases where communication links of licensed users are highly directional.

Table 6. Summary of DC percentages in relation to omni and directional antennae.

			DC Percentage		
			Channel '1'	Channel '12'	2.4–2.5 GHz
	Omni-directional	-	7.4013	11.5071	4.6449
Antenna Type	Directional	A1	3.7515	12.9776	3.9191
		A2	4.3158	12.6841	4.1967
		A3	5.1859	8.2525	3.1760

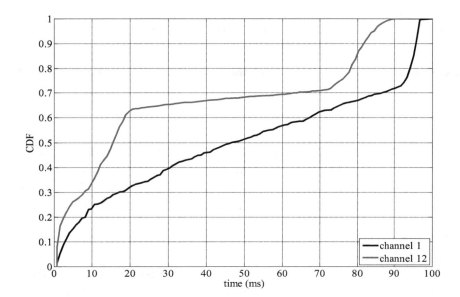

Figure 12. Empirical CDF of channels using OR combining technique.

Table 7. KS distance for fitted distribution.

CDF	Channel 1	Channel 12
GP	0.1610	0.1771
WB	**0.1459**	0.1654
EX	0.1480	0.1801
GM	0.1477	0.1651
LN	0.1751	**0.1473**

The opportunistic spectrum access will play an important role in future wireless networks like Internet of Things, device-to-device, 5G, and drones assisted, to improve spectrum utilization and overcome the spectrum demands from emerging applications in smart cities, connected health and retail sector, to count a few. Due to the wide spread availability for 2.4 GHz WLAN signals, our work will be a step forward for unlicensed users from emerging applications to access the spectrum without causing interference. The reported set of measurements and the numerical simple distributions to model the occupancy will allow users to access the spectrum without performing periodic spectrum sensing. In addition, our findings relating the impact of time resolutions and angle of arrival will be useful to properly plan spectrum occupancy measurements and build models in other wireless communication technologies.

7. Conclusions

Both omni-directional and directional high time resolution occupancy measurements were performed to model the distribution of the idle time window in the 2.4 GHz WLAN. It is found that distributions such as Gamma and lognormal can be used to model the idle state of a 2.4 GHz WLAN channel along with the generalized Pareto distribution. Moreover, previous reported measurements provide statistics of the idle time windows over the full span of a radio technology which does not provide the statistics per channel. Here, analysis is performed per channel, which shows that different concurrently measured channels in a radio technology may have different ITW statistics and their idle states can be modelled using different distributions. Both the time resolution of the SE and directional antennae can influence the state (idle/busy) of the signal and if not selected appropriately can produce longer idle time windows with the inability to detect short duration signals.

Author Contributions: This research was part of the A.A.C.'s PhD work at Durham University, under the supervision of S.S. A.A.C has conducted the experiments, data analysis and drafting of this manuscript under the guidelines of S.S.

Acknowledgments: The authors would like to thank the late S.M. Feeney for the design of the analogue circuits in the SE.

References

1. GSA. *The Road to 5G: Drivers, Applications, Requirements and Technical Development*; GSA: London, UK, 2015.
2. Sun, S.; Rappaport, T.S.; Shafi, M.; Tang, P.; Zhang, J.; Smith, P.J. Propagation Models and Performance Evaluation for 5G Millimeter-Wave Bands. *IEEE Trans. Veh. Technol.* **2018**, *67*, 8422–8439. [CrossRef]
3. Yadav, A.; Dobre, O.A. All Technologies Work Together for Good: A Glance at Future Mobile Networks. *IEEE Wirel. Commun.* **2018**, *25*, 10–16. [CrossRef]
4. Tandra, R.; Mishra, S.M.; Sahai, A. What is a Spectrum Hole and What Does it Take to Recognize One? *Proc. IEEE* **2009**, *97*, 824–848. [CrossRef]
5. Xu, W.; Trappe, W.; Zhang, Y. Channel Surfing: Defending Wireless Sensor Networks from Interference. In Proceedings of the 2007 6th International Symposium on Information Processing in Sensor Networks, Cambridge, MA, USA, 25–27 April 2007; pp. 499–508.
6. Geirhofer, S.; Tong, L.; Sadler, B.M. A Measurement-Based Model for Dynamic Spectrum Access in WLAN Channels. In Proceedings of the MILCOM 2006—2006 IEEE Military Communications Conference, Washington, DC, USA, 23–25 October 2006; pp. 1–7.
7. Haykin, S. Cognitive radio: Brain-empowered wireless communications. *IEEE J. Sel. Areas Commun.* **2005**, *23*, 201–220. [CrossRef]
8. Zhao, Q.; Sadler, B.M. A Survey of Dynamic Spectrum Access. *IEEE Signal Process. Mag.* **2007**, *24*, 79–89. [CrossRef]
9. Li, F.; Lam, K.; Meng, L.; Luo, H.; Wang, L. Trading-Based Dynamic Spectrum Access and Allocation in Cognitive Internet of Things. *IEEE Access* **2019**. [CrossRef]
10. Moon, B. Dynamic spectrum access for internet of things service in cognitive radio-enabled LPWANs. *Sensors* **2017**, *17*, 2818. [CrossRef] [PubMed]
11. Sharma, S.K.; Bogale, T.E.; Le, L.B.; Chatzinotas, S.; Wang, X.; Ottersten, B. Dynamic Spectrum Sharing in 5G Wireless Networks with Full-Duplex Technology: Recent Advances and Research Challenges. *IEEE Commun. Surv. Tutor.* **2018**, *20*, 674–707. [CrossRef]
12. Marotta, M.A.; Roveda Faganello, L.; Kist, M.; Bondan, L.; Wickboldt, J.A.; Zambenedetti Granville, L.; Rochol, J.; Bonato Both, C. Integrating dynamic spectrum access and device-to-device via cloud radio access networks and cognitive radio. *Int. J. Commun. Syst.* **2018**, *31*, e3698. [CrossRef]
13. Shen, F.; Ding, G.; Wang, Z.; Wu, Q. UAV-Based 3D spectrum sensing in spectrum-heterogeneous networks. *IEEE Trans. Veh. Technol.* **2019**, *68*, 5711–5722. [CrossRef]
14. Biggs, M.; Henley, A.; Clarkson, T. Occupancy analysis of the 2.4 GHz ISM band. *IEE Proc. Commun.* **2004**, *151*, 481–488. [CrossRef]

15. Arista Ramirez, D.A.; Cardenas-Juarez, M.; Pineda-Rico, U.; Arce, A.; Stevens-Navarro, E. Spectrum Occupancy Measurements in the Sub-6 GHz Band for Smart Spectrum Applications. In Proceedings of the 2018 IEEE 10th Latin-American Conference on Communications (LATINCOM), Guadalajara, Mexico, 14–16 November 2018; pp. 1–6.

16. Ayeni, A.A.; Faruk, N.; Bello, O.W.; Sowande, O.A.; Muhammad, M.Y. Spectrum Occupancy Measurements and Analysis in the 2.4–2.7 GHz Band in Urban and Rural Environments. *Int. J. Future Comput. Commun.* **2016**, *5*, 142–147. [CrossRef]

17. Cardenas-Juarez, M.; Diaz-Ibarra, M.A.; Pineda-Rico, U.; Arce, A.; Stevens-Navarro, E. On spectrum occupancy measurements at 2.4 GHz ISM band for cognitive radio applications. In Proceedings of the 2016 International Conference on Electronics, Communications and Computers (CONIELECOMP), Cholula, Mexico, 24–26 February 2016; pp. 25–31.

18. Cheema, A.A.; Salous, S. Digital FMCW for ultrawideband spectrum sensing. *Radio Sci.* **2016**, *51*, 1413–1420. [CrossRef]

19. Han, Y.; Wen, J.; Cabric, D.; Villasenor, J.D. Probabilistic Estimation of the Number of Frequency-Hopping Transmitters. *IEEE Trans. Wirel. Commun.* **2011**, *10*, 3232–3240. [CrossRef]

20. Geirhofer, S.; Tong, L.; Sadler, B.M. Cognitive Radios for Dynamic Spectrum Access—Dynamic Spectrum Access in the Time Domain: Modeling and Exploiting White Space. *IEEE Commun. Mag.* **2007**, *45*, 66–72. [CrossRef]

21. Stabellini, L. Quantifying and Modeling Spectrum Opportunities in a Real Wireless Environment. In Proceedings of the 2010 IEEE Wireless Communication and Networking Conference, Sydney, Australia, 18–21 April 2010; pp. 1–6.

22. López-Benítez, M.; Casadevall, F. Time-Dimension Models of Spectrum Usage for the Analysis, Design, and Simulation of Cognitive Radio Networks. *IEEE Trans. Veh. Technol.* **2013**, *62*, 2091–2104. [CrossRef]

23. Wellens, M.; Riihijärvi, J.; Mähönen, P. Empirical time and frequency domain models of spectrum use. *Phys. Commun.* **2009**, *2*, 10–32. [CrossRef]

24. Gupta, A.; Agarwal, S.; De, S. A New Spectrum Occupancy Model for 802.11 WLAN Traffic. *IEEE Commun. Lett.* **2016**, *20*, 2550–2553. [CrossRef]

25. Matinmikko, M.; Mustonen, M.; Höyhtyä, M.; Rauma, T.; Sarvanko, H.; Mämmelä, A. Distributed and directional spectrum occupancy measurements in the 2.4 GHz ISM band. In Proceedings of the 2010 7th International Symposium on Wireless Communication Systems, York, UK, 19–22 September 2010; pp. 676–980.

26. Wang, Z.; Salous, S. Spectrum Occupancy Statistics and Time Series Models for Cognitive Radio. *J. Signal Process. Syst.* **2011**, *62*, 145–155. [CrossRef]

27. Matinmikko, M.; Mustonen, M.; Höyhtyä, M.; Rauma, T.; Sarvanko, H.; Mämmelä, A. Cooperative spectrum occupancy measurements in the 2.4 GHz ISM band. In Proceedings of the 2010 3rd International Symposium on Applied Sciences in Biomedical and Communication Technologies (ISABEL 2010), Rome, Italy, 7–10 November 2010; pp. 1–5.

28. Urkowitz, H. Energy detection of unknown deterministic signals. *Proc. IEEE* **1967**, *55*, 523–531. [CrossRef]

Permissions

The contributors of this book come from diverse backgrounds, making this book a truly international effort. This book will bring forth new frontiers with its revolutionizing research information and detailed analysis of the nascent developments around the world.

We would like to thank all the contributing authors for lending their expertise to make the book truly unique. They have played a crucial role in the development of this book. Without their invaluable contributions this book wouldn't have been possible. They have made vital efforts to compile up to date information on the varied aspects of this subject to make this book a valuable addition to the collection of many professionals and students.

This book was conceptualized with the vision of imparting up-to-date information and advanced data in this field. To ensure the same, a matchless editorial board was set up. Every individual on the board went through rigorous rounds of assessment to prove their worth. After which they invested a large part of their time researching and compiling the most relevant data for our readers.

The editorial board has been involved in producing this book since its inception. They have spent rigorous hours researching and exploring the diverse topics which have resulted in the successful publishing of this book. They have passed on their knowledge of decades through this book. To expedite this challenging task, the publisher supported the team at every step. A small team of assistant editors was also appointed to further simplify the editing procedure and attain best results for the readers.

Apart from the editorial board, the designing team has also invested a significant amount of their time in understanding the subject and creating the most relevant covers. They scrutinized every image to scout for the most suitable representation of the subject and create an appropriate cover for the book.

The publishing team has been an ardent support to the editorial, designing and production team. Their endless efforts to recruit the best for this project, has resulted in the accomplishment of this book. They are a veteran in the field of academics and their pool of knowledge is as vast as their experience in printing. Their expertise and guidance has proved useful at every step. Their uncompromising quality standards have made this book an exceptional effort. Their encouragement from time to time has been an inspiration for everyone.

The publisher and the editorial board hope that this book will prove to be a valuable piece of knowledge for researchers, students, practitioners and scholars across the globe.

List of Contributors

Dong-Myeong Kim, Dongmin Kim, Hang-Geun Jeong and Donggu Im
Division of Electronic Engineering, Jeonbuk National University, Jollabuk-do 561-756, Korea

Azita Goudarzi, Mohammad Mahdi Honari and Rashid Mirzavand
IWT lab, University of Alberta, Edmonton, AB T6G 2R3, Canada

Dimitrios I. Lialios, Constantinos L. Zekios and Stavros V. Georgakopoulos
College of Engineering & Computing, Florida International University, Miami, FL 33174, USA

Nikolaos Ntetsikas, Konstantinos D. Paschaloudis and George A. Kyriacou
Department of Electrical & Computer Engineering, Democritus University of Thrace, 67100 Xanthi, Greece

Mahnoor Khalid, Syeda Iffat Naqvi, Fawad, Muhammad Jamil Khan and Yasar Amin
ACTSENA Research Group, Department of Telecommunication Engineering, University of Engineering and Technology, Taxila, Punjab 47050, Pakistan

Niamat Hussain
Department of Computer and Communication Engineering, Chungbuk National University, Cheongju 28644, Korea

MuhibUr Rahman
Department of Electrical Engineering, Polytechnique Montreal, Montreal, QC H3T 1J4, Canada

Seyed Sajad Mirjavadi
Department of Mechanical and Industrial Engineering, College of Engineering, Qatar University, 2713 Doha, Qatar

Kentaro Saito, Qiwei Fan, Nopphon Keerativoranan and Jun-ichi Takada
School of Environment and Society, Tokyo Institute of Technology, Tokyo 152-8550, Japan

Yong Chen
School of Physics and Electronic Electrical Engineering, Huaiyin Normal University, Huaian 223300, China

Gege Lu, Shiyan Wang and Jianpeng Wang
Ministerial Key Laboratory of JGMT, Nanjing University of Science and Technology, Nanjing 210094, China

Tayyaba Khan and Adeel Akram
Department of Telecommunication Engineering, University of Engineering and Technology, Taxila 47050, Pakistan

Yasar Amin
Department of Telecommunication Engineering, University of Engineering and Technology, Taxila 47050, Pakistan
Department of Electronic Systems, Royal Institute of Technology (KTH), Isafjordsgatan 26, SE 16440 Stockholm, Sweden

Hannu Tenhunen
Department of Electronic Systems, Royal Institute of Technology (KTH), Isafjordsgatan 26, SE 16440 Stockholm, Sweden
Department of Information Technology, TUCS, University of Turku, Turku 20520, Finland

Georgia E. Athanasiadou, Panagiotis Fytampanis, Dimitra A. Zarbouti and George V. Tsoulos
Wireless & Mobile Communications Lab, Department of Informatics and Telecommunications, University of Peloponnese, 22131 Tripolis, Greece

Panagiotis K. Gkonis
General Department, National and Kapodistrian University of Athens, Sterea Ellada, 34400 Dirfies Messapies, Greece

Dimitra I. Kaklamani
Intelligent Communications and Broadband Networks Laboratory, School of Electrical and Computer Engineering, National Technical University of Athens, 9 Heroon Polytechneiou str, Zografou, 15780 Athens, Greece

Ronis Maximidis and A. Bart Smolders
Department of Electrical Engineering, Electromagnetics Group, Eindhoven University of Technology, 5600 MB Eindhoven, The Netherlands

Diego Caratelli
Department of Electrical Engineering, Electromagnetics Group, Eindhoven University of Technology, 5600 MB Eindhoven, The Netherlands
Department of Research and Development, The Antenna Company, 5656 AE Eindhoven, The Netherlands

Giovanni Toso
Antenna and Sub-Millimeter Waves Section, European Space Agency, ESA/ESTEC, 2200 AG Noordwijk, The Netherlands

Zhuohang Zhang and Zhongming Pan
College of Intelligence Science and Technology, National University of Defense Technology, Changsha 410073, China

Nikolaos Nomikos
Department of Information and Communication Systems Engineering, University of the Aegean, 83200 Samos, Greece

Panagiotis Trakadas, Antonios Hatziefremidis and Stamatis Voliotis
General Department, National and Kapodistrian University of Athens, 34400 Psahna, Greece

Daniyal Ali Sehrai, Saad Hassan Kiani and Fazal Muhammad
Department of Electrical Engineering, City University of Science and Information Technology, Peshawar 25000, Pakistan

Mujeeb Abdullah
Department of Computer Science, Bacha Khan University, Charsadda 24420, Pakistan

Ahsan Altaf
Department of Electrical Engineering, Istanbul Medipol University, Istanbul 34083, Turkey

Muhammad Tufail
Department of Mechatronics Engineering, University of Engineering and Technology, Peshawar 25000, Pakistan

Muhammad Irfan and Saifur Rahman
Electrical Engineering Department, College of Engineering, Najran University Saudi Arabia, Najran 61441, Saudi Arabia

Adam Glowacz
Department of Automatic, Control and Robotics, AGH University of Science and Technology, 30-059 Krakow, Poland

Carolina Gouveia and José Vieira
Instituto de Telecomunicações, 3810-193 Aveiro, Portugal Departamento de Eletrónica, Telecomunicações e Informática, Universidade de Aveiro, 3810-193 Aveiro, Portugal

Caroline Loss
FibEnTech Research Unit, Universidade da Beira Interior, 6200-001 Covilhã, Portugal

Pedro Pinho
Instituto de Telecomunicações, 3810-193 Aveiro, Portugal Departamento de Engenharia Eletrónica, Telecomunicações e de Computadores, Instituto Superior de Engenharia de Lisboa, 1959-007 Lisboa, Portugal

Chai-Eu Guan and Takafumi Fujimoto
Graduate School of Engineering, Nagasaki University, Nagasaki 852-8521, Japan

Adnan Ahmad Cheema
School of Engineering, Ulster University, Jordanstown BT37 0QB, UK

Sana Salous
Department of Engineering, Durham University, Durham DH1 3LE, UK

Index